Advances in Mechanics and Mathematics

Volume 46

Advances in Continuum Mechanics

Series Editors
Paolo Maria Mariano, University of Florence
Richard D. James, University of Minnesota
Constantine Dafermos, Brown University

More information about this subseries at https://link.springer.com/bookseries/16097

Paolo Maria Mariano
Editor

Variational Views
in Mechanics

Editor
Paolo Maria Mariano
DICEA
University of Florence
Firenze, Italy

ISSN 1571-8689 ISSN 1876-9896 (electronic)
Advances in Mechanics and Mathematics
ISSN 2524-4639 ISSN 2524-4647
Advances in Continuum Mechanics
ISBN 978-3-030-90050-2 ISBN 978-3-030-90051-9 (eBook)
https://doi.org/10.1007/978-3-030-90051-9

Mathematics Subject Classification: 49-xx

© The Editor(s) (if applicable) and The Author(s), under exclusive license to Springer Nature Switzerland AG 2021, corrected publication 2022
This work is subject to copyright. All rights are solely and exclusively licensed by the Publisher, whether the whole or part of the material is concerned, specifically the rights of translation, reprinting, reuse of illustrations, recitation, broadcasting, reproduction on microfilms or in any other physical way, and transmission or information storage and retrieval, electronic adaptation, computer software, or by similar or dissimilar methodology now known or hereafter developed.
The use of general descriptive names, registered names, trademarks, service marks, etc. in this publication does not imply, even in the absence of a specific statement, that such names are exempt from the relevant protective laws and regulations and therefore free for general use.
The publisher, the authors, and the editors are safe to assume that the advice and information in this book are believed to be true and accurate at the date of publication. Neither the publisher nor the authors or the editors give a warranty, expressed or implied, with respect to the material contained herein or for any errors or omissions that may have been made. The publisher remains neutral with regard to jurisdictional claims in published maps and institutional affiliations.

This book is published under the imprint Birkhäuser, www.birkhauser-science.com by the registered company Springer Nature Switzerland AG
The registered company address is: Gewerbestrasse 11, 6330 Cham, Switzerland

Preface

In the title of a 1756 paper read before the Academy of Berlin and published later,[1] Leonhard Euler coined the syntagma *Calculus of Variations*, shortly after receiving a letter by Joseph-Louis Lagrange, dated August 12, 1755. In his correspondence, the younger Italian mathematician, born in Turin as Giuseppe Luigi Lagrangia, indicated a method for the determination of minima and maxima of functionals, a method bringing to what we commonly call the Euler-Lagrange equations.

Lagrange had been inspired by Euler's 1744 paper dealing with the isoperimetric problem.[2] Euler's view on the topic had been influenced by Newton's geometric approach. From his side, Lagrange derived anew Euler's results not resting on geometric intuition, simplifying them, and offering a more general view. Then, he wrote to Euler a first letter, which remained unanswered. He insisted. Euler answered a second letter, the one received in August: "Your solution to the isoperimetric problem," he wrote, "leaves nothing to be desired." His answer was sufficient to secure Lagrange a teaching position (as a substitute of the lecturer in mathematics) at the Royal School of Artillery in Turin, on September 26, 1755. It was the starting point of a brilliant career that would bring him first to the Academy of Berlin in 1766, under the auspices of Euler himself and d'Alembert, with a subsequent move in 1787 to the Académie des Sciences in Paris.

In his letter, Euler also wrote that "the importance of the matter has led me to outline, with the aid of your light, an analytical solution to which I will give no publicity until you yourself have published the whole of your research, so that I do not take away any part of the glory that is due to you." Euler was a leading mathematician. Lagrange was 19 years old.

However, these ideas did not emerge *ex nihilo*. Optimal problems interested even ancient Greeks (Hero and Pappus of Alessandria in the first century), although they did not have even rough tools of infinitesimal calculus. Isaac Newton, Jacob

---

[1] Euler L., Elementa calculi variationum, *Novi comment. acad. sc. Petrop.*, 10 (1764), 1766, 51–93.
[2] Euler L., *Methodus inveniendi lineas curvas maximi minimive proprietate gaudentes, sive solution problematis isoperimetrici latissimo sensu accepti*, 1744, Lausanne.

Bernoulli, and Pierre de Fermat, among others, had analogous interests. Lagrange's main merit was perhaps to recognize that Euler's 1744 results could imply a (say) new type of differential calculus, and he opened its construction, although rigor in the matter began to be clear later, after Karl Weierstrass' and Carl Gustav Jacobi's work, so in the nineteenth century. They found sufficient conditions for an extremum of a functional. More specifically, Weierstrass introduced the concept of a field of extremals, the excess function, a style renowned for rigor, and criticized the Dirichlet principle showing that not necessarily a variational problem with energy bounded from below has a minimum.[3]

Between 1900 and 1904, David Hilbert addressed new light on the issue: he proved that the Dirichlet principle is a way to find existence of a curve of minimum length connecting two points on a surface and solved the Laplace equation with boundary conditions in the plane.[4] Also, in formulating his 19th and 20th problems in the list discussed first in the 1900 International Congress of Mathematicians in Paris,[5] Hilbert opened a new program in calculus of variations, which has grown as a prominent sector of mathematical analysis per se. In his 1910 book on this matter, Jacques Hadamard wrote, in fact, that it was "a first chapter of functional calculus, whose development will without doubt be one of the first tasks in the analysis of the future."[6]

Also, Hilbert considered the calculus of variations at the core of analysis—with a role similar to that of the alphabet for reading and writing—and placed (in a sense) mechanics *within* calculus of variations.[7] That view portrayed the calculus of variations as a field with theoretical and aesthetic motivations per se.[8] However, and at variance in a sense, besides specific problems such as the isoperimetric or fastest descent ones, the former dating back to Dido's trick leading to the foundation of Carthage, according to the Roman historian Pompeius Trogus, calculus of variations essentially grabs the roots of its interest beyond pure mathematics on our common belief that nature minimizes energy at constant entropy and maximizes entropy at constant energy, a belief not falsified so far by common experiments. Already Pierre

---

[3] Weierstrass K., Vorlesungen über Variationsrechnung, *Math. Werke*, Bd. 7., 1927, Akademische Verlagsgesellschaft, Leipzig.

[4] Hilbert D. (1900), Über das Dirichletsche Prinzip, *Jahresberich des Deutschen Mathematicker-Vereinigung*, **59**, 1900, 161–186.
Hilbert D., Über das Dirichletsche Prinzip, *Mathematische Annalen*, **59**, 1904, 161–186.

[5] 19th Problem: *Are the solutions of regular problems in calculus of variations always analytics?*.
20th Problem: *Do variational problems with certain boundary conditions admit solutions?*

[6] Hadamard J., *Leçons sur le calcul des variationsç*, Tome I (seul paru). La variation première et les conditions du premier ordre. Les conditions de l'extremum libre, edited by M. Fréchet, Cours du Collège de France, 1910, Librairie Scientifique A. Hermann et Fils.

[7] Giaquinta M., Hilbert e il calcolo delle variazioni, *Le Matematiche*, **LV**, 2000, 47–58.

[8] In his treatise, Jacques Hadamard had a view of this type, also pursued later by Gilbert Ames Bliss with his 1946 treatise *Lectures on Calculus of Variations*, where he exposed systematically essential results known at that times.

de Fermat believed that nature operates by means and ways that are "easiest and fastest."

Problems requiring recourse to calculus of variations techniques to be tackled emerge in several sectors, even in social sciences, but—I repeat—above all in mechanics (be it classical, quantum, or relativistic), condensed matter physics, and chemistry, among others. The technological need of objects with some optimal properties—for example, shape, strength, and conductivity—under some constraints or need of optimal control of processes further motivates in engineering the interest for techniques pertaining to the calculus of variations.

Looking at mechanics from a theoretical side, we realize that the motion of a three-dimensional rigid body can be viewed as a geodetic curve (the one with minimal length) over the special orthogonal group, while incompressible perfect fluids move along geodetic paths over the special group of diffeomorphisms. We can also aim at controlling optimally the motion of certain systems, be them multi-rigid-bodies with flexible mutual constraints or continua suffering distributed strain, as, for example, rods are. We may ask to find the minimal energy of an atom, a molecule, a thin film, or we may tackle optimality questions connected with chemical reactions, or we aim at printing and connecting microstructures, in order to obtain an artifact with some optimized properties, what we call a metamaterial. Also, we may be interested in optimal transportation of mass, charges, and their like.

Energy minimization characterizes equilibrium configurations. When coupled with appropriate monotonicity conditions mimicking irreversible behavior, such a minimization procedure may allow us to describe classes of (rate-independent) dissipative processes, such as plastic flows, damage, some phase transitions, or nucleation and grow of fractures, by adapting Ennio De Giorgi's construction of minimizing movements. The idea is to partition the time interval into finitely many sub-intervals, presuming to go from the state at instant t_k to the one at t_{k+1} by minimizing some functional onto an appropriate function class. For example, in fracture processes, a crack path can be viewed as coinciding with the jump set of special bounded variation functions (SBV) or the support of varifolds.[9] The two possibilities are not equivalent: when we choose just deformations in SBV, imposing that they are one-to-one and preserve almost everywhere the orientation of volumes outside the jump set, we are thinking essentially about fractures that open or undergo relative slip between their margins, while when we make use of varifolds, we may describe fractures which have a portion of the margins in contact, although no material bonds intervene between them.

This and other examples of phenomena suggesting recourse to a variational view to be described open often-challenging analytical problems. Tackling them drives the evolution of this rich sector of mathematical analysis, possibly indicating

---

[9] SBV is a class of functions with derivative a measure having absolutely continuous component with respect to the Lebesgue volume n-dimensional measure, and a nonsingular part concentrated over a $\mathcal{H}^{n-1}$-measurable set, with $\mathcal{H}^{n-1}$ the $(n-1)$-dimensional Hausdorff measure. Varifolds are vector-valued measures admitting a generalized notion of curvature.

connections with other sectors, as it happened in the analysis of parabolic partial differential equations, now connected with calculus of variations in the optimal transportation theory.

Already a beginner in mathematical analysis knows the way of finding minima, maxima, and stationary points of a differentiable real function on $[a, b] \subset \mathbb{R}$ through the evaluation of zeros for the first derivative and the analysis of second derivative, when available. The classical approach to calculus of variations has been the proper extension of that standard method to functions defined on functional spaces rather than just on $\mathbb{R}^n$. The derivative becomes a variation obtained through appropriate test functions. The Euler-Lagrange equations determine necessary conditions for a function u to be an extremal for a functional $\mathcal{F}$ with values given, for example, by

$$\mathcal{F}(u) := \int_\Omega F(x, u(x), Du(x)) \, d\mu(x),$$

where F is a scalar density assumed to be differentiable with respect to its entries, and μ a volume measure over Ω, a smooth open set in $\mathbb{R}^n$. When F is convex and the pertinent Euler-Lagrange equations admit unique solution u, we are sure that $\mathcal{F}$ attains its minimum value at u. Otherwise, once proving existence of solutions for the Euler-Lagrange equations, we should evaluate the second variation of $\mathcal{F}$ over them. When Ω is an interval in $\mathbb{R}$, the pertinent Euler-Lagrange equations have ordinary character, with pertinent boundary data, and we do not always find conditions assuring existence of their solutions. Beyond one-dimensional ground space, the Euler-Lagrange equations are partial, with pertinent difficulties.

Around the end of nineteenth century, Bernhard Riemann suggested to reverse the view on minimization of functionals along a path already used (in a sense implicitly) by Carl Friedrich Gauss and William Thompson Lord (1st Baron) Kelvin. If, with X a topological space from now on, we are able to find a minimum for

$$\mathcal{F} : X \to \mathbb{R} \cup \{\infty\}$$

by looking just at minimizing sequences, we have, in turn, a solution of the pertinent Euler-Lagrange equations in some sense, that is, depending on the regularity showed by the minimum. This is (roughly) what we call *direct method* in calculus of variations, explored by many scholars, starting from David Hilbert, who formulated related questions in his 1900 program, which has already been mentioned.

Such an approach emerges from what we do on functions on $\mathbb{R}$. In fact, to prove that a continuous real function defined on a compact set $K \in \mathbb{R}^n$ attains its minimum value, first we take a minimizing sequence $\{x_j\}$ such that

$$f(x_j) \to \inf_{x \in K} f(x)$$

as $j \to \infty$. In K there exists a converging subsequence $\{x_j\}$ and the continuity of f implies $\lim_{j \to \infty} f(x_j) = f(x)$. Although working on minimizing sequences is a key idea, the version of this procedure in infinite-dimensional spaces cannot be reached straight away. In fact, consider, for example, $\mathcal{F}$ to be such that, for $u \in L^2(\Omega, d\mu)$, whenever $\|u_j \to u\|_{L^2}$ as $j \to +\infty$, $\mathcal{F}(u_j) \to \mathcal{F}(u)$, that is, $\mathcal{F}$ is strongly continuous. If we choose K to be now the unit ball

$$K := \left\{ u \in L^2(\Omega, d\mu) : \|u\|_{L^2} \leq 1 \right\}$$

as a putative set for finding the minimum of $\mathcal{F}$, although K is closed and bounded, we do not necessarily find a convergent subsequence $\{u_j\}$ in K. If we look at weak convergence, we find that every sequence in K has a weakly convergent subsequence. However, $\mathcal{F}$ is not necessarily weakly continuous. In other words, the more we relax the notion of convergence, the less likely $\mathcal{F}$ is continuous on the pertinent sequences. Things may be adjusted when $\mathcal{F}$ is such that

$$\liminf_{j \to \infty} \mathcal{F}(u_j) \geq \mathcal{F}(u)$$

as $u_j \rightharpoonup u$. In this case, we say that $\mathcal{F}$ is weakly lower semicontinuous—remarkably, Leonida Tonelli established first in 1920 necessary and sufficient conditions of lower semicontinuity for a functional defined on a one-dimensional space. Thus, if $\mathcal{F}$ is lower semicontinuous, for $\{u_j\}$ a minimizing sequence in the sense that

$$\mathcal{F}(u_j) \to \inf \{\mathcal{F}(u) : u \in C\} =: \gamma,$$

there exists a subsequence $\{u_j\}$ such that $u_j \rightharpoonup u$, so that

$$\gamma = \lim_{j \to \infty} \mathcal{F}(u_j) \geq \mathcal{F}(u) \geq \gamma,$$

that is, $\mathcal{F}(u) = \gamma$. Instead of thinking in sequential terms (convergence of sequences being not necessarily weak), we can speak of lower semicontinuity for a functional $\mathcal{F}: X \to \mathbb{R} \cup \{+\infty\}$, if for any $t \in \mathbb{R}$ the set

$$\mathcal{F}_t := \{u \in X : \mathcal{F}(u) > t\}$$

is open. A functional lower semicontinuous in this topological sense is also so in sequential terms. The opposite is true if every point of X admits a countable fundamental system of neighborhoods.

In analyzing a functional class in terms of the direct method, a key point is to have at disposal a lower semicontinuity result. In the academic year 1968/1969, in the unpublished notes of a course held in Rome at the "Istituto Nazionale di Alta Matematica," Ennio De Giorgi presented the first proof of the (sequential) lower semicontinuity of the functional

$$\mathcal{F}(u, v) = \int_\Omega F(x, u(x), v(x)) \, dx,$$

with respect to strong convergence of u and weak convergence of v, under assumption that the density $F(x, s, \xi)$ is jointly continuous with respect to the three variables entering it and convex in ξ (e.g., the L^p norm is a lower semicontinuous functional). Such a result opened new paths in calculus of variations. In the subsequent body of semicontinuity theorems, obtained under various assumptions, a specific result involving the Dirichlet functional,

$$\mathcal{F}(u) = \frac{1}{2} \int_\Omega |Du|^2 \, dx,$$

in which we take u as a map with values on a finite-dimensional differentiable manifold $\mathcal{M}$ not embedded into a linear space, and assumed to be Riemannian and complete, plays a role in the general model-building framework for the mechanics of complex materials, a format in which descriptors of the material microstructure are manifold-valued maps.

Another idea pertaining to calculus of variations, which emerges as a useful (at time essential) tool for justifying rigorously some mechanical models, is the notion of Γ-convergence, also due to De Giorgi. Such a notion of convergence may decisively help in analyses that aim at justifying schemes involving dimensional reduction from thick material layers to thin films or the passage from a discrete (atomic-scale) representation of matter to a continuum view.

A sketch of the idea goes as follows: Imagine having a sequence of functionals and a companion sequence of minimizers. A question is to find conditions assuring that the limiting function is a minimum for the limiting functional, if any. Consider, in fact, functionals $\mathcal{F}_\varepsilon : X_\varepsilon \to \mathbb{R} \cup \{\infty\}$ and a sequence $\{\min \mathcal{F}_\varepsilon(u_\varepsilon) : u_\varepsilon \in X_\varepsilon\}$, which we assume to be equi-coercive, that is, there exists a pre-compact minimizing sequence such that $\mathcal{F}_\varepsilon(u_\varepsilon) \leq \inf \mathcal{F}_\varepsilon + o(1)$, and also $u_\varepsilon \to u_0$, as $\varepsilon \to 0$, with u_0 solution to $\{\min \mathcal{F}_0(u_0) : u_0 \in X_0\}$. We call $\mathcal{F}_0$ the Γ-limit of $\mathcal{F}_\varepsilon$ when two conditions are satisfied: (i) for every $u \in X_0$ and every $u_\varepsilon \to u$ we have

$$\mathcal{F}(u) \leq \liminf_{\varepsilon \to 0} \mathcal{F}_\varepsilon(u_\varepsilon);$$

(ii) there exists a sequence $\bar{u}_\varepsilon \to u_0$ for every $u_0 \in X_0$ such that

$$\inf \mathcal{F}_0 \geq \limsup_{\varepsilon \to 0} \inf \mathcal{F}_\varepsilon.$$

Γ-convergence and equi-coerciveness imply convergence of minimum problems.

Again, the choice of convergence is crucial: a weaker convergence, with many converging sequences, makes equi-coerciveness easier to fulfill, but at the same time makes the lim inf inequality more difficult to hold. Often, an appropriate choice is strong convergence in L^p spaces. Connected with the selection of convergence is the companion choice of energy scaling to assure equi-coerciveness.

Analytical problems in calculus of variations are manifold and faceted. They exceed largely the brief incomplete sketch above. Also, besides purely analytical questions, when we look at the world around us with the aim of interpreting it qualitatively and quantitatively, we meet recurrently phenomena offering themselves as a playground for calculus of variations or suggesting further analytical problems in a fruitful mathematical field.

This miscellany offers a partial, although variegate, view on problems in mechanics that can be well-analyzed by means of variational techniques. Topics range from topology optimization to identification of material properties, optimal control, plastic flows, gradient polyconvexity, obstacle problems, quasi-monotonicity, and waves. These chapters offer results opening views on further possible research work. Also, they are examples of how foundational knowledge and command of appropriate mathematical techniques may address us towards applications going out of the rut and indicating, as such, possible new scientific and technological paths.

Preliminary versions of these chapters have been included in a special issue of the *Journal of Optimization Theory and Applications* (vol. 184, issue 1, 2020, 1–314) on *Calculus of Variations in Mechanics and Related Fields*. For some of the original papers, type and amount of variations leading to the new versions have been so extended to justify changes in the titles and, when appropriate, in the list of authors.

I thank the former editor-in-chief of the *Journal of Optimization Theory and Applications*, Prof. Franco Giannessi, who promoted that special issue; his successor, Prof. Tamás Terlaky, who gave permission to print modified versions of the papers; and all staff at Birkhäuser's and Springer's offices who supported and pushed ahead the whole project.

Firenze, Italy   Paolo Maria Mariano

The original version of this book was revised: Revised book has been uploaded to Springerlink. The correction to this book is available at https://doi.org/10.1007/978-3-030-90051-9_12.

Contents

Numerical Study of Microstructures in Multiwell Problems
in Linear Elasticity 1
Sergio Conti and Georg Dolzmann

Surface Shear Waves in a Functionally Graded Half-Space 31
Andrey Sarychev, Alexander Shuvalov, and Marco Spadini

Modeling of Microstructures in a Cosserat Continuum Using
Relaxed Energies: Analytical and Numerical Aspects 57
Muhammad Sabeel Khan and Klaus Hackl

The Polar-Isogeometric Method for the Simultaneous
Optimization of Shape and Material Properties of Anisotropic
Shell Structures 89
Christian Fourcade, Paolo Vannucci, Dosso Felix Kpadonou,
and Paul de Nazelle

Gradient Polyconvexity and Modeling of Shape Memory Alloys 133
Martin Horák, Martin Kružík, Petr Pelech, and Anja Schlömerkemper

Placement of an Obstacle for Optimizing the Fundamental
Eigenvalue of Divergence Form Elliptic Operators 157
Anisa M. H. Chorwadwala and Souvik Roy

Quasi-Monotonicity Formulas for Classical Obstacle Problems
with Sobolev Coefficients and Applications 185
Matteo Focardi, Francesco Geraci, and Emanuele Spadaro

Optimal Feedback for Structures Controlled by Hydraulic
Semi-active Dampers 205
Ido Halperin, Grigory Agranovich, and Yuri Ribakov

Multi-Displacement Requirement in a Topology Optimization Algorithm Based on Non-uniform Rational Basis Spline Hyper-Surfaces 223
Marco Montemurro, Thibaut Rodriguez, Paul Le Texier, and Jérôme Pailhès

Anti-plane Shear in Hyperelasticity 259
Jendrik Voss, Herbert Baaser, Robert J. Martin, and Patrizio Neff

Identification of Diffusion Properties of Polymer-Matrix Composite Materials with Complex Texture 289
Marianne Beringhier, Marco Gigliotti, and Paolo Vannucci

Correction to: Variational Views in Mechanics C1

Numerical Study of Microstructures in Multiwell Problems in Linear Elasticity

Sergio Conti and Georg Dolzmann

1 Introduction

The mathematical analysis of diffusionless phase transformations in elastic solids, in particular austenite–martensite transformations, has inspired a lot of research starting from the seminal papers [5, 6, 13] and subsequently also led to a new approach to the variational modelling in nonlinear plasticity starting in [10, 28]. Mechanically, these systems are characterized by the spontaneous formation of finely oscillating microstructures, which are difficult to study numerically. From the viewpoint of analysis, the challenges are related to questions of lower semicontinuity in the vector-valued calculus of variations and to the notion of quasiconvexity in the sense of Morrey [27]. The analytical investigations also inspired the search for efficient algorithms for the numerical solution of the corresponding variational problems. The key observation that establishes the link between microstructures experimentally observed in elasto-plastic materials, and the mathematical description in nonlinear elasto-plasticity is that the lack of quasiconvexity in the variational formulation may lead to oscillating minimizing sequences, which are hard to capture with numerical schemes based on energy minimization with descent methods and hard to resolve explicitly in mesh-based numerical approximations. Whereas the details of the microstructure can often be ignored, its presence is crucial in order to understand macroscopic material properties, such as the energetics and the stress–strain response.

S. Conti (✉)
Institut für Angewandte Mathematik, Universität Bonn, Bonn, Germany
e-mail: sergio.conti@uni-bonn.de

G. Dolzmann
Fakultät für Mathematik, Universität Regensburg, Regensburg, Germany
e-mail: georg.dolzmann@ur.de

© The Author(s), under exclusive license to Springer Nature Switzerland AG 2021
P. M. Mariano (ed.), *Variational Views in Mechanics*, Advances in Continuum Mechanics 46, https://doi.org/10.1007/978-3-030-90051-9_1

One of the main tools used in the mathematical analysis both from the analytical and the numerical point of view is the concept of relaxation [19], which transforms a variational problem I that fails to be lower semicontinuous into an associated variational problem I^{rel} that is lower semicontinuous in the appropriate topology. The relaxed variational problem I^{rel} is defined so that minimizing sequences for I converge to minimizers of I^{rel}, and, vice versa, minimizers of I^{rel} are limits of minimizing sequences for I. Therefore, a study of the minimizers of I^{rel} permits to understand the asymptotic behavior of minimizing sequences of I and to efficiently study the effective macroscopic material behavior without resolving the details of the microstructure. This approach is, however, only useful if one can characterize I^{rel} explicitly, which typically involves finding an explicit formula for the quasiconvex envelope of the energy density entering I, see Sect. 2 for the precise definition.

In [16], we presented our approach to the numerical approximation of relaxed variational problems in the framework of nonlinear elasto-plasticity with a focus on microstructures in single-slip finite elasto-plasticity. In this chapter, we focus on the implementation of variational problems in the context of linear elasticity, and we verify the excellent performance of the proposed scheme in the context of variational models related to austenite–martensite transformations for which some relaxed variational problems have been characterized analytically. In Sect. 2, we introduce the relevant notions of convexity, state preliminary results, and describe common strategies that are useful in the search for relaxed variational models. A detailed description of the available results and an illustration of the approaches described in Sect. 2 are contained in Sect. 3. The algorithm itself is briefly described in Sect. 4, and results from numerical experiments are presented in Sect. 5.

Notation We denote the space real $n \times n$-matrices by $\mathbb{R}^{n \times n}$ and the space of symmetric matrices by $\mathbb{R}^{n \times n}_{\text{sym}}$ and define the orthogonal projection $\pi_n \colon \mathbb{R}^{n \times n} \to \mathbb{R}^{n \times n}_{\text{sym}}$ by $F \mapsto \pi_n(F) = (F + F^T)/2$. We denote by $\langle F, G \rangle = \operatorname{tr} F^T G$ the scalar product in $\mathbb{R}^{n \times n}$ and use the same for its subspace $\mathbb{R}^{n \times n}_{\text{sym}}$.

2 Fundamental Notions of Convexity and Relaxed Energy Densities

The theory of relaxation is based on the notion of quasiconvexity which, despite of its importance in the vector-valued calculus of variations, is poorly understood. Therefore, other notions of convexity, which provide necessary and sufficient conditions for quasiconvexity, play an important role in the mathematical analysis. We first introduce these notions, and then we sketch two widely used ideas to characterize relaxed energies.

2.1 Notions of Convexity and Their Relations

In this chapter, we focus on models for elastic behavior under the assumption that the elastic response is governed by linear elasticity. Therefore, we recall the fundamental notions of convexity in this setting. The relation with the classical notions in nonlinear elasticity [19] is established through the concatenation with the projection onto symmetric matrices. More precisely, a function $f\colon \mathbb{R}^{n\times n}_{\text{sym}} \to \mathbb{R}$ is said to be symmetric rank-one convex (quasiconvex, polyconvex) if the function $\widehat{f}\colon \mathbb{R}^{n\times n} \to \mathbb{R}$ given by $F \mapsto f(\pi_n(F))$ is rank-one convex (quasiconvex, polyconvex), see [9] for a detailed discussion of this and related concepts. This identification induces the following definitions.

Definition 1 Two matrices $\epsilon, \eta \in \mathbb{R}^{n\times n}_{\text{sym}}$ are said to be (symmetrically) compatible if there exist $a, b \in \mathbb{R}^n$ with

$$\epsilon - \eta = \frac{1}{2}(a \otimes b + b \otimes a) =: a \odot b,$$

where $(a \otimes b)_{ij} = a_i b_j$ for $i, j = 1, \ldots, n$.

Lemma 1 ([24, Lemma 4.1]) *A matrix* $\epsilon \in \mathbb{R}^{n\times n}_{\text{sym}}$, $\epsilon \neq 0$, *can be written in the form* $\epsilon = a \odot b$ *with* $a, b \in \mathbb{R}^n$, *if and only if either*

(a) ϵ *has rank one or*
(b) ϵ *has rank two and its nonzero eigenvalues have opposite sign.*

Definition 2 Suppose that $f\colon \mathbb{R}^{n\times n}_{\text{sym}} \to \mathbb{R}$ is given. Then,

1. f is said to be symmetric rank-one convex if for all symmetrically compatible ϵ, $\eta \in \mathbb{R}^{n\times n}_{\text{sym}}$ and all $\lambda \in [0, 1]$, the inequality

$$f(\lambda\epsilon + (1-\lambda)\eta) \leq \lambda f(\epsilon) + (1-\lambda)f(\eta)$$

   holds.
2. f is said to be symmetric quasiconvex if for all $\epsilon \in \mathbb{R}^{n\times n}_{\text{sym}}$ and all test functions $\phi \in C_c^\infty((0,1)^n; \mathbb{R}^n)$, the inequality

$$\int_{(0,1)^n} f(\epsilon)\,\mathrm{d}x \leq \int_{(0,1)^n} f(\epsilon + \pi_n(\nabla\phi))\,\mathrm{d}x$$

   holds.
3. f is said to be symmetric polyconvex if there exists a convex function $g\colon \mathbb{R}^{\tau(n)} \to \mathbb{R}$ such that for all $\xi \in \mathbb{R}^{n\times n}$, the identity

$$f(\pi_n(\xi)) = g(M(\xi))$$

holds, where $M \in \mathbb{R}^{\tau(n)}$ denotes the vector of all minors of ξ. For example, for $n = 2$, we have $\tau(2) = 5$ and $M(\xi) = (\xi, \det \xi)$, and for $n = 3$, we have $\tau(3) = 19$ and $M(\xi) = (\xi, \operatorname{cof} \xi, \det \xi)$.

Remark 1 For every function $f: \mathbb{R}^{n \times n}_{\text{sym}} \to \mathbb{R}$, the implications

$$f \text{ convex} \Rightarrow f \text{ symmetric polyconvex} \Rightarrow f \text{ symmetric quasiconvex}$$
$$\Rightarrow f \text{ symmetric rank-one convex}$$

hold. The definition of quasiconvexity does not depend on the domain of integration, and the cube $(0, 1)^n$ can be replaced by any open domain with $\mathcal{L}^n(\partial \Omega) = 0$, see [7].

Remark 2 Since our applications concern two-dimensional models, we note that there are in this case significant differences between linear and nonlinear elasticity. In fact, the determinant is a null-Lagrangian or polyaffine, but the map $F \mapsto \det F$ (seen as a map from $\mathbb{R}^{n \times n}_{\text{sym}}$ to $\mathbb{R}$) is not symmetric quasiconvex. In fact, $\det(\pi_2(\xi)) = \det \xi - \frac{1}{2}(\xi_{12} - \xi_{21})^2$, and for any $\phi \in C_c^\infty((0, 1)^2; \mathbb{R}^2)$,

$$\int_{(0,1)^2} \det \left(\frac{1}{2}(\nabla \phi + (\nabla \phi)^T)\right) \mathrm{d}x = \int_{(0,1)^2} \left[\det \nabla \phi - \frac{1}{2}(\partial_1 \phi^{(2)} - \partial_2 \phi^{(1)})^2\right] \mathrm{d}x.$$

These formulas also show that $F \mapsto -\det F$ is symmetric polyconvex. In general, the quasiconvex functions presented in [31] fail to be symmetric quasiconvex, see [9] for a detailed discussion and proofs.

Remark 3 One can show that a function $f: \mathbb{R}^{2 \times 2}_{\text{sym}} \to \mathbb{R}$ is symmetric polyconvex if and only if there exists a convex function $g: \mathbb{R}^{2 \times 2}_{\text{sym}} \times \mathbb{R} \to \mathbb{R}$ nonincreasing in its second argument with $f(\epsilon) = g(\epsilon, \det \epsilon)$, see [9, Theorem 4.1].

Definition 3 Suppose that $f: \mathbb{R}^{n \times n}_{\text{sym}} \to \mathbb{R}$ is given. The symmetric rank-one convex (quasiconvex, polyconvex) envelope of f is defined by

$$\widetilde{f} = \sup\{g \leq f: g \text{ symmetric rank-one convex (quasiconvex, polyconvex)}\}.$$

One uses the notation $\widetilde{f} = f^{rc}$, $\widetilde{f} = f^{qc}$, and $\widetilde{f} = f^{pc}$ in each of the three cases. The convex envelope of f is denoted by f^{**}.

In view of Remark 1, the polyconvex envelope is a lower bound and the rank-one convex envelope is an upper bound for the quasiconvex envelope of a finite-valued function, respectively, $f^{pc} \leq f^{qc} \leq f^{rc}$. The quasiconvex envelope is also referred to as the relaxation, the relaxed or the macroscopic energy density. In practice, it is a difficult problem to find an explicit formula for the relaxation, and this has been achieved only in a few special cases, see, for example, [11, 17, 23, 24, 30].

2.2 Strategies for the Characterization of Relaxed Energies

Since symmetric rank-one convexity is a necessary condition and since symmetric polyconvexity is a sufficient condition for symmetric quasiconvexity, most approaches for the characterization of the relaxed energy density f^{qc} are based on upper bounds inspired by symmetric rank-one convexity and lower bounds inspired by symmetric polyconvexity. In view of the organization of the necessary calculations involved in the characterization of a relaxed energy, one can envision two strategies.

2.2.1 Construction of an Upper Bound and a Matching Lower Bound

The common approach consists in first finding by explicit constructions an upper bound for the relaxation and then verifying that this upper bound is at the same time a lower bound. More precisely, this can be achieved by calculating by successive relaxation along symmetric rank-one lines an upper bound for the symmetric rank-one convex envelope and by showing that this bound is symmetric polyconvex. To implement this scheme, one defines a sequence of functions $f^{(k)}$ for $k \in \mathbb{N}_0$ in the following way: set

$$f^{(0)}: \mathbb{R}^{n \times n}_{\text{sym}} \to \mathbb{R}, \quad \epsilon \mapsto f^{(0)}(\epsilon) = f(\epsilon)$$

and inductively $f^{(k+1)}: \mathbb{R}^{n \times n}_{\text{sym}} \to \mathbb{R}$ for $k \in \mathbb{N}_0$ by

$$f^{(k+1)}(\epsilon) = \inf\{\lambda f^{(k)}(\xi) + (1-\lambda) f^{(k)}(\eta):$$
$$\lambda \in [0,1], \ \xi, \eta \text{ compatible}, \ \epsilon = \lambda \xi + (1-\lambda)\eta\}.$$

Then, $f^{(k)} \geq W^{rc}$ for all k. If one can find a closed expression for small k, then one can check, for example, based on Remark 3, whether $f^{(k)}$ is symmetric polyconvex. If this is the case, then $f^{rc} = f^{qc} = f^{pc}$ and a characterization has been found.

We illustrate this approach in Sect. 3.3 for the compatible and the incompatible case of the two-well problem in two dimensions.

2.2.2 Construction of a Lower Bound and a Matching Upper Bound

In some applications, the translation method, which originates in the theory of homogenization and of optimal bounds, has been applied successfully as an alternative to the polyconvexity of the lower bound. A general discussion of the translation methods and various examples can be found in [20], where it is also shown that in some cases the translation method can provide better lower bounds than polyconvexity.

The translation method considers that for a (symmetric) quasiconvex function q the inequality $f - q \geq (f - q)^{**}$, which leads to the bound

$$f \geq (f - q)^{**} + q.$$

The right-hand side is (symmetric) quasiconvex as the sum of a convex and a (symmetric) quasiconvex function and by construction a lower bound for f^{qc}. Since this function is a lower bound for all q (symmetric) quasiconvex, one can try to find an optimal q, which may depend on a specific argument as well. One of the main difficulties is a lack of a good description of all (symmetric) quasiconvex functions. In applications, one considers a subset of this set which consists of all null-Lagrangians and all quadratic (symmetric) rank-one convex functions. As null-Lagrangians are linear combinations of minors [18, 19], and no nontrivial minor can be written in terms of the symmetric part of the matrix alone, no nontrivial null-Lagrangians exist [9], and one has only (symmetric) quadratic rank-one convex functions as a standard choice for the translation method. We illustrate this approach in Sect. 3.1.3 following [12] for the two-well problem in two dimensions, where the function $q(\epsilon) = -\det(\epsilon)$ is chosen as symmetric rank-one convex translation.

2.2.3 Concept of the Algorithm

Our algorithm addresses the approximation problem by providing an upper bound for the symmetric rank-one convex envelope of the energy. Given suitable growth conditions that are often met in problems in elasticity, successive relaxation along symmetric rank-one lines is known to converge to the symmetric rank-one convex envelope [25, Eq. (5.15)–(5.16)], see also [19, Th. 6.10].

Proposition 1 *Suppose that $f \colon \mathbb{R}^{n \times n}_{\mathrm{sym}} \to \mathbb{R}$ is given and that there exists a symmetric rank-one convex function $m \colon \mathbb{R}^{n \times n}_{\mathrm{sym}} \to \mathbb{R}$ with $m \leq f$ on $\mathbb{R}^{n \times n}_{\mathrm{sym}}$. Define the function $f^{(k)}$, $k \in \mathbb{N}_0$, as in Sect. 2.2.1. Then, $\lim_{k \to \infty} f^{(k)} = f^{rc}$.*

Proof By definition, $f^{(k+1)} \leq f^{(k)}$. Let $g := \inf_k f^{(k)} = \lim_{k \to \infty} f^{(k)}$, where the limit is interpreted pointwise. If h is a symmetric rank-one convex function and $h \leq f$, then (by induction) $h \leq f^{(k)}$ for all k. In particular, $m \leq g$ and g is real-valued. By the same argument, $f^{rc} \leq g$. Furthermore, from its definition, one checks that g is symmetric rank-one convex, therefore $g \leq f^{rc}$.

3 The N-Well Problem in Linear Elasticity

A model problem that arises, for example, in models for materials undergoing solid-to-solid phase transformations concerns the N-well problem in linear elasticity, where $N \in \mathbb{N}$ is the number of phases in the low-temperature phase. We use this model in order to illustrate the general concepts described in Sect. 2.2 and to verify the efficiency of the proposed algorithm for relaxation in the linearized setting, see [8] for details of the modelling. The relaxation of this stored energy has attracted a lot of attention and was first investigated for two phases in the context of optimal bounds in homogenization and later in the context of relaxation of energies, see [1–4, 22, 26]. The special case of two phases in $\mathbb{R}^{n \times n}$ with equal elastic moduli was investigated in [24, 30] based, in a first step, on the characterization of the relaxation with fixed volume fractions and then, in a second step, on the minimization in the volume fractions. We mention that even in the two-dimensional case, no general results are available for more than two phases. However, if the elastic moduli are equal and the transformation strains are pairwise strain compatible, then the relaxed energy coincides with the convex envelope, see [8, Result 12.1, p. 215]. Moreover, three phases in two dimensions were investigated based on a tensor of geometric parameters in [21, 29], and bounds in the three-dimensional setting were derived in [32]. In more general situations, the algorithm described in this chapter offers a unique opportunity to investigate the structure of the phase diagram.

To fix notation, suppose that an elastic material undergoes a phase transformation in the solid state and that the low temperature phase has N different phases which are characterized by the symmetric stress-free transformation strains $\epsilon_i^\top \in \mathbb{R}^{n \times n}_{\text{sym}}$ and the elastic moduli α_i, symmetric and positive definite tensors of fourth order. If the minimum of the energy in $\epsilon_i^\top$ is denoted by w_i, then the free energy for the ith phase is given by

$$W_i(\epsilon) = \frac{1}{2} \langle \alpha_i (\epsilon - \epsilon_i^\top), \epsilon - \epsilon_i^\top \rangle + w_i,$$

and the system is governed by the minimum of these energies,

$$W(\epsilon) = \min_{i=1,\ldots,N} W_i(\epsilon_i).$$

3.1 Chenchiah–Bhattacharya Relaxation Result in Two Dimensions

Since we present numerical simulations with the implementation of our algorithm for two phases in two dimensions only, we recall the main results in [12] in the case of cubic elasticity in two dimensions using the same notation for the convenience of the reader. Here, cubic elasticity refers to the situation that the elastic behavior of the material depends on three parameters, the bulk modulus κ, the diagonal shear modulus μ, and the off-diagonal shear modulus η. The theoretical results are used to evaluate the efficiency of the proposed algorithm.

3.1.1 Definitions in the Two-Dimensional Case

The orthogonal projections $\Lambda_h, \Lambda_d, \Lambda_o \colon \mathbb{R}^{2\times 2}_{\text{sym}} \to \mathbb{R}^{2\times 2}_{\text{sym}}$ onto the associated linear spaces are defined by

$$\Lambda_h \epsilon = \frac{\epsilon_{11} + \epsilon_{22}}{2}(e_1 \otimes e_1 + e_2 \otimes e_2), \quad R(\Lambda_h) = \text{span}\{e_1 \otimes e_1 + e_2 \otimes e_2\},$$

$$\Lambda_d \epsilon = \frac{\epsilon_{11} - \epsilon_{22}}{2}(e_1 \otimes e_1 - e_2 \otimes e_2), \quad R(\Lambda_d) = \text{span}\{e_1 \otimes e_1 - e_2 \otimes e_2\},$$

$$\Lambda_o \epsilon = \epsilon_{12}(e_1 \otimes e_2 + e_2 \otimes e_1), \quad R(\Lambda_o) = \text{span}\{e_1 \otimes e_2 + e_2 \otimes e_1\},$$

where R denotes the range of the operators and e_1, e_2 the standard basis in $\mathbb{R}^2$. Thus, $\Lambda_h + \Lambda_d + \Lambda_o = \text{Id}$ is the identity on $\mathbb{R}^{2\times 2}_{\text{sym}}$, and the elastic tensor $\alpha = \alpha(\kappa, \mu, \eta)$ can be viewed as a mapping

$$\alpha \colon \mathbb{R}^{2\times 2}_{\text{sym}} \to \mathbb{R}^{2\times 2}_{\text{sym}}, \quad \epsilon \mapsto (2\kappa \Lambda_h + 2\mu \Lambda_d + 2\eta \Lambda_o)\epsilon. \tag{1}$$

We assume $\kappa, \mu, \eta > 0$. Finally, denote by $T \colon \mathbb{R}^{n\times n}_{\text{sym}} \to \mathbb{R}^{n\times n}_{\text{sym}}$ the mapping

$$\epsilon \mapsto T\epsilon = \epsilon - (\text{tr}\,\epsilon)\text{Id} = (-\Lambda_h + \Lambda_d + \Lambda_o)(\epsilon),$$

which represents the quadratic form $-2\det(\cdot)$, that is, for a symmetric matrix ϵ, $T\epsilon = -\text{cof}\,\epsilon$ and $\langle T\epsilon, \epsilon\rangle = -2\det \epsilon$.

3.1.2 Minimization with Fixed Volume Fractions

The translation method has close connections to the theory of homogenization, and the adaption of the method is based on the idea that one tries in a first step to find the minimal energy under the assumption that the energies W_1 and W_2 are used on fixed volume fractions λ_1 and λ_2, respectively. In view of Remark 1, we fix $\Omega = (0, 1)^2$. The subsets of Ω, on which one uses W_1 and W_2, are defined by a phase function $\chi \in L^\infty(\Omega; \{0, 1\}^2)$ with $\chi_1 \chi_2 = 0$ and $\chi_1 + \chi_2 = 1$, that is, the subset on which W_i is used is given by $\chi_i = 1$ and $\chi_{i+1} = 0$, where one computes indices modulo 2. The average of χ on Ω is denoted by $\langle \chi \rangle \in \mathbb{R}^2$, and we assume $\langle \chi \rangle \notin \{e_1, e_2\}$. Fix $\bar{\epsilon} \in \mathbb{R}^{2\times 2}_{\text{sym}}$. The task is to find for fixed $\lambda \in [0, 1]^2$ with $\lambda_1 + \lambda_2 = 1$ the density (formula (1.6) in [12])

$$\overline{W}_\lambda(\bar{\epsilon}) = \inf_{\langle \chi \rangle = \lambda} \inf_{u|_{\partial\Omega} = \bar{\epsilon}x} \int_{(0,1)^2} \sum_{i=1}^{2} \chi_i(x) W_i(\epsilon(x))\,dx,$$

where $\epsilon = \frac{1}{2}(\nabla u + \nabla u^T)$. This is achieved by providing a lower bound based on a suitable translation of the energy with the determinant and a construction

Numerical Study of Microstructures in Linear Elasticity

which shows that the lower bound can be (asymptotically) realized by admissible microstructures.

3.1.3 The Lower Bound Based on Translation with the Determinant

In order to obtain the lower bound, one first rewrites the integral in the definition of $\overline{W}_\lambda(\bar{\epsilon})$ as

$$\int_{(0,1)^2} \sum_{i=1}^{2} \chi_i(x)[(W_i + \beta \det)(\epsilon) - \beta \det \epsilon] dx$$

$$= \sum_{i=1}^{2} \langle \chi_i \rangle \cdot \frac{1}{\langle \chi_i \rangle} \int_{\{\chi_i=1\}} (W_i + \beta \det)(\epsilon) dx - \beta \int_{(0,1)^2} \det \epsilon \, dx \, .$$

For $\beta \geq 0$, the function $\epsilon \mapsto -\beta \det \epsilon$ is symmetric polyconvex, so that the last term is not smaller than $-\beta \det \bar{\epsilon}$. If β is additionally chosen so that $W_i + \beta \det$ is convex for both $i = 1$ and $i = 2$, then one can apply Jensen's inequality in the first term. Letting $\epsilon_i = \langle \chi_i \epsilon \rangle / \langle \chi_i \rangle$, and using that the boundary data imply $\lambda_1 \epsilon_1 + \lambda_2 \epsilon_2 = \bar{\epsilon}$, one obtains (formula (3.2) in [12])

$$\overline{W}_\lambda(\bar{\epsilon}) \geq \max_{\substack{\beta \geq 0 \\ W_{1,2}+\beta \det \text{convex}}} \min_{\substack{\epsilon_1, \epsilon_2 \in \mathbb{R}^{2\times 2}_{\text{sym}} \\ \lambda_1 \epsilon_1 + \lambda_2 \epsilon_2 = \bar{\epsilon}}} \sum_{i=1}^{2} \lambda_i (W_i + \beta \det)(\epsilon_i) - \beta \det(\bar{\epsilon}). \quad (2)$$

We first determine the set of admissible β. First, we need $\beta \geq 0$. As $W_i + \beta \det$ is a polynomial of degree two, it is convex if and only if its quadratic part

$$\frac{1}{2} \langle \alpha_i \epsilon, \epsilon \rangle + \beta \det \epsilon = \frac{1}{2} \langle \alpha_i \epsilon, \epsilon \rangle - \frac{1}{2} \beta \langle T\epsilon, \epsilon \rangle$$

is positive semidefinite. Equivalently, the operator

$$S_i(\beta) := \alpha_i - \beta T = (2\kappa_i + \beta)\Lambda_h + (2\mu_i - \beta)\Lambda_d + (2\eta_i - \beta)\Lambda_o$$

needs to be positive definite for $i = 1$ and $i = 2$. Recalling that Λ_h, Λ_d, and Λ_o are orthogonal projections onto orthogonal subspaces, this is the same as $0 \leq \beta \leq \gamma := \gamma(\alpha_1, \alpha_2) := \min\{2\mu_1, 2\mu_2, 2\eta_1, 2\eta_2\}$.

We then minimize in ϵ_i. The expression in (2) is a maximum in β of

$$W_\lambda(\beta, \bar{\epsilon}) = \min_{\substack{\epsilon_1, \epsilon_2 \in \mathbb{R}^{2\times 2}_{\text{sym}} \\ \lambda_1 \epsilon_1 + \lambda_2 \epsilon_2 = \bar{\epsilon}}} \lambda_1 W_1(\epsilon_1) + \lambda_2 W_2(\epsilon_2) + \beta \lambda_1 \lambda_2 \det(\epsilon_2 - \epsilon_1) \, .$$

Writing $\epsilon_1 = \bar{\epsilon} - \lambda_2 \hat{\epsilon}$ and $\epsilon_2 = \bar{\epsilon} + \lambda_1 \hat{\epsilon}$, we obtain with $\hat{\epsilon}$ as independent variable

$$W_\lambda(\beta, \bar{\epsilon}) = \min_{\hat{\epsilon} \in \mathbb{R}^{2 \times 2}_{\text{sym}}} \lambda_1 W_1(\bar{\epsilon} - \lambda_2 \hat{\epsilon}) + \lambda_2 W_2(\bar{\epsilon} + \lambda_1 \hat{\epsilon}) + \beta \lambda_1 \lambda_2 \det(\hat{\epsilon}),$$

which can be rewritten, expanding the various terms, as

$$\lambda_1 W_1(\bar{\epsilon}) + \lambda_2 W_2(\bar{\epsilon}) + \lambda_1 \lambda_2 \langle \alpha_2(\bar{\epsilon} - \epsilon_2^T) - \alpha_1(\bar{\epsilon} - \epsilon_1^T), \hat{\epsilon} \rangle + \frac{\lambda_1 \lambda_2}{2} \langle (\lambda_2 \alpha_1 + \lambda_1 \alpha_2 - \beta T) \hat{\epsilon}, \hat{\epsilon} \rangle.$$

We minimize in $\hat{\epsilon}$. The coefficient of the quadratic term is $\lambda_2 S_1 + \lambda_1 S_2$ and hence strictly positive definite for $\beta \in [0, \gamma)$. In this range, the minimizer is

$$\hat{\epsilon} = \Delta \epsilon^*(\beta, \bar{\epsilon}) = (\lambda_2 \alpha_1 + \lambda_1 \alpha_2 - \beta T)^{-1}(\Delta(\alpha \epsilon^T) - (\Delta \alpha)\bar{\epsilon}), \tag{3}$$

where we write briefly $\Delta(x) = x_2 - x_1$, and one finds the optimal decomposition for $\bar{\epsilon}$ as $\bar{\epsilon} = \lambda_1 \epsilon_1^*(\beta, \bar{\epsilon}) + \lambda_2 \epsilon_2^*(\beta, \bar{\epsilon})$ with (formula (3.9) in [12])

$$\epsilon_1^*(\beta, \bar{\epsilon}) = \bar{\epsilon} - \lambda_2 \Delta \epsilon^* = (\lambda_2 \alpha_1 + \lambda_1 \alpha_2 - \beta T)^{-1}((\alpha_2 - \beta T)\bar{\epsilon} - \lambda_2 \Delta(\alpha \epsilon^T)), \tag{4a}$$

$$\epsilon_2^*(\beta, \bar{\epsilon}) = \bar{\epsilon} + \lambda_1 \Delta \epsilon^* = (\lambda_2 \alpha_1 + \lambda_1 \alpha_2 - \beta T)^{-1}((\alpha_1 - \beta T)\bar{\epsilon} + \lambda_1 \Delta(\alpha \epsilon^T)) \tag{4b}$$

and therefore an explicit formula for $W_\lambda(\beta, \bar{\epsilon})$. In particular, (3) implies that $\Delta \epsilon^*(\beta, \epsilon)$ is equal to zero on $[0, \gamma(\alpha_1, \alpha_2))$ if it has a zero in this interval since the matrix which depends on β is invertible on the entire interval.

It remains to maximize W_λ in β since one seeks the maximum of all lower bounds at fixed volume fractions. The important observation [12, Lemma 3.5] is that the map

$$[0, \gamma(\alpha_1, \alpha_2)) \ni \beta \mapsto W_\lambda(\beta, \bar{\epsilon})$$

satisfies for $\Delta \epsilon^*(\beta, \bar{\epsilon}) \neq 0$

$$\frac{\partial W_\lambda}{\partial \beta}(\beta, \bar{\epsilon}) = \lambda_1 \lambda_2 \det(\Delta \epsilon^*(\beta, \bar{\epsilon})), \quad \frac{\partial^2 W_\lambda}{\partial \beta^2}(\beta, \bar{\epsilon}) < 0$$

and is therefore strictly concave. Moreover,

$$\frac{\partial^2 W_\lambda}{\partial \beta^2}(\beta, \bar{\epsilon}) = \lambda_1 \lambda_2 \frac{\partial}{\partial \beta} \det(\Delta \epsilon^*(\beta, \bar{\epsilon})) < 0,$$

and the map $\beta \mapsto -\det(\Delta \epsilon^*(\beta, \bar{\epsilon}))$ is strictly increasing, that is, $\partial_\beta W_\lambda(\cdot, \bar{\epsilon})$ has at most one zero. If $\Delta \epsilon^*(\beta, \bar{\epsilon}) = 0$, then by (3), the function $W_\lambda(\cdot, \bar{\epsilon})$ is constant.

These observations lead to the four different regimes for the maximization of $W_\lambda(\cdot, \overline{\epsilon})$ with Regime 0 being the case with $\Delta \epsilon^* = 0$. Here, $W_\lambda(\cdot, \overline{\epsilon})$ is constant, and one may choose $\beta = 0$. The function $\partial_\beta W_\lambda(\cdot, \overline{\epsilon})$ has no zero, and the maximum is attained in a boundary point if the value of the strictly increasing function $-\det(\Delta \epsilon^*(\beta, \overline{\epsilon}))$ at $\beta = 0$ is positive, Regime I, or if its value at $\beta = \gamma(\alpha_1, \alpha_2)$ is negative, Regime III. In Regime I, $\partial_\beta W_\lambda$ is negative, and the maximum is attained at $\beta = 0$, in Regime III, $\partial_\beta W_\lambda$ is positive and the minimum is attained at $\beta = \gamma(\alpha_1, \alpha_2)$. In the remaining case, Regime II, there exists exactly one zero β_{II}, which provides a maximum for $W_\lambda(\cdot, \overline{\epsilon})$. The situation is summarized in [12, Theorem 2.1]. Define

$$\beta^*(\overline{\epsilon}) = \begin{cases} 0 & \text{if } -\det(\Delta \epsilon^*(\cdot, \overline{\epsilon})) \equiv 0 & \text{Regime 0}, \\ 0 & \text{if } -\det(\Delta \epsilon^*(0, \overline{\epsilon})) > 0 & \text{Regime I}, \\ \gamma(\alpha_1, \alpha_2) & \text{if } -\det(\Delta \epsilon^*(\gamma(\alpha_1, \alpha_2), \overline{\epsilon})) < 0 & \text{Regime III}, \\ \beta_{II} & \text{otherwise} & \text{Regime II}, \end{cases}$$

and the corresponding matrices

$$(\lambda_2 \alpha_1 + \lambda_1 \alpha_2 - \beta^*(\overline{\epsilon})T)\epsilon_1^*(\beta^*(\overline{\epsilon}), \overline{\epsilon}) = (\alpha_2 - \beta^*(\overline{\epsilon})T)\overline{\epsilon} - \lambda_2(\alpha_2 \epsilon_2^\top - \alpha_1 \epsilon_1^\top),$$
$$(\lambda_2 \alpha_1 + \lambda_1 \alpha_2 - \beta^* T(\overline{\epsilon}))\epsilon_2^*(\beta^*(\overline{\epsilon}), \overline{\epsilon}) = (\alpha_1 - \beta^*(\overline{\epsilon})T)\overline{\epsilon} + \lambda_1(\alpha_2 \epsilon_2^\top - \alpha_1 \epsilon_1^\top).$$

Note that the matrix on the left-hand side may not be invertible for $\beta^*(\overline{\epsilon}) = \gamma(\alpha_1, \alpha_2)$. With this notation in place, the expression for the relaxation at fixed volume fractions is given by

$$\overline{W}_\lambda(\overline{\epsilon}) = \sum_{i=1}^2 \lambda_i W_i(\epsilon_i^*(\beta^*(\overline{\epsilon}), \overline{\epsilon})) + \beta^*(\overline{\epsilon}) \lambda_1 \lambda_2 \det(\epsilon_2^*(\beta^*(\epsilon), \overline{\epsilon}) - \epsilon_1^*(\beta^*(\overline{\epsilon}), \overline{\epsilon})),$$

and a lower bound for the relaxation can be obtained by minimization in λ. It remains to verify that this lower bound is optimal in the sense that it can be realized by microstructures with the given volume fractions. Since the optimal strains ϵ_i^* are compatible in Regimes 0, I, and II, this has to be shown in Regime III and is accomplished in [12] with an explicit construction based on second-order laminates. For $\beta = \gamma = \gamma(\alpha_1, \alpha_2)$, at least one of the translated energies $W_i + \gamma \det$ is not strictly convex, and after relabeling the energies we may assume that this is true for $W_1 + \gamma \det$. Therefore, there exists a nontrivial element ϵ_n in the kernel of $\alpha_1 - \gamma T$, and from $\alpha_1 > 0$ and $\gamma > 0$, we obtain $\det \epsilon_n = -\frac{1}{2} \langle T \epsilon_n, \epsilon_n \rangle < 0$. Since in this regime $\det(\Delta \epsilon^*) > 0$, the function $z \mapsto \det(\Delta \epsilon^* + z \epsilon_n) = 0$ has two distinct roots $z_1 < 0 < z_2$. Furthermore, the map $z \mapsto (W_1 + \gamma \det)(\Delta \epsilon^* + z \epsilon_n)$ is affine.

We decompose ϵ_1^* along the direction ϵ_n into a convex combination of two matrices, see Fig. 1,

$$\epsilon^I = \epsilon_1^* - t_1 \epsilon_n, \quad \epsilon^{II} = \epsilon_1^* - t_2 \epsilon_n$$

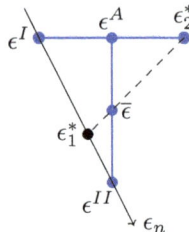

Fig. 1 Sketch of the second-order laminate constructed in Sect. 3.1 for regime III. The dashed line is the (typically not rank one) decomposition of $\bar{\epsilon}$ into ϵ_1^* and ϵ_2^* and the arrow the direction of ϵ_n, along which W_1 is affine

with $t_1 t_2 < 0$. To do so, we seek ϵ^I in such a way that it is rank one connected to ϵ_2^*,

$$\det(\epsilon^I - \epsilon_2^*) = \det(\epsilon_1^* - t_1\epsilon_n - \epsilon_2^*) = \det(\Delta\epsilon^* + t_1\epsilon_n),$$

and hence $t_1 = w_1 \in \{z_1, z_2\}$. The matrix ϵ^{II} is constructed in such a way that it is rank one connected to a convex combination with weight ρ of ϵ^I and ϵ_2^*, $\epsilon^A = \rho\epsilon^I + (1-\rho)\epsilon_2^*$,

$$\det(\epsilon^{II} - (\rho\epsilon^I + (1-\rho)\epsilon_2^*)) = (1-\rho)^2 \det\left(\Delta\epsilon^* + \frac{t_2 - \rho t_1}{1-\rho}\epsilon_n\right) = 0.$$

Therefore, $(t_2 - \rho t_1)/(1-\rho) = w_2 \in \{z_1, z_2\} \setminus \{w_1\}$ and

$$w_2 = \frac{t_2 - \rho t_1}{1-\rho} \quad \Leftrightarrow \quad t_2 = \rho w_1 + (1-\rho)w_2.$$

These rank-one directions are not parallel. Since $\bar{\epsilon} = \lambda_1\epsilon_1^* + \lambda_2\epsilon_2^*$, we need to determine $\rho \in (0, 1)$ and $\mu \in (0, 1)$ in such a way that

$$\bar{\epsilon} = \mu\epsilon^A + (1-\mu)\epsilon^{II} = \mu(1-\rho)\epsilon_2^* + (\mu\rho + (1-\mu))\epsilon_1^* - (\mu\rho t_1 + (1-\mu)t_2)\epsilon_n.$$

Thus, $\lambda_2 = \mu(1-\rho)$, $\lambda_1 = \mu\rho + 1 - \mu = 1 - \lambda_2$,

$$\mu\rho w_1 + (1-\mu)(\rho w_1 + (1-\rho)w_2) = \rho w_1 + (1-\mu)(1-\rho)w_2 = 0,$$

and hence

$$1 - \mu = -\frac{\rho}{1-\rho} \cdot \frac{w_1}{w_2}.$$

We now find

$$\lambda_1 = \mu\rho + 1 - \mu = \rho\left[1 + \frac{\rho}{1-\rho} \cdot \frac{w_1}{w_2}\right] - \frac{\rho}{1-\rho} \cdot \frac{w_1}{w_2} = \rho\left[1 - \frac{w_1}{w_2}\right],$$

and then

$$\rho = \lambda_1 \cdot \frac{w_2}{w_2 - w_1}, \quad \frac{w_1}{w_2} = \frac{\rho - \lambda_1}{\rho}, \quad 1 - \mu = \frac{\lambda_1 - \rho}{1 - \rho}, \quad \mu = \frac{1 - \lambda_1}{1 - \rho}.$$

Note that $\rho \in (0, \lambda_1)$. Finally,

$$t_1 = w_1, \quad t_2 = \rho w_1 + (1 - \rho) w_2 = w_2 \left(\rho \cdot \frac{\rho - \lambda_1}{\rho} + 1 - \rho \right) = w_2 (1 - \lambda_1),$$

and since $w_1 w_2 < 0$, ϵ_1^* is a convex combination of two matrices with weights

$$\frac{-t_1}{t_2 - t_1} = \frac{\lambda_1 - \rho}{\lambda_1 (1 - \rho)}, \quad \frac{t_2}{t_2 - t_1} = \frac{(1 - \lambda_1)\rho}{\lambda_1 (1 - \rho)}.$$

For $w_1 = z_2$ and $w_2 = z_1$, this is exactly the construction reported in [12], which contains a typo in the definition of ρ after [12, Lemma 3.12].

It remains to verify that this second-order laminate coincides with the lower bound on the energy obtained through translation [12, Proof of Theorem 3.10]. By definition,

$$f^{(2)} \leq \frac{\lambda_1 - \rho}{1 - \rho} W_1(\epsilon^{II}) + \frac{\lambda_2}{1 - \rho} (\rho W_1(\epsilon^{I}) + (1 - \rho) W_2(\epsilon_2^*))$$

$$= \lambda_1 \left[\frac{\lambda_1 - \rho}{\lambda_1 (1 - \rho)} (W_1 + \gamma \det)(\epsilon^{II}) + \frac{\lambda_2 \rho}{\lambda_1 (1 - \rho)} (W_1 + \gamma \det)(\epsilon^{I}) \right]$$

$$+ \lambda_2 W_2(\epsilon_2^*) - \gamma \left[\frac{\lambda_1 - \rho}{1 - \rho} \det(\epsilon^{II}) + \frac{\lambda_2 \rho}{1 - \rho} \det(\epsilon^{I}) \right].$$

The translated energy $W_1 + \gamma \det$ is affine along the direction ϵ_n, and the first expression in the square brackets is equal to $(W_1 + \gamma \det)(\epsilon_1^*)$. Moreover, the determinant is affine along rank-one directions, and hence

$$\det(\overline{\epsilon}) = \frac{\lambda_1 - \rho}{1 - \rho} \det(\epsilon^{II}) + \frac{1 - \lambda_1}{1 - \rho} ((1 - \rho) \det(\epsilon_2^*) + \rho \det(\epsilon^{I}))$$

$$= \left[\frac{\lambda_1 - \rho}{1 - \rho} \det(\epsilon^{II}) + \frac{\lambda_2 \rho}{1 - \rho} \det(\epsilon^{I}) \right] + \lambda_2 \det(\epsilon_2^*).$$

These identities lead to

$$f^{(2)} \leq \lambda_1 W(\epsilon_1^*) + \gamma \lambda_1 \det(\epsilon_1^*) + \lambda_2 W(\epsilon_2^*) + \gamma \lambda_2 \det(\epsilon_2^*) - \gamma \det(\overline{\epsilon}).$$

The same calculation that leads starting with (2) to the formula for $W_{\lambda,\beta}$ implies that the right-hand side is equal to the translated energy. We remark that the same type of laminate is identified by the numerical simulation, see Sect. 5 below.

3.2 Kohn's Relaxation Result with Equal Moduli

Kohn [24] discusses a derivation based both on Fourier analysis and on a translation approach with quadratic functions under the assumption that both phases have the same elastic moduli. Let W be given by

$$W(\xi) = \min\{W_1(\xi), W_2(\xi)\},$$

where each phase is a quadratic function of the linear stain ξ,

$$W_i(\xi) = \frac{1}{2} \langle \alpha(\xi - \epsilon_i^\top), \xi - \epsilon_i^\top \rangle + w_i, \quad i = 1, 2.$$

Here, $\epsilon_i^\top$ are the eigenstrains of the two phases and w_i their free energies. Fix $\theta \in [0, 1]$, the volume fraction of phase one described by W_1, see Sect. 3.1.2 for a detailed discussion. The fact that the two phases have the same elastic moduli permits to use Fourier transformation in the definition of the quasiconvex envelope. Using this approach, Kohn [24] has shown that the relaxation Q_θ with fixed volume fraction θ is

$$Q_\theta W(\xi) = \theta W_1(\xi) + (1-\theta) W_2(\xi) - \frac{\theta(1-\theta)}{2} g$$

with

$$g = \max_{|k|=1} \left| \pi_{\alpha^{1/2} V(k)} \alpha^{1/2} (\epsilon_1^\top - \epsilon_2^\top) \right|^2, \tag{5}$$

where π_V denotes the orthogonal projection onto the subspace $V \subset \mathbb{R}^{n \times n}$, and where for $k \in \mathbb{R}^n$, $k \neq 0$, the subspace $V(k)$ is given by

$$V(k) = \{k \otimes v + v \otimes k, \ v \in \mathbb{R}^n\}.$$

An optimization in θ leads to an explicit formula for the relaxed energy.

Theorem 1 ([24, Theorem 3.5]) *The symmetric quasiconvex relaxation of the two-well energy with equal elastic moduli in n dimensions is given for strains $\xi \in \mathbb{R}^{n \times n}_{\text{sym}}$ by the following formulas with g defined in (5):*

1. *if ξ satisfies $W_1(\xi) - W_2(\xi) - \frac{g}{2} \geq 0$, then the optimal value of θ is zero and*

$$W^{qc}(\xi) = W_2(\xi);$$

2. *if ξ satisfies $W_1(\xi) - W_2(\xi) + \frac{g}{2} \leq 0$, then the optimal value of θ is one and*

$$W^{qc}(\xi) = W_1(\xi);$$

3. otherwise, if $|W_1(\xi) - W_2(\xi)| \leq \frac{g}{2}$, then the optimal value of θ is

$$\theta_* = \frac{1}{g}\left(W_2(\xi) - W_1(\xi) + \frac{g}{2}\right),$$

and the value of the relaxed energy is

$$W^{qc}(\xi) = W_2(\xi) - \frac{1}{2g}\left(W_2(\xi) - W_1(\xi) + \frac{g}{2}\right)^2.$$

If ϵ_1^T and ϵ_2^T are compatible, then W^{qc} equals the convex envelope of W [24, Sec. 4]; otherwise, W^{qc} is not convex and one obtains a closed formula if g can be calculated explicitly. This is the case for isotropic elasticity,

$$\alpha\xi = \kappa(\operatorname{tr}\xi)\operatorname{Id} + 2\mu\xi^D, \quad \xi^D = \xi - \frac{1}{n}(\operatorname{tr}\xi)\operatorname{Id}$$

with bulk modulus κ and shear modulus μ. For $n = 2$, this corresponds to $\alpha = 2\kappa\Lambda_h + 2\mu\Lambda_d + 2\mu\Lambda_o$. If $\epsilon_1^T - \epsilon_2^T = \eta_0\operatorname{Id}$, then [24, Prop. 4.4]

$$g = c\eta_0^2 \quad \text{with } c = \frac{\kappa^2 n^3}{\kappa n + 2(n-1)\mu}. \tag{6}$$

In two dimensions, formulas depending on the eigenvalues η_1 and η_2 of $\eta = \epsilon_1^T - \epsilon_1^T$ are available [24, Prop. 4.5]. If $\eta_1\eta_2 \leq 0$, then ϵ_1^T and ϵ_2^T are compatible, and

$$g = |\alpha^{1/2}(\epsilon_1^T - \epsilon_2^T)|^2 = (\kappa - \mu)(\eta_1 + \eta_2)^2 + 2\mu(\eta_1^2 + \eta_2^2). \tag{7}$$

Otherwise, ϵ_1^T and ϵ_2^T are incompatible and

$$g = \frac{\mu^2}{\kappa + \mu}\left(\frac{\kappa}{\mu}|\eta_1 + \eta_2| + |\eta_1 - \eta_2|\right)^2.$$

3.3 Examples

We illustrate the calculation of relaxed energies in two prototypical cases with $\alpha_1 = \alpha_2 = \operatorname{Id}$, that is, using the notation of [12], for

$$\kappa_1 = \kappa_2 = \mu_1 = \mu_2 = \eta_1 = \eta_2 = \frac{1}{2}.$$

In this case, $\gamma(\alpha_1, \alpha_2) = 1$. We fix the average $\bar{\epsilon}$ and discuss briefly both the compatible and the incompatible cases.

3.3.1 The Compatible Case with $\epsilon_1^\top = \text{diag}(1, -1)$ and $\epsilon_2^\top = \text{diag}(-1, 1)$

We use this case to illustrate the general strategy formulated in Sect. 2.2.1 and provide a closed formula for $f^{(1)}$ based on the geometry in the subspace of diagonal matrices, see Fig. 2, and then compare the result with the formulas due to Kohn and Chenchiah and Bhattacharya, respectively. The key observation is that $\Lambda_d \epsilon_i^\top = \epsilon_i^\top$ for $i = 1, 2$, so that the energy can be written as

$$W(\epsilon) = \min\left\{\frac{1}{2}|\epsilon - \epsilon_1^\top|^2, \frac{1}{2}|\epsilon - \epsilon_2^\top|^2\right\}$$

$$= \frac{1}{2}\min\left\{|\Lambda_d\epsilon - \epsilon_1^\top|^2, |\Lambda_d\epsilon - \epsilon_2^\top|^2\right\} + \frac{1}{2}|(\Lambda_h + \Lambda_o)\epsilon|^2 = f_1(\Lambda_d\epsilon) + f_2(\epsilon),$$

that is, as the sum of the nonconvex function f_1, which depends only on the projection Λ_d of ϵ, and the convex function f_2, which depends only on the other two. The image of Λ_d, which is the direction $\Delta\epsilon^\top = \epsilon_2^\top - \epsilon_1^\top$, is a symmetric rank-

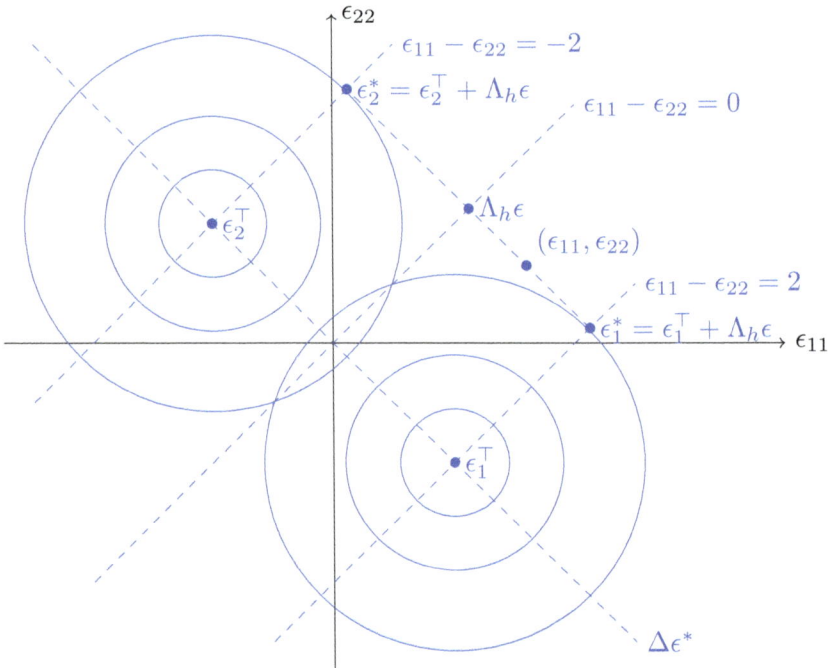

Fig. 2 Construction of the relaxation of the two-well problem in the compatible case shown in the plane of all diagonal matrices: the level sets of the functions W_1 and W_2 in the definition of the energy W are the families of circles centered at $\epsilon_1^\top$ and $\epsilon_2^\top$, respectively. Along the line through $\Lambda_h\epsilon$ with direction $\Delta\epsilon^\top$, the energy W_i is minimal in matrix ϵ_i^*, and the function $\widetilde{W}$ constructed as a candidate for W^{rc} is constant along this line with $\widetilde{W}(\epsilon) = W_i(\epsilon_i^*)$

one direction. Therefore, W^{rc} is not larger than $f_1^{**} + f_2$. As the latter function is convex, we obtain that all envelopes of W coincide with $f_1^{**} + f_2$. The geometry is illustrated in Fig. 2 in the set of symmetric matrices and shows three different regions: for $\epsilon_{11} - \epsilon_{22} \geq 2$, $W = W_1$ is convex, for $\epsilon_{11} - \epsilon_{22} \leq -2$, $W = W_2$ is convex, and for $-2 \leq \epsilon_{11} - \epsilon_{22} \leq 2$, the energy fails to be convex. The level sets of the energies W_1 and W_2 are circles centered at $\epsilon_1^\top$ and $\epsilon_2^\top$, and the energy is minimal in the two points on the line $t \mapsto \epsilon + t\Delta\epsilon^\top$, which are tangential to these circles. This construction leads for $-2 \leq \epsilon_{11} - \epsilon_{22} \leq 2$ to

$$\epsilon_i^* = \epsilon_i^\top + \Lambda_h \epsilon + \Lambda_o \epsilon, \quad W_i(\epsilon_i^*) = \frac{1}{2}|\Lambda_h \epsilon|^2 + \frac{1}{2}|\Lambda_o \epsilon|^2,$$

a convex function. One obtains

$$W^{qc}(\epsilon) = \begin{cases} W_1(\epsilon), & \text{if } \epsilon_{11} - \epsilon_{22} \geq 2, \\ W_i(\Lambda_h \epsilon + \epsilon_i^\top + \Lambda_o \epsilon), & \text{if } -2 \leq \epsilon_{11} - \epsilon_{22} \leq 2, \\ W_2(\epsilon), & \text{if } \epsilon_{11} - \epsilon_{22} \leq -2. \end{cases}$$

Being the sum of the convex functions f_1^{**} and f_2, this is convex and continuous.

Finally, we compare with the formulas in [24] and [12]. According to (7), $g = |\epsilon_1^\top - \epsilon_2^\top|^2 = 8$, and the relaxed energy is given for

$$|W_1(\epsilon) - W_2(\epsilon)| \leq \frac{g}{2} \quad \Leftrightarrow \quad |\langle \epsilon, \epsilon_1^\top - \epsilon_2^\top \rangle| \leq 4 \quad \Leftrightarrow \quad |\epsilon_{11} - \epsilon_{22}| \leq 2$$

by Kohn's formula

$$W_2(\epsilon) - \frac{1}{2g}(W_2(\epsilon) - W_1(\epsilon) + \frac{g}{2})^2$$

$$= \frac{1}{2}|\epsilon|^2 - \frac{1}{2}\langle \frac{\epsilon_1^\top - \epsilon_2^\top}{\|\epsilon_1^\top - \epsilon_2^\top\|}, \epsilon \rangle^2 = \frac{1}{2}|\Lambda_h \epsilon|^2 + \frac{1}{2}|\Lambda_o \epsilon|^2.$$

Using the notation in [12], $\alpha_1 = \alpha_2 = \text{Id}$, $\Delta\alpha = 0$, $\Delta(\alpha\epsilon^\top) = \text{diag}(-2, 2)$, $\gamma = 1$, and for $\beta \in [0, \gamma)$, the optimal ϵ_i^* for the translated energy are given by

$$\epsilon_1^* = \bar{\epsilon} - \frac{\lambda_2}{1-\beta}\text{diag}(-2, 2), \quad \epsilon_2^* = \bar{\epsilon} + \frac{\lambda_1}{1-\beta}\text{diag}(-2, 2).$$

A short calculation leads to

$$\lambda_1 W_1(\epsilon_1^*) + \lambda_2 W_2(\epsilon_2^*) = \frac{1}{2}|\bar{\epsilon}|^2 + 4\frac{\lambda_1\lambda_2}{(1-\beta)^2}(\lambda_1 + \lambda_2) + 1 + (\lambda_2 - \lambda_1)\langle \bar{\epsilon}, \epsilon_1^\top \rangle$$

$$+ 2\frac{\lambda_1\lambda_2}{1-\beta}\langle \text{diag}(-2, 2), \epsilon_1^\top \rangle$$

and

$$\beta\lambda_1\lambda_2 \det(\epsilon_2^* - \epsilon_1^*) = \beta\lambda_1\lambda_2 \det\left(\frac{\lambda_1}{1-\beta}\operatorname{diag}(-2,2) + \frac{\lambda_2}{1-\beta}\operatorname{diag}(-2,2)\right)$$

$$= \beta\lambda_1\lambda_2 \frac{1}{(1-\beta)^2} \det(\operatorname{diag}(-2,2)) = -\frac{4\beta\lambda_1\lambda_2}{(1-\beta)^2},$$

so that we obtain an explicit formula for the relaxed energy,

$$W_\lambda(\beta, \bar{\epsilon}) = \frac{1}{2}|\bar{\epsilon}|^2 - \frac{4\lambda_1\lambda_2}{1-\beta} + 1 + (\lambda_2 - \lambda_1)\langle\bar{\epsilon}, \epsilon_1^T\rangle.$$

One needs to compute first the maximum in $\beta \in [0, 1)$ and then the minimum in $\lambda_1 + \lambda_2 = 1$ to obtain the relaxed energy. The maximum in β is attained at $\beta = 0$, which means that in this situation the translation does not help to identify good constructions or microstructures. We find

$$\lambda_1 = \frac{1}{2} + \frac{1}{4}(\epsilon_{11} - \epsilon_{22}),$$

and λ_1 is an admissible choice as long as $|\epsilon_{11} - \epsilon_{22}| \leq 2$. All three expressions for the relaxed energy are equal.

3.3.2 The Incompatible Case with $\epsilon_1^T = -\operatorname{Id}$ and $\epsilon_2^T = \operatorname{Id}$

To find an upper bound, we construct as in the compatible case the approximation $f^{(1)}$. Fix $\epsilon \in \mathbb{R}^{2\times 2}_{\text{sym}}, a \in \mathbb{S}^1$, and consider the symmetric rank-one line $t \mapsto \epsilon + ta \otimes a$ with parameters $t_1 < 0 < t_2$, that is,

$$\epsilon = \frac{t_2}{t_2 - t_1}(\epsilon + t_1 a \otimes a) - \frac{t_1}{t_2 - t_1}(\epsilon + t_2 a \otimes a),$$

and find the minimum of

$$\frac{t_2}{t_2 - t_1} W_1(\epsilon + t_1 a \otimes a) - \frac{t_1}{t_2 - t_1} W_2(\epsilon + t_2 a \otimes a)$$

$$= \frac{|\epsilon|^2}{2} + 1 + \frac{t_2 + t_1}{t_2 - t_1}\operatorname{tr}\epsilon + \frac{2t_1 t_2}{t_2 - t_1} - \frac{1}{2}t_1 t_2,$$

in a, t_1, and t_2. This expression is in fact independent of a and needs to be minimized in $t_1 \leq 0 \leq t_2$. Since minimization along a symmetric rank-one line corresponds to a convexification along this line, the derivatives of the energies corresponding to the parameters t_1 and t_2 must coincide

$$\langle \epsilon - \epsilon_1^\top + t_1 a \otimes a, a \otimes a \rangle = \langle \epsilon - \epsilon_2^\top + t_2 a \otimes a, a \otimes a \rangle,$$

which gives $t_2 = t_1 + 2$. Inserting in the previous expression yields $t_1 = -(\operatorname{tr}\epsilon + 1)$, and recalling that $|\epsilon|^2 + 2\det\epsilon = (\operatorname{tr}\epsilon)^2$,

$$f^{(1)}(\epsilon) = -\det\epsilon + \frac{1}{2}.$$

The choice of t_1 is admissible if $t_1 = -(\operatorname{tr}\epsilon + 1) \le 0$ and $t_2 = -\operatorname{tr}\epsilon + 1 \ge 0$, and this leads to the restriction $|\operatorname{tr}\epsilon| \le 1$. We compute

$$W_1(\epsilon) = \frac{1}{2}|\epsilon|^2 - \langle \epsilon, \epsilon_1^\top \rangle + \frac{1}{2}|\epsilon_1^\top|^2 = \frac{1}{2}|\epsilon|^2 + \operatorname{tr}\epsilon + 1,$$

which, for $\operatorname{tr}\epsilon = -1$, is the same as $\frac{1}{2}|\epsilon|^2 - \frac{1}{2}(\operatorname{tr}\epsilon)^2 + \frac{1}{2}$. Therefore, $f^{(1)}$ is continuous. In view of Remark 3, we represent $f^{(1)}$ as

$$f^{(1)}(\epsilon) = -\det\epsilon + \begin{cases} \frac{1}{2}(\operatorname{tr}\epsilon)^2 + \operatorname{tr}\epsilon + 1, & \text{if } \operatorname{tr}\epsilon \le -1, \\ \frac{1}{2}, & \text{if } -1 \le \operatorname{tr}\epsilon \le 1, \\ \frac{1}{2}(\operatorname{tr}\epsilon)^2 - \operatorname{tr}\epsilon + 1, & \text{if } \operatorname{tr}\epsilon \le -1. \end{cases}$$

Therefore, $f^{(1)}$ is a convex function of ϵ and $\det\epsilon$ which is nonincreasing in $\det\epsilon$ and therefore symmetric polyconvex.

Also, in this case, we compare with [24] and [12]. According to (6) with $\kappa = \mu = 1/2$, the parameter g is given by $c = 1$ and $g = \eta_0^2 c = 4$, and relaxation is present for $|W_1(\epsilon) - W_2(\epsilon)| \le g/2$, that is, for $|\operatorname{tr}\epsilon| \le 1$. The relaxed energy is

$$W_K^{qc}(\epsilon) = W_2(\epsilon) - \frac{1}{2g}(W_2(\epsilon) - W_1(\epsilon) + \frac{g}{2})^2 = \frac{1}{2}|\epsilon|^2 - \frac{1}{2}(\operatorname{tr}\epsilon)^2 + \frac{1}{2}.$$

To compare with the results in [12], recall $\Delta(\alpha\epsilon^\top) = 2\mathrm{Id}$, $\Delta\alpha = 0$, and the translations are strictly convex for $0 \le \beta < \gamma(\alpha_1, \alpha_2) = 1$. The fourth-order tensor for the computation of the optimal strains is

$$(\lambda_2\alpha_1 + \lambda_1\alpha_2 - \beta T)^{-1} = \frac{1}{1+\beta}\Lambda_h + \frac{1}{1-\beta}\Lambda_d + \frac{1}{1-\beta}\Lambda_o.$$

Invertibility is lost for $\beta = \gamma(\alpha_1, \alpha_2)$. The optimal strains are (see (4a)–(4b))

$$\epsilon_1^*(\beta, \bar\epsilon) = \bar\epsilon - \frac{2\lambda_2}{1+\beta}\mathrm{Id}, \quad \epsilon_2^*(\beta, \bar\epsilon) = \bar\epsilon + \frac{2\lambda_1}{1+\beta}\mathrm{Id}.$$

Following the same calculations as in the compatible case, one finds

$$W_\lambda(\beta, \bar{\epsilon}) = \frac{1}{2}|\bar{\epsilon}|^2 - \frac{4\lambda_1\lambda_2}{1+\beta} + (\lambda_2 - \lambda_1)\langle\bar{\epsilon}, \epsilon_1^\top\rangle + 1.$$

The maximum is attained at $\beta = \gamma = 1$ with

$$\overline{W}_\lambda(\bar{\epsilon}) = \frac{1}{2}|\bar{\epsilon}|^2 - 2\lambda_1\lambda_2 + (\lambda_2 - \lambda_1)\langle\bar{\epsilon}, \epsilon_1^\top\rangle + 1.$$

Since $\lambda_2 = 1 - \lambda_1$, the optimal value for λ_1 is $\lambda_1 = 1/2 + (1/2)\langle\bar{\epsilon}, \epsilon_1^\top\rangle$. This value is admissible if $-1 \leq \mathrm{tr}\,\bar{\epsilon} \leq 1$, otherwise $\lambda_1 \in \{0, 1\}$, and no microstructure is formed. We also note that the derivative of the function $W_\lambda(\beta, \bar{\epsilon})$ is given by (formula (3.13) in [12])

$$\frac{\partial}{\partial \beta} W_\lambda(\beta, 0) = -\lambda_1\lambda_2 \phi(\Delta\epsilon^*(\beta, 0)) = \lambda_1\lambda_2 \det(\Delta\epsilon^*(\beta, 0)) = \frac{4\lambda_1\lambda_2}{(1+\beta)^2}.$$

This expression has a limit as $\beta \to \gamma(\alpha_1, \alpha_2)$ and $\bar{\epsilon}$ is in Regime III since

$$\phi(\Delta\epsilon^*(\gamma(\alpha_1, \alpha_2), \bar{\epsilon})) = -\det(\Delta\epsilon^*(\gamma(\alpha_1, \alpha_2), \bar{\epsilon})) < 0.$$

4 Numerical Relaxation

We employ a numerical algorithm for the determination of lamination convex envelopes that was first presented in [15], and that has been successfully applied to the geometrically nonlinear two-well problem [14], to microstructure in nematic elastomers [15] and to models in finite crystal plasticity [16]. Whereas all these applications to finite elasticity and plasticity were based on a version of the algorithm that incorporates $SO(2)$-invariance, in the present study, we developed a version adapted to geometrically linear elasticity, which incorporates invariance under linearized rotations. We first present the key ideas of the algorithm.

The rank-one convex envelope f^{rc} is approximated numerically using Proposition 1. In the numerical implementation, a small bound k on the largest order of lamination is used, such as $k = 3$. If one can verify that $f^{(k-1)} = f^{(k)}$, then automatically also $f^{rc} = f^{(k-1)}$. Each $f^{(h)}$, $h = 1, \ldots, k$, is determined by computing the optimal laminate of order h. In turn, laminates are determined by a mixture of a global search, with careful storage of information from previous searches, and a local optimization, which is fast and is performed using the Polak–Ribière conjugate gradient algorithm. The global search, which aims at producing good starting points for the local optimization, is the crucial part of the algorithm. It is based on the idea of storing information on laminates once it has been acquired and of transferring this information to neighboring parts of phase space. This permits to obtain good starting conditions for the local optimization also in regions of phase space where the energy is locally rank-one convex and leads to substantially more accurate results than if one were to use random initial conditions, as was demonstrated for a model problem in [15].

Information about the laminates is stored using a hierarchical structure in phase space, which automatically refines in the regions which are explored during the numerical study. This makes crucial use of the variational nature of the problem, which permits to easily compare the quality of two laminates with the same barycenter. Specifically, whenever a good laminate is available for some strain $\bar\epsilon$, then the program rounds it to some $\bar\epsilon_\delta$, projecting each of the components of $\bar\epsilon$ onto $\delta\mathbb{Z}$ for some fixed small δ and then checks if a node for $\bar\epsilon_\delta$ is present and generates it if not. If the node is not present, the new laminate is stored; if the node was already present, and a laminate of the same order was already stored, then the energies of the two are compared; and if the old one has a higher energy than the new one, then it is replaced.

If a good laminate for a strain $\bar\epsilon$ is required, then the algorithm rounds to $\bar\epsilon_\delta$ and then tries to locate it in the data structure. If it is not present, it is searched at higher level virtual grids, composed of cells represented by matrices of the form $2^i \delta \mathbb{Z}$, up to some maximum level. The local optimization is then performed starting with the best initial condition that has been found and the result stored.

Periodically, during the optimization, a systematic improvement of the stored data on laminates is performed. This consists of four steps, which are performed for each node for which information has been stored. In each step, the old laminate is replaced by the new one, if the energy of the new one is smaller. First, a local search is started starting from a few random values of the lamination parameters. Second, a local search is started starting from the values of the neighboring cells. Here, neighboring is interpreted in three senses: cells at the same scale which share at least a vertex with the one considered, cells at the next smaller scale which are contained in the one considered, and the cell at the next higher scale which contains the one considered. Third, the algorithm takes the optimal first-order laminate $\lambda \delta_{\epsilon_1} + (1-\lambda)\delta_{\epsilon_2}$ stored at this cell and tries to improve it with laminates centered in ϵ_1 and ϵ_2, automatically generating higher order laminates.

This data improvement step is rather time consuming, and in practice it is convenient to store the result on disk for later use. It permits to obtain a good approximation of laminates also in regions where the energy is locally (but not globally) rank-one convex, since microstructures are transferred from neighboring regions. The fact that no new grid elements are created guarantees that memory storage (and run time) does not explode during the improvement procedure. The transfer of information between one virtual grid and those at higher and lower levels leads to a very rapid transfer of information over large distances in phase space and also over regions that are not populated at the lower levels. We refer to [15] for further details on this algorithm.

Our numerical code has in all cases we investigated reported that the second-order lamination convex envelope $f^{(2)}$ is rank-one convex, and in particular no matrices have been found where laminates of order three or higher are needed.

5 Numerical Results

As a first verification of the proposed algorithm, we investigated the special situations discussed in Sect. 3.3. The algorithm computes correctly the relaxed energies and finds optimal first-order laminates in both situations, see Fig. 3 for the simulation in the incompatible case.

In view of the discussion in Sect. 3, we identify a situation in which second-order laminates are needed for the relaxation at fixed volume fraction and use our algorithm to investigate whether the full relaxation, which also involves the minimization in the volume fractions, requires second-order laminates. The chemical energy w_i of the ith phase is of no importance if the minimization is carried out at fixed volume fractions since it contributes a constant energy $\lambda_1 w_1 + \lambda_2 w_2$. However, this expression is an affine translation if the minimization in λ is included, and therefore we fix in the following $w_1 = w_2 = 0$.

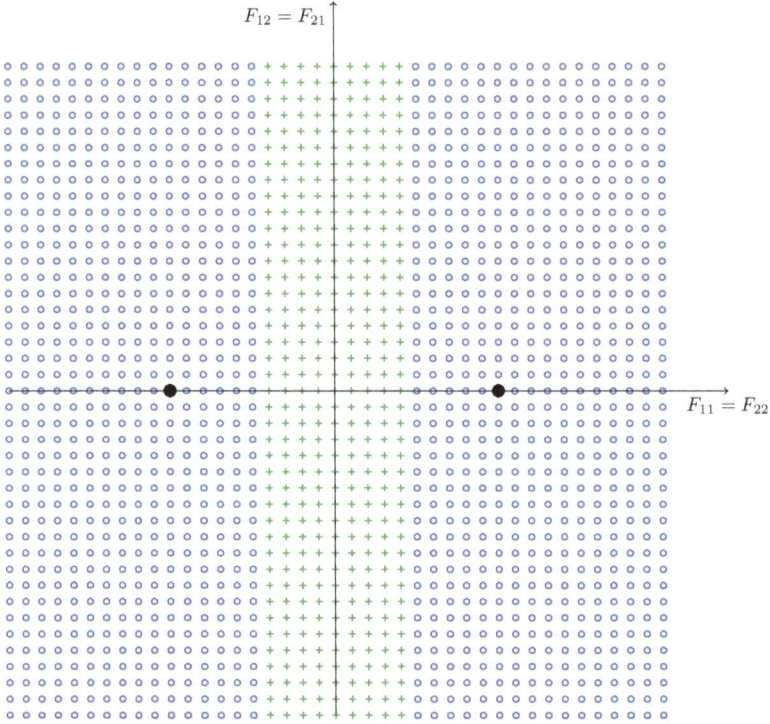

Fig. 3 Numerical phase diagram in the plane of matrices of the form $\begin{pmatrix} x & y \\ y & x \end{pmatrix}$ for the model with equal moduli discussed in Sect. 3.3, which is the same as the model of Sect. 5 with the parameters in (8). The relaxed energy coincides with the unrelaxed energy for $|\operatorname{tr}\epsilon| \geq 1$ and can be obtained with a first-order laminated for $|\operatorname{tr}\epsilon| < 1$. The bullets mark the two minima of the energy

We choose two materials with positive moduli and elasticity tensors $\alpha_1 = \alpha(\kappa_1, \mu_1, \eta_1)$ and $\alpha_2 = \alpha(\kappa_2, \mu_2, \eta_2)$. In a first set of computations, we assume

$$\kappa_1 = \kappa_2 = \frac{1}{2}, \quad \mu_1 = \mu_2 = \frac{1}{2}, \quad \eta_1 = \eta_2 = \frac{1}{2}, \quad w_1 = w_2 = 0, \quad (8)$$

corresponding to the simpler, equal-moduli case analyzed by Kohn and discussed in Sect. 3.2. In a second set of computations, we use different elastic moduli,

$$\kappa_1 = \kappa_2 = \frac{1}{2}, \quad \mu_1 = \mu_2 = \frac{1}{2}, \quad \frac{1}{10} = \eta_2 < \eta_1 = 1, \quad w_1 = w_2 = 0. \quad (9)$$

Figure 4 shows a typical numerical phase diagram that can be obtained with the algorithm. We represent areas in which the relaxed and the unrelaxed energies

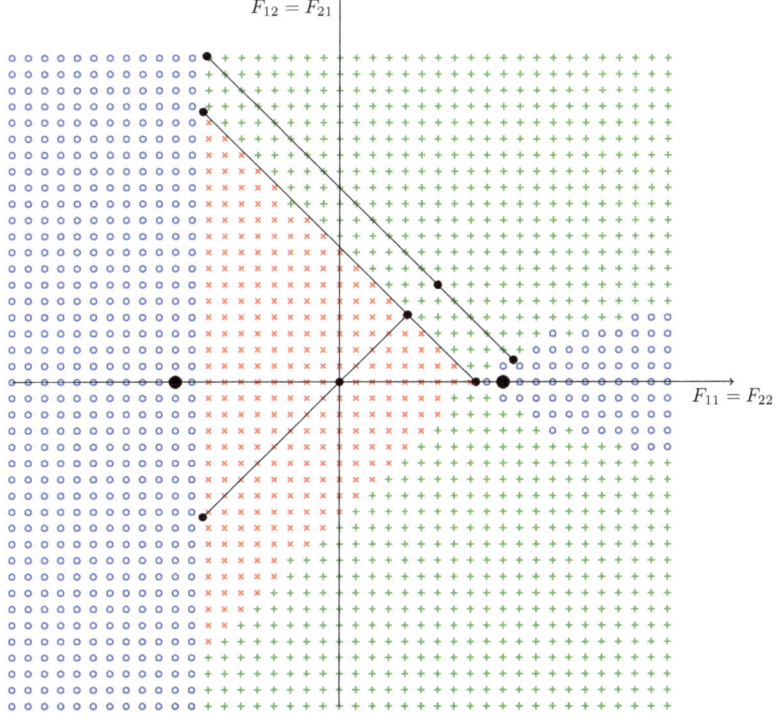

Fig. 4 Numerical phase diagram with the parameters in (9) in the plane of matrices of the form $\begin{pmatrix} x & y \\ y & x \end{pmatrix}$. The two larger dots mark the transformation strains, $\epsilon_{1,2}^T = \pm \mathrm{Id}$. Three distinct regions emerge in which the relaxed energy coincides with the energy (blue area) in which it can be calculated with first-order (green area) and second-order (red area) laminates, respectively. The solid lines indicate the laminates computed numerically for the matrices 0 and $\begin{pmatrix} 0.6 & 0.6 \\ 0.6 & 0.6 \end{pmatrix}$. They are contained, within numerical precision, in this plane. The optimal laminate in the region in which second-order laminates emerge is supported on three points as predicted in [12], see Sect. 3.3

coincide with blue dots, regions in which the relaxed energy is obtained with first-order laminates with green dots, and regions with second-order laminates with red dots, respectively. Optimal laminates of first and second order are shown as well. The numerical results are in prefect agreement with the theory developed in [12] and allow us to make predictions about the full relaxation of the energy, not only the relaxation at fixed volume fractions.

Figure 5 shows a different cut through the three-dimensional space $\mathbb{R}^{2\times 2}_{\mathrm{sym}}$, $\begin{pmatrix} x & y \\ y & x+0.5 \end{pmatrix}$. The general structure of the phase diagram is similar, showing stability on a scale which is smaller than the distance between the two minima of the energy.

The effect of the relaxation with first- and second-order laminates, that is, numerical approximations of the functions $f^{(1)}$ and $f^{(2)}$, is demonstrated in Fig. 6. The line plotted in this figure corresponds to the $y = 0$ line in Fig. 4. The curve for first-order laminates is not symmetric since $\eta_1 \neq \eta_2$, and symmetry is recovered for second-order laminates. Figure 7 shows a different cut through phase space.

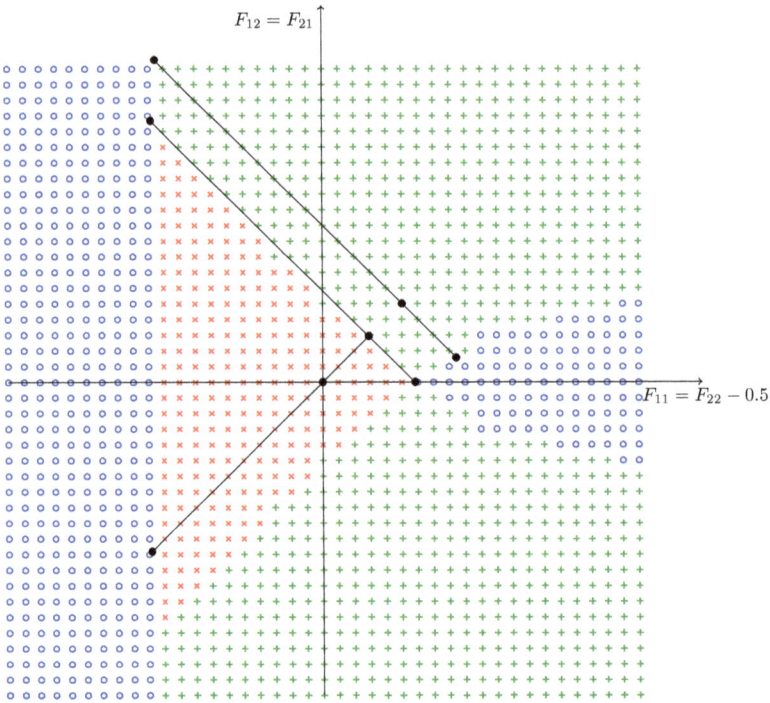

Fig. 5 Phase diagram in the plane $\begin{pmatrix} x & y \\ y & x+0.5 \end{pmatrix}$, and decomposition of the matrix $\begin{pmatrix} 0 & 0 \\ 0 & 0.5 \end{pmatrix}$, for the parameters in (9). The laminate turns out to be, to numerical precision, completely supported in the plane displayed in the picture

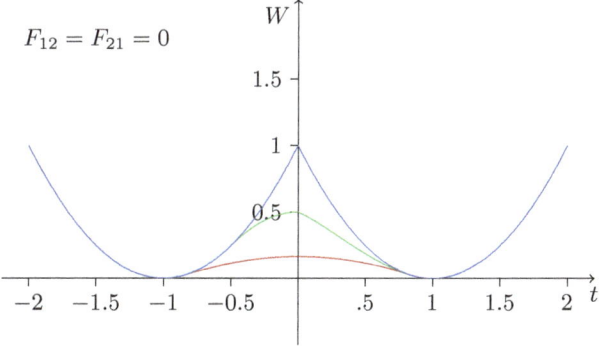

Fig. 6 Numerical approximation of the relaxed energy with parameters as in Eq. (9) with first-order (green curve) and second-order (red curve) laminates, along the rank-two line $F = \text{diag}(t, t)$. The blue curve represents the unrelaxed energy

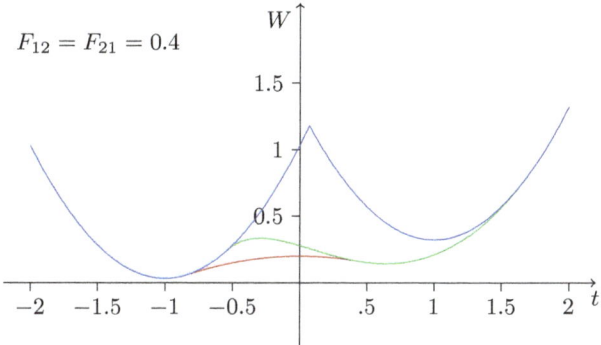

Fig. 7 Numerical approximation of the relaxed energy with parameters as in Eq. (9) with first-order (green curve) and second-order (red curve) laminates, along the rank-two line $\begin{pmatrix} t & 0.4 \\ 0.4 & t \end{pmatrix}$. The blue curve represents the unrelaxed energy

We finally consider the macroscopic mechanical response of the system, in the sense of a stress–strain curve. We focus for simplicity on the rank-one line $\text{diag}(t, 0)$. In Fig. 8, we show the response for the equal-moduli case. The full relaxation is obtained with first-order laminates, which make the stress–strain monotone. Figure 9 shows the corresponding result for the case of unequal moduli. We see that the first lamination convex envelope $f^{(1)}$ still has a non-monotone dependence of strain on t. Indeed, in this case, second-order laminates are needed to obtain the relaxation of the energy. This is also apparent from the corresponding plots of the energies, which are shown in Fig. 10.

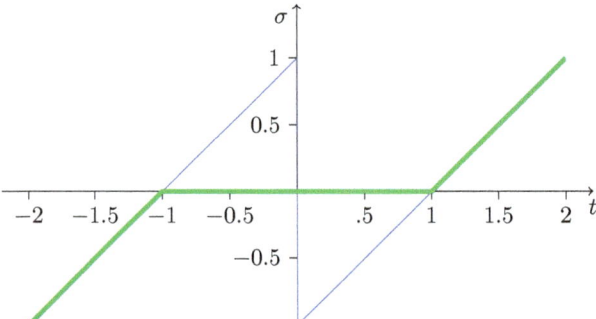

Fig. 8 Numerical approximation of the stress–strain curve for a uniaxial deformation along $F = t e_1 \otimes e_1$ and the material parameters of Eq. (8). The blue curve corresponds to the unrelaxed energy and the green one to the first lamination convex envelope which for this curve to numerical precision coincides with the relaxed energy

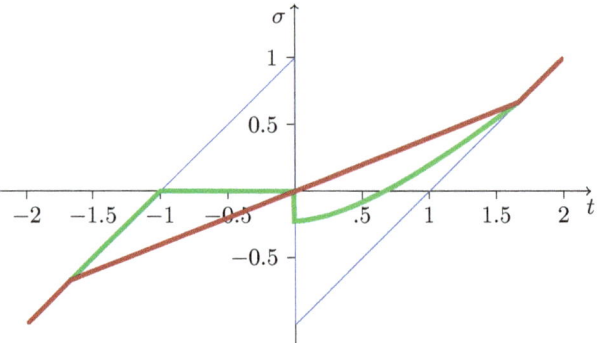

Fig. 9 Numerical approximation of the stress–strain curve for a uniaxial deformation along $F = t e_1 \otimes e_1$ and the material parameters of Eq. (9). The blue curve corresponds to the unrelaxed energy, the green one to the first lamination convex envelope, and the red one to the second lamination convex envelope, which to numerical precision coincides with the relaxed energy

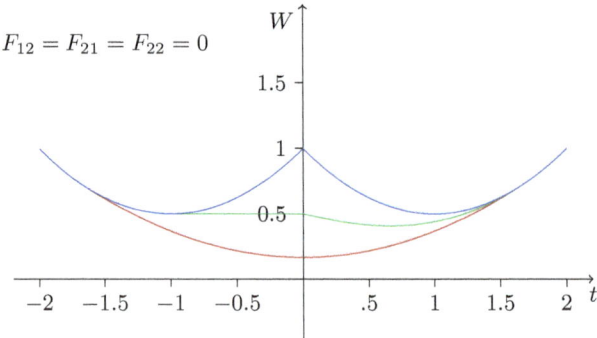

Fig. 10 Energy along the same line as Fig. 9

6 Conclusions

We presented an extension of the algorithm for the numerical computation of relaxed energies to the case if linear elasticity. As in the case of nonlinear elasticity, we obtained numerical phase diagrams that allow excellent predictions about the structure of the relaxed energy in situations in which analytical formulas are not available.

At variance with explicit numerical relaxation formulas, our approach does not depend on the details of the functional form of the energy chosen. In particular, it would immediately apply to non-quadratic energies, to problems with three or more wells, or to situations where the different minima do not have the same energy. An extension to three dimensions would be very interesting but is probably challenging, as the local minimization along rank-one lines involves more dimensions, and in particular the phase space, in which one has to store good laminates, would have six instead of three dimensions.

This algorithm also offers a unique opportunity for an integration with a finite element simulation and thus an efficient simulation of relaxed variational problems in solid mechanics.

Acknowledgments This work was partially supported by the Deutsche Forschungsgemeinschaft (DFG German Research Foundation) through SFB 1060, *"The mathematics of emergent effects,"* project 211504053, and SPP 2256 *"Variational Methods for Predicting Complex Phenomena in Engineering Structures and Materials,"* projects 441211072 and project 441468770.

References

1. Allaire, G., Kohn, R.V.: Explicit optimal bounds on the elastic energy of a two-phase composite in two space dimensions. Quart. Appl. Math. **51**(4), 675–699 (1993). https://doi.org/10.1090/qam/1247434
2. Allaire, G., Kohn, R.V.: Optimal bounds on the effective behavior of a mixture of two well-ordered elastic materials. Quart. Appl. Math. **51**(4), 643–674 (1993). https://doi.org/10.1090/qam/1247433
3. Allaire, G., Kohn, R.V.: Optimal lower bounds on the elastic energy of a composite made from two non-well-ordered isotropic materials. Quart. Appl. Math. **52**(2), 311–333 (1994). https://doi.org/10.1090/qam/1276240
4. Allaire, G., Lods, V.: Minimizers for a double-well problem with affine boundary conditions. Proc. Roy. Soc. Edinburgh Sect. A **129**(3), 439–466 (1999). https://doi.org/10.1017/S0308210500021454
5. Ball, J.M., James, R.D.: Fine phase mixtures as minimizers of the energy. Arch. Ration. Mech. Analy. **100**, 13–52 (1987)
6. Ball, J.M., James, R.D.: Proposed experimental tests of a theory of fine microstructure and the two-well problem. Phil. Trans. R. Soc. Lond. A **338**, 389–450 (1992)
7. Ball, J.M., Murat, F.: $W^{1,p}$-quasiconvexity and variational problems for multiple integrals. J. Funct. Anal. **58**(3), 225–253 (1984). https://doi.org/10.1016/0022-1236(84)90041-7
8. Bhattacharya, K.: Microstructure of martensite. In: Oxford Series on Materials Modelling. Oxford University Press, Oxford (2003)

9. Boussaid, O., Kreisbeck, C., Schlömerkemper, A.: Characterizations of symmetric polyconvexity. Arch. Ration. Mech. Anal. **234**(1), 417–451 (2019). https://doi.org/10.1007/s00205-019-01395-4
10. Carstensen, C., Hackl, K., Mielke, A.: Non-convex potentials and microstructures in finite-strain plasticity. R. Soc. Lond. Proc. Ser. A **458**(2018), 299–317 (2002)
11. Cesana, P., DeSimone, A.: Quasiconvex envelopes of energies for nematic elastomers in the small strain regime and applications. J. Mech. Phys. Solids **59**(4), 787–803 (2011)
12. Chenchiah, I.V., Bhattacharya, K.: The relaxation of two-well energies with possibly unequal moduli. Arch. Ration. Mech. Anal. **187**(3), 409–479 (2008). https://doi.org/10.1007/s00205-007-0075-3
13. Chipot, M., Kinderlehrer, D.: Equilibrium configurations of crystals. Arch. Rational Mech. Anal. **103**, 237–277 (1988)
14. Conti, S., Dolzmann, G.: Relaxation of a model energy for the cubic to tetragonal phase transformation in two dimensions. Math. Models. Methods App. Sci. **24**, 2929–2942 (2014). https://doi.org/10.1142/S0218202514500419
15. Conti, S., Dolzmann, G.: An adaptive relaxation algorithm for multiscale problems and application to nematic elastomers. J. Mech. Phys. Solids **113**, 126–143 (2018). https://doi.org/10.1016/j.jmps.2018.02.001
16. Conti, S., Dolzmann, G.: Numerical study of microstructures in single-slip finite elastoplasticity. J. Optim. Theory Appl. **184**, 43–60 (2020). https://doi.org/10.1007/s10957-018-01460-0
17. Conti, S., Ortiz, M.: Dislocation microstructures and the effective behavior of single crystals. Arch. Rat. Mech. Anal. **176**, 103–147 (2005). https://doi.org/10.1007/s00205-004-0353-2
18. Conti, S., Dolzmann, G., Kirchheim, B., Müller, S.: Sufficient conditions for the validity of the Cauchy-Born rule close to $SO(n)$. J. Eur. Math. Soc. (JEMS) **8**, 515–530 (2006). https://doi.org/10.4171/JEMS/65
19. Dacorogna, B.: Direct Methods in the Calculus of Variations, vol. 78. Springer, Berlin (2007)
20. Firoozye, N.B.: Optimal use of the translation method and relaxations of variational problems. Comm. Pure Appl. Math. **44**(6), 643–678 (1991). https://doi.org/10.1002/cpa.3160440603
21. Firoozye, N.B., Kohn, R.V.: Geometric parameters and the relaxation of multiwell energies. In: Microstructure and Phase Transition. IMA Volumes in Mathematics and Its Applications, vol. 54, pp. 85–109. Springer, New York (1993). https://doi.org/10.1007/978-1-4613-8360-4_6
22. Grabovsky, Y.: Bounds and extremal microstructures for two-component composites: a unified treatment based on the translation method. Proc. Roy. Soc. London Ser. A **452**(1947), 919–944 (1996). https://doi.org/10.1098/rspa.1996.0046
23. Khan, M.S., Hackl, K.: Modeling of microstructures in a Cosserat continuum using relaxed energies. In: Trends in Applications of Mathematics to Mechanics. Springer INdAM Series, vol. 27, pp. 103–125. Springer, Cham (2018)
24. Kohn, R.V.: The relaxation of a double-well energy. Contin. Mech. Thermodyn. **3**(3), 193–236 (1991). https://doi.org/10.1007/BF01135336
25. Kohn, R.V., Strang, G.: Optimal design and relaxation of variational problems. II. Comm. Pure Appl. Math. **39**, 139–182 (1986). https://doi.org/10.1002/cpa.3160390202
26. Lurie, K.A., Cherkaev, A.V.: On a certain variational problem of phase equilibrium. In: Material Instabilities in Continuum Mechanics (Edinburgh, 1985–1986), pp. 257–268. Oxford Science Publlication, Oxford University Press, New York (1988)
27. Morrey, Jr., C.B.: Quasi-convexity and the lower semicontinuity of multiple integrals. Pacific J. Math. **2**, 25–53 (1952)
28. Ortiz, M., Repetto, E.A.: Nonconvex energy minimization and dislocation structures in ductile single crystals. J. Mech. Phys. Solids **47**(2), 397–462 (1999)
29. Palombaro, M., Smyshlyaev, V.P.: Relaxation of three solenoidal wells and characterization of extremal three-phase H-measures. Arch. Ration. Mech. Anal. **194**(3), 775–722 (2009). https://doi.org/10.1007/s00205-008-0204-7

30. Pipkin, A.C.: Elastic materials with two preferred states. Quart. J. Mech. Appl. Math. **44**(1), 1–15 (1991). https://doi.org/10.1093/qjmam/44.1.1
31. Šverák, V.: New examples of quasiconvex functions. Arch. Rational Mech. Anal. **119**(4), 293–300 (1992). https://doi.org/10.1007/BF01837111
32. Tang, Q., Zhang, K.: Bounds for effective strains of geometrically linear elastic multiwell model. J. Math. Anal. Appl. **339**(2), 1264–1276 (2008). https://doi.org/10.1016/j.jmaa.2007.07.051

Surface Shear Waves in a Functionally Graded Half-Space

Andrey Sarychev, Alexander Shuvalov, and Marco Spadini

1 Introduction

The paper deals with the problem of solvability of the bi-parametric Sturm-Liouville equation on a half-line with the Neumann initial condition and the condition of vanishing at infinity. In the physical context, this is formally a problem of existence of time-harmonic surface shear waves propagating with a given frequency ω and the tangential wave number k along the traction-free boundary of a functionally graded semi-infinite medium with continuously depth-dependent density and shear modulus. Surface acoustic waves are relevant to various fields of physics and applications [7]. The model of functionally graded media implies the materials with a continuous spatial variation of their properties which occurs for natural reasons or is purposely manufactured in order to realize a desired physical behavior [6, 15].

It is known that the boundary-value problem in hand does not admit a solution for a generic couple (ω, k); in fact, solutions may exist on a certain a union of a number of eigencurves $\omega(k)$ [3, Ch.6], which are referred to as dispersion branches in the context of physics. The purpose of the present paper is to study the conditions for existence of the surface shear waves and in particular to characterize the admissible (ω, k) pairs.

The situation is elementary when $\rho(y)$ and $\mu(y)$ are constant (in fact, then the shear surface wave does not exist). There is ample literature regarding the case of $\rho(y)$ and $\mu(y)$ being periodic functions, see [18] and bibliography therein. Much work on the shear surface waves has been done for the settings where $\rho(y)$ and

---

A. Sarychev · M. Spadini (✉)
DiMaI, Università di Firenze, Firenze, Italy
e-mail: andrey.sarychev@unifi.it; marco.spadini@unifi.it

A. Shuvalov
Université de Bordeaux, Talence, France
e-mail: alexander.shuvalov@u-bordeaux.fr

$\mu(y)$ are constant everywhere below a finite depth or they are described by one of those particular functions of y which render the shear-wave equation explicitly solvable; another well-elaborated approach is related to the WKB asymptotic expansion restricted to high ω and k, see, e.g., [1, 19, 20]. In contrast to the above settings, we address the general case where ω, k may take any values, and $\rho(y)$ and $\mu(y)$ are arbitrary continuous functions for $y \geq 0$.

The paper is organized as follows. Section 2 describes the problem setting. Section 3 contains the auxiliary results. In Sect. 4 we formulate the corresponding parametric Sturm-Liouville problem on a half-line and introduce the assumptions on the material coefficients $\rho(y)$, $\mu(y)$. Section 5 presents the formulations of the main results which are the criteria for non-existence of surface waves (Theorem 5.1), for the existence of $N(k)$ surface-wave solutions with $N(k) \to \infty$ as $k \to \infty$ (Theorem 5.2), and for an interesting possibility of the existence of infinite number of solutions $N(k) = \infty$ for any given k (Theorem 5.3). The criteria for these three options are specified in terms of explicit conditions on the functions $\rho(y)$ and $\mu(y)$. Section 6 provides the proofs of the above theorems. Sections 7 and 8 contain research outlooks and conclusions.

2 Mathematical Setting of the Problem

Consider an elastically isotropic half-space a density ρ and a shear modulus μ. Let the axis Y be orthogonal to the planar half-space boundary and directed into its depth. The axis X may be taken along an arbitrary orientation in the boundary plane (if the medium is elastically monoclinic, then XY is supposed to be a plane of crystallographic symmetry).

The vector equation of 2D linear isotropic elastodynamics in the absence of sources splits into two independent equations, one of which is a scalar equation

$$\partial_x \left(\mu \partial_x \hat{u} \right) + \partial_y \left(\mu \partial_y \hat{u} \right) = \rho \partial_{tt} \hat{u}, \tag{1}$$

describing the shear acoustic wave with a mechanical displacement $\hat{u}(x, y, t)$ induced in the direction orthogonal to the plane XY.

We assume the half-space to be made of functionally graded material, so that ρ and μ depend on one coordinate y (further assumptions are introduced in Sects. 3 and 4). Under this condition, we seek the solution of (1) in the form

$$\hat{u}(x, y, t) = u(y) e^{i(kx - \omega t)}, \tag{2}$$

where ω and k are real positive frequency and wave number. Then Eq. (1) reduces to the ordinary second-order differential equation

$$\left(\mu(y) u'(y) \right)' + \left(\omega^2 \rho(y) - k^2 \mu(y) \right) u(y) = 0. \tag{3}$$

The sought non-zero function $u(y)$ is supposed to satisfy the Neumann boundary condition at the half-space surface $y = 0$:

$$u'(y)|_{y=0} = 0, \qquad (4)$$

which implies that the half-space surface is mechanically unloaded. Aiming at finding the shear surface waves in the given medium, we supplement (4) with the condition of decay at infinite depth:

$$\lim_{y \to \infty} u(y) = 0. \qquad (5)$$

As it was mentioned in the Introduction, the boundary-value problem (3)–(5) admits the surface-wave solutions only for the particular values of the parameters ω and k, which form curves (dispersion branches) in the ωk-plane. Moreover, the functions $\rho(y)$ and $\mu(y)$ allowing existence of the shear surface waves are also restricted by certain conditions. Our goal is to infer these conditions and to characterize the corresponding set of dispersion branches $\omega(k)$.

3 Second-Order Linear Ordinary Differential Equation on a Half-Line: Auxiliary Results

3.1 Second-Order Linear Equation

Equation (3) is a particular type of the second-order linear differential equation

$$\left(\mu(y)u'(y)\right)' + \gamma(y)u = 0 \qquad (6)$$

defined on a half-line $[0, +\infty[$.

Assumption 3.1 *We assume from now on that the function* $\mu(s) \geq \underline{\mu} > 0$ *on* $[0, +\infty[$, *is continuous on* $[0, +\infty[$ *and admits a finite limit* $\lim_{s \to \infty} \mu(s) = \mu_\infty > 0$.

The following substitution of the independent variable

$$\tau(y) = \int_0^y (\mu(s))^{-1} ds \qquad (7)$$

is invertible ($\tau(y)$ is strictly growing) and satisfies the relation: $\frac{d}{d\tau} = \mu \frac{d}{dy}$.

By Assumption 3.1, the functions $\mu(s)$, $(\mu(s))^{-1}$ are both bounded on $[0, +\infty[$ and therefore the function $\tau(y)$ and its inverse $y(\tau)$ are Lipschitzian. Besides, $\int_0^{+\infty} (\mu(s))^{-1} ds = \infty$, i.e., $\tau(y)$ is Lipschitzian homeomorphism of $[0, +\infty[$ onto $[0, +\infty[$.

This substitution transforms (6) into the standard form

$$\frac{d^2\bar{u}}{d\tau^2} + \bar{\gamma}(\tau)\bar{u}(\tau) = 0, \tag{8}$$

where $\bar{u}(\tau) = u(y(\tau))$ and $\bar{\gamma}(\tau) = \mu(y(\tau))\gamma(y(\tau))$.

Another form of (6) is its representation as a system of first-order differential equations for the variables $u(y)$, $w(y) = \mu(y)u'(y)$:

$$u'(y) = \frac{w(y)}{\mu(y)}, \quad w'(y) = -\gamma(y)u(y), \tag{9}$$

or in the matrix form

$$Z'(y) = \frac{dZ}{dy} = C(y)Z(y), \tag{10}$$

$$Z = \begin{pmatrix} w \\ u \end{pmatrix}, \quad C(y) = \begin{pmatrix} 0 & -\gamma(y) \\ (\mu(y))^{-1} & 0 \end{pmatrix}. \tag{11}$$

Performing substitution (7), we transform (10) into the system for the function $\bar{Z}(\tau) = Z(y(\tau))$

$$\frac{d\bar{Z}}{d\tau} = \bar{C}(\tau)\bar{Z}(\tau), \quad \bar{C}(\tau) = \begin{pmatrix} 0 & -\bar{\gamma}(\tau) \\ 1 & 0 \end{pmatrix}. \tag{12}$$

We concentrate for a moment on the asymptotic properties of the solutions of (6), (8), (10), (12) at infinity.

3.2 Asymptotic Properties of Solutions for $y \to +\infty$

The matrix of the coefficients $C(y)$ of the system (10) for each y is traceless, hence, by the Liouville formula, the Wronskian of a fundamental system of solutions is constant in y. This precludes a possibility of having two independent solutions which would both tend to zero at infinity.

Important characteristics of the asymptotics of the system at infinity are determined by the limit of the coefficient matrix for $y \to +\infty$ (if it exists):

$$C_\infty = \lim_{y \to +\infty} C(y) = \begin{pmatrix} 0 & -\gamma_\infty \\ (\mu_\infty)^{-1} & 0 \end{pmatrix},$$

where $\mu_\infty = \lim_{y \to +\infty} \mu(y)$, $\gamma_\infty = \lim_{y \to +\infty} \gamma(y)$.

Whenever $\det C_\infty = \gamma_\infty(\mu_\infty)^{-1} > 0$, or, equivalently, $\gamma_\infty > 0$, the eigenvalues of C_∞ are purely imaginary and one can conclude (see Proposition 3.2 below) the non-existence of a solution of system (10) with $\lim_{y \to +\infty} u(y) = 0$. If on the contrary $\det C_\infty < 0$, then the eigenvalues of C_∞ are real numbers of opposite signs and the existence of a solution of (10) with $\lim_{y \to \infty} u(y) = 0$ is guaranteed under some additional conditions on the function $\gamma(y)$. Note that $\det \bar{C}_\infty = \mu_\infty^2 \det C_\infty$ and therefore a similar conclusion holds for the solutions of system (12).

Let us introduce linear space $\mathcal{G}$ of the coefficients $\gamma(y)$ of Eq. (6) as a space of functions $\gamma(y) = \gamma_\infty + \beta(y)$, with γ_∞ being a constant and $\beta(y)$ a continuous function on $[0, +\infty[$ such that:

$$\lim_{y \to +\infty} \beta(y) = 0, \tag{13}$$

$$\int_0^{+\infty} |\beta(y)| dy < \infty. \tag{14}$$

Evidently $\lim_{y \to \infty} \gamma(y) = \gamma_\infty$.
Introduce in $\mathcal{G}$ the norm

$$\|\gamma(\cdot)\|_{01} = |\gamma_\infty| + \|\beta(\cdot)\|_{C^0} + \|\beta(\cdot)\|_{L_1}. \tag{15}$$

For each $y_0 \in [0, +\infty[$ we introduce a subset $\mathcal{G}^-(y_0) \subset \mathcal{G}$ consisting of the functions $\gamma(y) = \gamma_\infty + \beta(y)$, for which $\gamma_\infty < 0$ and $\gamma_\infty + \beta(y) < 0$ on $[y_0, +\infty[$. Similarly, for each $y_0 \in [0, +\infty[$ we define $\mathcal{G}^+(y_0) \subset \mathcal{G}$ consisting of the functions $\gamma(y) = \gamma_\infty + \beta(y)$, for which $\gamma_\infty > 0$ and $\gamma_\infty + \beta(y) > 0$ on $[y_0, +\infty[$. Both $\mathcal{G}^-(y_0)$ and $\mathcal{G}^+(y_0)$ are open subsets of $\mathcal{G}$ in the above introduced norm. It is easy to verify that substitution (7) transforms the space $\mathcal{G}$ into itself and transforms the sets $\mathcal{G}^-(y_0)$, $\mathcal{G}^+(y_0)$ into $\mathcal{G}^-(\tau(y_0))$, $\mathcal{G}^+(\tau(y_0))$, correspondingly.

The first classical result concerns the so-called non-elliptic case for Eq. (8), where the coefficient $\bar{\gamma}(\cdot) \in \mathcal{G}^-(\tau_0)$.

Proposition 3.1 (See [5, §6.12]) *Consider the equation*

$$u''(\tau) + \bar{\gamma}(\tau)u = u''(\tau) + \left(-\lambda^2 + \beta(\tau)\right)u = 0, \quad \lambda > 0. \tag{16}$$

Assume $\beta(\tau)$ to be continuous and to satisfy (13). *Then there exist two solutions $u_\lambda(\tau), u_{-\lambda}(\tau)$ of Eq.* (16) *and $\tau_0 \geq 0$ such that $\forall \tau \geq \tau_0$:*

$$c_2' \exp\left[\lambda \tau - d_1' \int_{\tau_0}^\tau |\beta(\theta)| d\theta\right] \leq u_\lambda(\tau) \leq c_1' \exp\left[\lambda \tau + d_1' \int_{\tau_0}^\tau |\beta(\theta)| d\theta\right], \tag{17}$$

$$c_2 \exp\left[-\lambda \tau - d_1 \int_{\tau_0}^\tau |\beta(\theta)| d\theta\right] \leq u_{-\lambda}(\tau) \leq c_1 \exp\left[-\lambda \tau + d_1 \int_{\tau_0}^\tau |\beta(\theta)| d\theta\right], \tag{18}$$

where $c_1, c_2, d_1, c_1', c_2', d_1'$ are constant.

Corollary 3.1 (See [14, §XI.9]) *Assume the assumptions of Proposition 3.1 to hold and $\beta(\cdot)$ to satisfy (14). Then the solutions $u_\lambda, u_{-\lambda}$ satisfy*

$$u_\lambda \sim \frac{u'_\lambda}{\lambda} \sim e^{\lambda\tau}, \quad u_{-\lambda} \sim -\frac{u'_{-\lambda}}{\lambda} \sim e^{-\lambda\tau}$$

as $\tau \to +\infty$.

Corollary 3.2 *For each $\tilde{\gamma}(\cdot)$ sufficiently close to $\bar{\gamma}(\cdot)$ in the norm (15) the equation*

$$u''(\tau) + \tilde{\gamma}(\tau)u(\tau) = 0$$

has a decaying solution.

Next we pass on to the elliptic case (see [14, §XI.8]; Corollary 8.1), where the coefficient $\bar{\gamma}(\cdot) \in \mathcal{G}^+(y_0)$.

Proposition 3.2 *Consider the equation*

$$u''(\tau) + \bar{\gamma}(\tau)u = u''(\tau) + \left(\lambda^2 + \beta(\tau)\right)u = 0, \ \lambda > 0 \quad (19)$$

with $\bar{\gamma}(\cdot) \in \mathcal{G}^+(y_0)$. Then for any real a, b there is a unique solution of equation (19) with the asymptotics

$$u(\tau) = (a + o(1))\cos\lambda\tau + (b + o(1))\sin\lambda\tau, \quad (20)$$
$$u'(\tau) = (-\lambda a + o(1))\sin\lambda\tau + (\lambda b + o(1))\cos\lambda\tau,$$

as $\tau \to +\infty$.

3.3 Prüfer's Angle Coordinate

We consider Prüfer's coordinates (see [3, 14]):

$$r = (u^2 + \mu^2 u'^2)^{\frac{1}{2}} = (u^2 + w^2)^{\frac{1}{2}}, \ \varphi = \text{Arctg}\,\frac{u}{w}, \quad (21)$$

where again $w = \mu u'$. For the vector-function Z introduced in (11), we denote φ by Arg Z (the choice of a continuous branch is done in a standard way). In coordinates (21) system (6) takes the form:

$$r' = \left(\mu^{-1}(y) - \gamma(y)\right) r \sin\varphi \cos\varphi, \ \varphi' = \gamma(y)\sin^2\varphi + \mu^{-1}(y)\cos^2\varphi; \quad (22)$$

note that the second equation is decoupled from the first one.

We list some facts concerning the evolution of Arg $Z(y)$. Recall that $\mu(y)$ in Eqs. (6) and (22) meets Assumption 3.1.

Proposition 3.3

i) *If $\gamma(y) \geq 0$ (respectively, $\gamma(y) > 0$) on an interval, then for a solution $Z(y)$ of (9) Prüfer's angle variable $\varphi = \text{Arg } Z$ is non-decreasing (increasing) on the interval.*

ii) *If $\gamma(y) < 0$ on an interval I, then the first and the third quadrants—Arg $Z \in\,]0, \pi/2[$ and Arg $Z \in\,]\pi, 3\pi/2[$—are invariant for system (9) on I.*

iii) *For any $\gamma(y)$ there is a kind of weakened monotonicity for Arg Z: if Arg $Z(\tilde{y}) > m\pi$, then Arg $Z(y) > m\pi$ for any $y > \tilde{y}$.*

Property i) follows from (22). So does property ii) since, according to (22), $\varphi'(\pi m) > 0$ and $\varphi'(\pi/2 + \pi m) < 0$ for negative γ. Property iii) follows from the fact that in (22) $\varphi'(m\pi) = \mu^{-1}(m\pi) > 0$.

3.4 Oscillatory Equations

Second-order linear differential equation is called *oscillatory* [14, §XI.5] on $[0, +\infty[$ when its every solution has infinite number of zeros on $[0, +\infty[$, or equivalently the set of zeros of any solution has no upper limit, or equivalently for every solution its Prüfer's coordinate Arg Z (see the previous Subsection) satisfies

$$\limsup_{y \to +\infty} \text{Arg } Z(y) = +\infty.$$

An obvious example of oscillatory equation is (19) (cf. Proposition 3.2).

We are interested in the conditions under which Eq. (19) with vanishing λ or, the same, Eq. (8) are oscillatory. The result can be found in [14, §XI.5], [13, Ch.2,§6]; we formulate it for Eq. (8).

Proposition 3.4 *Let $\bar{\gamma}(\cdot)$ in Eq. (8) be continuous of bounded variation on every interval $[0, T]$. Let $\bar{\gamma}(\tau) > 0$ on some interval $[\tau_0, +\infty[$, and*

$$\int_{\tau_0}^{+\infty} (\bar{\gamma}(\tau))^{1/2}\, d\tau = +\infty, \tag{23}$$

$$\int_{\tau_0}^{T} (\bar{\gamma}(\tau))^{-1} |d\gamma(\tau)| = o\left(\int_{\tau_0}^{T} (\bar{\gamma}(\tau))^{1/2}\, d\tau \right), \text{ as } T \to +\infty. \tag{24}$$

Then Eq. (8) is oscillatory.

Below we deal with the functions $\bar{\gamma}(\tau)$ which tend to 0 as $\tau \to +\infty$. In the following Proposition we describe some classes of such functions which satisfy conditions (23), (24).

Proposition 3.5 *Positive function* $\bar{\gamma}(\tau)$, $\tau \in [1, +\infty]$ *satisfies conditions* (23)–(24), *whenever any of the following conditions hold:*

1. $\bar{\gamma}(\tau)$ *is monotonously decreasing and there exist* $c > 0$, $\delta \in]0, 2[$, $\tau_0 \geq 1$ *such that*

$$\bar{\gamma}(\tau) \geq \frac{c}{\tau^{2-\delta}}, \ \forall \tau \in [\tau_0, +\infty[; \tag{25}$$

2. *function* $\bar{\gamma}(\tau)$ *is continuously differentiable on* $[1, +\infty[$, *satisfies* (25) *and*

$$\bar{\gamma}'(\tau) = o\left((\bar{\gamma}(\tau))^{3/2}\right), \ as \ \tau \to +\infty. \tag{26}$$

Indeed, if the condition 1. holds, then (23) immediately follows from (25). Under the same condition we have $|d\bar{\gamma}(\tau)| = -d\bar{\gamma}(\tau)$, and hence

$$\int_{\tau_0}^{T} (\bar{\gamma}(\tau))^{-1} |d\gamma(\tau)| = \ln \bar{\gamma}(\tau_0) - \ln \bar{\gamma}(T) \leq$$

$$\leq \ln \bar{\gamma}(\tau_0) - \ln \frac{c}{T^{2-\delta}} = \ln \bar{\gamma}(\tau_0) - \ln c + (2-\delta) \ln T,$$

while by (25)

$$\int_{\tau_0}^{T} (\bar{\gamma}(\tau))^{1/2} d\tau \geq c^{1/2} \int_{\tau_0}^{T} \frac{d\tau}{\tau^{1-\delta/2}} = c^{1/2} \left(T^{\delta/2} - \tau_0^{\delta/2}\right),$$

wherefrom (24) follows. A typical example of the function which satisfies the condition 1. would be $\bar{\gamma}(\tau) = \frac{c}{\tau^{2-\delta}}$ with $c > 0$, $\delta \in (0, 2)$.

Condition (26) appears in [14, §XI, Corollary 5.3]; once it is satisfied, one gets

$$|d\bar{\gamma}(\tau)| = o\left((\bar{\gamma}(\tau))^{3/2} d\tau\right) \Rightarrow \frac{|d\bar{\gamma}(\tau)|}{\bar{\gamma}(\tau)} = o\left((\bar{\gamma}(\tau))^{1/2} d\tau\right),$$

wherefrom (24) follows.

An example of a nonmonotonous function which satisfies the condition 2 of Proposition 3.4 could be $\bar{\gamma}(\tau) = \frac{a(\tau)}{\tau^{2-\delta}}$, where $a'(\tau) = \frac{\cos \tau}{\tau^{1-\delta/4}}$ and an additive constant a_0 in the expression $a(\tau) = a_0 + \int_1^{\tau} \frac{\cos \xi}{\xi^{1-\delta/4}} d\xi$ is chosen in such a way that $a(\tau) > 0$ and $\lim_{\tau \to +\infty} a(\tau) = a_\infty > 0$, so that (23) holds. Non-monotonicity of $\bar{\gamma}(\tau) = \frac{a(\tau)}{\tau^{2-\delta}}$, as well as condition (26) for it, can be checked by direct computation.

3.5 Hamiltonian Form

The system (10) can be given the following Hamiltonian form

$$u' = \frac{\partial H}{\partial w} = \frac{w}{\mu(y)}, \quad w' = -\frac{\partial H}{\partial u} = -\gamma(y)u \qquad (27)$$

with the Hamiltonian

$$H = \frac{1}{2}\left(\frac{w^2}{\mu(y)} + \gamma(y)u^2\right).$$

We denote by $\vec{h}$ the (Hamiltonian) vector field at the right-hand side of (27).

As it is well known, Eq. (6) follows from a variational principle, i.e., (6) is the Euler-Lagrange equation which represents necessary minimality condition for a variational problem

$$\int_0^{+\infty} \mu(y)\left(u'(y)\right)^2 - \gamma(y)\left(u(y)\right)^2 dy \to \min$$

with appropriate boundary conditions. The Hamiltonian form of the minimality condition for the variational problem is precisely (27).

For Prüfer's angle $\varphi = \text{Arctan}\left(\frac{u}{w}\right)$, there holds

$$\varphi' = \frac{-w'u + wu'}{u^2 + w^2} = \frac{\gamma u^2 + w^2/\mu}{u^2 + w^2} = \frac{2H}{u^2 + w^2}.$$

The last equation is equivalent to the differential equations (22) for Prüfer's coordinate φ.

Remark 3.1 A simple but useful (see [2]) computation is provided by the derivation of $u(y)w(y)$ along the trajectories of Hamiltonian system (27):

$$\frac{d}{dy}(uw) = \partial_{\vec{h}}(uw) = \left(\partial_{\vec{h}}u\right)w + u\left(\partial_{\vec{h}}w\right) = -\gamma u^2 + \frac{w^2}{\mu}, \qquad (28)$$

wherefrom it follows, among other things, that uw is non-decreasing (respectively increasing) on the intervals where $\gamma(y) \leq 0$ (respectively $\gamma(y) < 0$).

Proposition 3.3 and Remark 3.1 allow us to arrive at a conclusion on qualitative behavior of solutions on an interval where $\gamma(\tau) < 0$ in (16). According to Proposition 3.1, there is a decaying solution along which (according to Remark 3.1) uw grows. Hence the solution approaches the origin either in the second or in the fourth quadrants where $uw < 0$.

Proposition 3.6 *Let $\bar{\gamma}(\tau)$ meet the assumptions of Proposition 3.1 with $\bar{\gamma}(\tau) < 0$ for $\tau \in [\tau_0, +\infty[$. Then the decaying solutions $\pm u(\tau)$ of (16) correspond to the solutions $\pm Z(\tau)$ of (9) with $\text{Arg } Z(\tau) \in [\pi/2, \pi]$ and $\text{Arg}(-Z)(\tau) \in [3\pi/2, 2\pi]$ for $\tau \in [\tau_0, +\infty[$.*

Other solutions which start in the same quadrants escape to either the first or the third quadrant, which, according to Proposition 3.3, are invariant for (16) whenever $\gamma(\tau) < 0$. According to Remark 3.1, the product uw (positive in these quadrants) grows along the respective trajectories, which tend to infinity.

3.6 Sturmian Properties of Trajectories

We provide few results from the Sturm theory. First result is classical [3, 13], [14, Ch. X,XI] and follows directly from the second equation (22).

Proposition 3.7 (Comparison Result) *Consider a pair of second-order equations*

$$\big(\mu(y)u'(y)\big)' + \gamma(y)u = 0, \quad \big(\mu(y)u'(y)\big)' + \tilde{\gamma}(y)u(y) = 0, \tag{29}$$

where $\mu(y)$ meets Assumption 3.1 and

$$\tilde{\gamma}(y) \geq \gamma(y), \ \forall y \in [y_0, +\infty[.$$

If for $y_1 \geq y_0$ and a pair of vector solutions $Z = \begin{pmatrix} w \\ u \end{pmatrix}$, $\tilde{Z} = \begin{pmatrix} \tilde{w} \\ \tilde{u} \end{pmatrix}$ of the first and the second equations (29)

$$\text{Arg } \tilde{Z}(y_1) = \text{Arg } Z(y_1),$$

then

$$\forall y \geq y_1 : \text{Arg } \tilde{Z}(y) \geq \text{Arg } Z(y)$$

and

$$\forall y \in [y_0, y_1] : \text{Arg } \tilde{Z}(y) \leq \text{Arg } Z(y). \tag{30}$$

We provide analogue of the comparison result (in particular, of relation (30)) for the decaying solutions of (29) at $y_1 = +\infty$ with a (short) proof.

Proposition 3.8 (Comparison Result for Decaying Solutions on a Half-Line)
Consider the pair of second-order equations (29) with the coefficient $\mu(y)$ meeting Assumption 3.1 and with $\gamma(y)$, $\tilde{\gamma}(y)$ belonging to $\mathcal{G}^-(y_0)$. Let

$$0 > \tilde{\gamma}(y) \geq \gamma(y), \quad \forall y \in [y_0, +\infty[. \tag{31}$$

If (in the notation of Proposition 3.7) $Z, \tilde{Z}$ are the decaying solutions of equations (29),

$$\operatorname{Arg} \tilde{Z}(y) \leq \operatorname{Arg} Z(y), \quad \forall y \geq y_0. \tag{32}$$

Proof Without loss of generality, we may assume $\mu(y) \equiv 1$; otherwise we perform substitution (7) of the independent variable which preserves relation (31) for the coefficients.

By (31) and (28), the functions uw and $\tilde{u}\tilde{w}$ are increasing on $[y_0, +\infty[$. As long as the limits of these functions at $+\infty$ are null, we conclude that $(uw)(y) < 0$, $(\tilde{u}\tilde{w})(y) < 0$ on $[y_0, +\infty[$ and then without loss of generality we may assume that $u(y), \tilde{u}(y)$ are positive, while $w(y), \tilde{w}(y)$ are negative on $[y_0, +\infty[$.

Denote $\tilde{\gamma}(y) - \gamma(y)$ by $\Delta\gamma(y)$ and represent the second one of the equations (29) as

$$\tilde{u}'' + \gamma(y)\tilde{u} = -\Delta\gamma(y)\tilde{u}; \tag{33}$$

$\Delta\gamma(y) > 0$ by (31).

Applying the integral form of the Lagrange identity (or Green's formula, see [14, §XI.2]) to the respective vector solutions $Z = \begin{pmatrix} w \\ u \end{pmatrix}$, $\tilde{Z} = \begin{pmatrix} \tilde{w} \\ \tilde{u} \end{pmatrix}$ of Eqs. (29), of which the second one is written as (33), we conclude:

$$\forall y \geq y_0: \ (u\tilde{w} - w\tilde{u})|_y^{+\infty} = \int_y^{+\infty} -\Delta\gamma(s)\tilde{u}(s)u(s)ds < 0.$$

Given that $(u\tilde{w} - w\tilde{u})$ vanishes at $+\infty$, we obtain:

$$\forall y \geq y_0: \ -u(y)\tilde{w}(y) + w(y)\tilde{u}(y) = \int_y^{+\infty} -\Delta\gamma(s)\tilde{u}(s)u(s)ds < 0. \tag{34}$$

Dividing the inequality in (34) by the positive value $w(y)\tilde{w}(y)$, we get

$$\forall y \geq y_0: \ \frac{\tilde{u}(y)}{\tilde{w}(y)} \leq \frac{u(y)}{w(y)},$$

wherefrom (32) follows. □

We establish the continuous dependence of decaying solutions on the coefficient $\gamma(\cdot)$ in $\|\cdot\|_{01}$-norm.

Proposition 3.9 (Continuous Dependence of Decaying Solutions on the Right-Hand Side) *Consider Eqs. (29). Let $\gamma(\cdot) = -\lambda^2 + \beta(\cdot) \in \mathcal{G}_{y_0}^-$ for some $y_0 \in [0, +\infty[$. Then, for any $\tilde{\gamma}(\cdot) = -\tilde{\lambda}^2 + \tilde{\beta}(\cdot)$ sufficiently close to $\gamma(\cdot)$ in $\|\cdot\|_{01}$-norm:*

i) *both Eqs. (29) possess the decaying vector solutions $Z(\cdot), \tilde{Z}(\cdot)$ with Arg Z, Arg $\tilde{Z} \in [\pi/2, \pi]$;*
ii) *for each $y \in [y_0, +\infty[$*

$$\left|\operatorname{Arg} \tilde{Z}(y) - \operatorname{Arg} Z(y)\right| \to 0, \text{ as } \|\tilde{\gamma}(\cdot) - \gamma(\cdot)\|_{01} \to 0.$$

Proof Again one may proceed under the assumption $\mu(y) \equiv 1$.

i) Any $\tilde{\gamma}(\cdot)$ sufficiently close to $\gamma(\cdot)$ in $\|\cdot\|_{01}$-norm belongs to $\mathcal{G}_{y_0}^-$, which is open with respect to the norm. The existence of the decaying solutions $Z(y), \tilde{Z}(y)$ follows from Corollary 3.2. Since both γ and $\tilde{\gamma}$ are negative on $[y_0, +\infty[$, we conclude by Proposition 3.6 that Arg $Z(y)$ and Arg $\tilde{Z}(y)$ lie in $[\pi/2, \pi]$ for $y \in [y_0, +\infty[$. This implies that for $s \in [y_0, +\infty[$ $w(s), \tilde{w}(s)$ are negative, while $u(s), \tilde{u}(\tau)$ are positive and by (9) decrease.
ii) Recall that $\Delta \gamma(\cdot) = \tilde{\gamma}(\cdot) - \gamma(\cdot)$. Invoking the equality in (34) and dividing it by $-u(y)\tilde{u}(y)$, we get

$$\frac{\tilde{w}(y)}{\tilde{u}(y)} - \frac{w(y)}{u(y)} = \int_y^{+\infty} \Delta\gamma(s) \frac{\tilde{u}(s)}{\tilde{u}(y)} \frac{u(s)}{u(y)} ds = \int_y^{+\infty} \Delta\gamma(s) v(s) \tilde{v}(s) d\tau, \quad (35)$$

where $v(s) = \frac{\tilde{u}(s)}{\tilde{u}(y)}$, $\tilde{v}(s) = \frac{u(s)}{u(y)}$ are the solutions of the first and second equation (29), which are normalized by the condition: $v(y) = \tilde{v}(y) = 1$.

By the aforesaid $v(s), \tilde{v}(s)$ decrease; hence

$$v(s) \leq 1, \tilde{v}(s) \leq 1, \text{ for } s \geq y. \quad (36)$$

According to Proposition 3.1, there exist $c_1, d_1 > 0, s_0 > y$ such that

$$v(s) \leq c_1 \exp\left(-\lambda s + d_1 \int_{s_0}^s |\beta(\sigma)| d\sigma\right), \forall s > s_0. \quad (37)$$

From the proof of the Proposition (see [5, §6.12, §2.6]), it follows that one can choose any $c_1 > 1$, a sufficiently large d_1 in (37) and then choose s_0 such that $d_1 \sup_{s \geq s_0} |\beta(s)| < \lambda$. The same holds for the second one of equations (29).

For each $\tilde{\gamma}$ from a small neighborhood of γ in $\|\cdot\|_{01}$-norm, $\tilde{\lambda}$ and λ as well as $\sup_{\tau \geq \tau_0} |\beta(\tau)|$ and $\sup_{\tau \geq \tau_0} |\tilde{\beta}(\tau)|$ are close. Thus, one can choose common c_1, d_1, τ_0 for all the equations with the coefficient $\tilde{\gamma}$ from the neighborhood.

Besides, there is a common upper bound B for the corresponding norms $\|\tilde{\beta}(\cdot)\|_{L_1}$. Then, by (36) and (37),

$$\left|\int_y^{+\infty} \Delta\gamma(s)v(s)\tilde{v}(s)ds\right| \leq \int_y^{s_0} |\Delta\gamma(s)|\,ds + c_1^2 e^{2d_1 B} \int_{s_0}^{\infty} e^{-\lambda s} |\Delta\gamma(s)|\,ds \tag{38}$$

with the right-hand side tending to 0 as $\|\Delta\gamma(s)\|_{01} \to 0$.

Note that $\mathrm{Arg}\, Z = \mathrm{Arccot}\, \frac{w(y)}{u(y)}$, $\mathrm{Arg}\, \tilde{Z} = \mathrm{Arccot}\, \frac{\tilde{w}(y)}{\tilde{u}(y)}$ and since the function $z \mapsto \mathrm{Arccot}\, z$ is Lipschitzian with constant 1, it follows that

$$\left|\mathrm{Arg}\, Z(y) - \mathrm{Arg}\, \tilde{Z}(y)\right| = \left|\mathrm{Arccot}\, \frac{w(y)}{u(y)} - \mathrm{Arccot}\, \frac{\tilde{w}(y)}{\tilde{u}(y)}\right| \leq \left|\frac{w(y)}{u(y)} - \frac{\tilde{w}(y)}{\tilde{u}(y)}\right|$$

and then, by (35) and (38), the left-hand side tends to 0 as $\|\Delta\gamma(\tau)\|_{01} \to 0$. □

4 Existence of Surface Shear Waves and Parametric Sturm-Liouville Problem

We come back to Eq. (3) and simplify the notations putting $\Omega = \omega^2$, $K = k^2$, $A = (K, \Omega)$,

$$\gamma_A(y) = \Omega\rho(y) - K\mu(y), \tag{39}$$

thus arriving at the equation

$$(\mu(y)u'(y))' + \gamma_A(y)u(y) = 0 \tag{40}$$

with the vector parameter A.

We are interested in the solutions, which satisfy simultaneously the condition at infinity (5) and the traction-free boundary condition (4). In other words, we are dealing with parametric Sturm-Liouville problem on a half-line for Eq. (40) with the boundary conditions (4)–(5) and the parameters Ω, K.

Let us introduce the vector-function $a(y) = (\rho(y), \mu(y))$, which characterizes our medium.

Assumption 4.1 (Positivity, Lipschitz Continuity, and Limit at Infinity) *The function $a(y) = (\rho(y), \mu(y))$ is Lipschitz continuous on $[0, +\infty[$; its components admit positive values. There exists a finite limit*

$$\lim_{y \to +\infty} a(y) = a_\infty,\ a_\infty = (\rho_\infty, \mu_\infty),\ \rho_\infty > 0,\ \mu_\infty > 0.$$

It follows immediately that $\rho(y), \mu(y)$ are bounded from below by positive constants on $[0, +\infty[$.

Assumption 4.2 (Integral Boundedness) *The function*

$$\hat{a}(y) = (\hat{\rho}(y), \hat{\mu}(y)) = a(y) - a(+\infty) = (\rho(y) - \rho_\infty, \mu(y) - \mu_\infty)$$

is integrable on $[0, +\infty[: \int_0^\infty |\hat{\rho}(y)| + |\hat{\mu}(y)| \, dy < \infty.$

We consider the ratio $\frac{\mu(y)}{\rho(y)}$, which may be given a physical meaning of the squared "local" phase velocity of the shear wave at the depth y. By our assumptions, $\frac{\mu(y)}{\rho(y)}$ is bounded positive function on $[0, +\infty[$ and $\lim_{y \to \infty} \frac{\mu(y)}{\rho(y)} = \frac{\mu_\infty}{\rho_\infty}$.

The following assumption states that this function attains either local maximum or local minimum at infinity.

Assumption 4.3 (Sign Permanence at Infinity) *There exists an interval* $]\bar{y}, +\infty[$ *such that either*

$$\frac{\mu(y)}{\rho(y)} < \frac{\mu_\infty}{\rho_\infty}, \quad \forall y \in]\bar{y}, +\infty[, \tag{41}$$

or

$$\frac{\mu(y)}{\rho(y)} > \frac{\mu_\infty}{\rho_\infty}, \quad \forall y \in]\bar{y}, +\infty[. \tag{42}$$

Graphs (a) and (b) in Fig. 1 provide examples of the functions $\frac{\mu(y)}{\rho(y)}$ which satisfy (41) or (42) correspondingly; graph 1c at the same Figure does not satisfy any of the two conditions.

The vector of parameters $A = (K, \Omega)$ have positive components. Assumptions 4.1–4.2 imply the following properties of the function $\gamma_A(y)$ defined by (39).

Lemma 4.1 *Under Assumptions 4.1 and 4.2, for any $A \in \mathbb{R}_+^2$ and $\gamma_A(\cdot)$ defined by (39) there holds:*

1. $\gamma_A(y) - \gamma_A(+\infty) \xrightarrow{y \to +\infty} 0;$
2. $\int_0^\infty |\gamma_A(y) - \gamma_A(+\infty)| dy < \infty;$
3. *if $\gamma_A(y)$ admits positive values, then so does $\gamma_{A'}(y)$ with any A' sufficiently close to A;*
4. *for each $A = (K, \Omega)$ with $\frac{\Omega}{K} < \frac{\mu_\infty}{\rho_\infty}$, there exists an interval $[y_-, +\infty[$ on which $\gamma_A(y) < 0$.*

The proofs of 1.–4. are elementary and we just sketch the last one. By hypothesis $\Omega \rho_\infty - K \mu_\infty < 0$ and by Assumption 4.1, $\gamma_A(y) = \Omega \rho(y) - K \mu(y) < 0$ for y from some neighborhood $[y_-, +\infty[$ of the infinity.

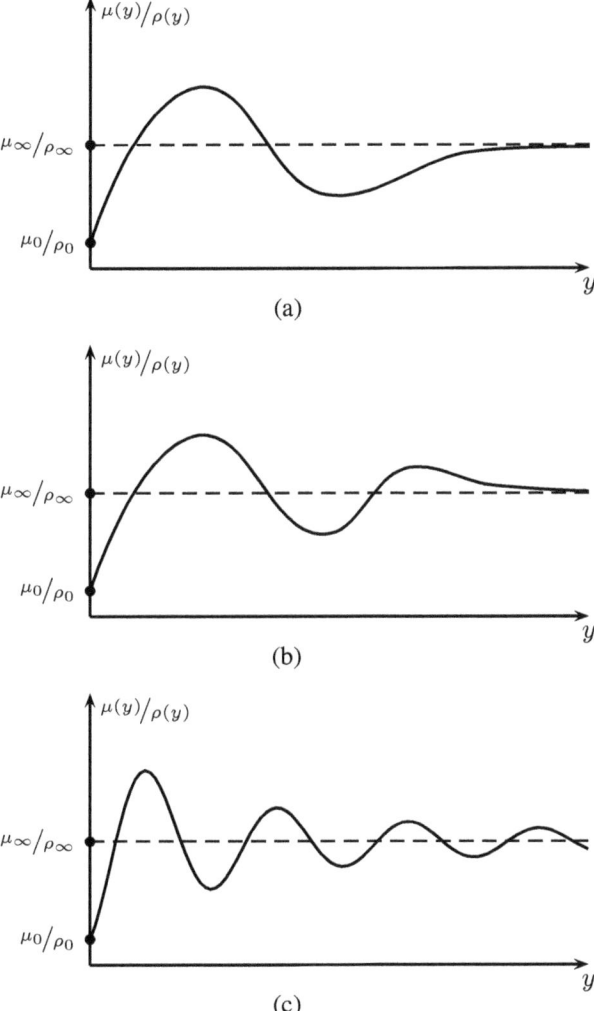

Fig. 1 Graphs of $\frac{\mu(y)}{\rho(y)}$ with different types of behavior at infinity

Performing the substitution of the independent variable

$$y \mapsto \tau(y) = \int_0^y (\mu(s))^{-1} \, ds$$

(cf. (7)) we bring Eq. (40) to the form:

$$\frac{d^2 \bar{u}}{d\tau^2} + \bar{\gamma}_A(\tau) \bar{u}(\tau) = 0, \tag{43}$$

where

$$\bar{\gamma}_A(\tau) = \Omega \bar{\rho}(\tau) - K\bar{\mu}(\tau), \tag{44}$$

$$\bar{\rho}(\tau) = \mu(y(\tau))\rho(y(\tau)), \ \bar{\mu}(\tau) = (\mu(y(\tau)))^2, \ \bar{u}(\tau) = u(y(\tau)), \tag{45}$$

and $y(\tau)$ is the inverse function. Note that $\bar{\mu}(0) = \mu(0)$, $\bar{\rho}(0) = \rho(0)$ and

$$\bar{\mu}(+\infty) = \mu_\infty, \ \bar{\rho}(+\infty) = \rho_\infty, \ \bar{\gamma}_A(+\infty) = \Omega\mu_\infty\rho_\infty - K\mu_\infty^2.$$

We wish to check that Assumptions 4.1–4.3 remain valid after this substitution.

Proposition 4.1 *Let Assumptions 4.1–4.3 hold for the functions $\rho(y), \mu(y)$. Then the same Assumptions hold (after an appropriate change of notation) for the functions $\bar{\rho}(\tau), \bar{\mu}(\tau)$.*

Proof By Assumption 4.1, $\tau(y)$ defined by (7) is Lipschitzian homeomorphism of $[0, +\infty[$ onto itself. Hence the functions $\bar{\mu}, \bar{\rho}$ defined by (45) are bounded, Lipschitzian and have finite limits at infinity, i.e., Assumption 4.1 is valid for them.

By (43), (45)

$$\int_0^{+\infty} |\bar{\rho}(\tau) - \bar{\rho}(+\infty)| \, d\tau = \int_0^{+\infty} |\rho(y(\tau))\mu(y(\tau)) - \rho_\infty\mu_\infty| \, d\tau =$$

$$= \int_0^{+\infty} |\rho(y)\mu(y) - \rho_\infty\mu_\infty| \, (\mu(y))^{-1} dy \le \int_0^{+\infty} |\rho(y) - \rho_\infty| dy +$$

$$+ \int_0^{+\infty} \rho_\infty |\mu(y) - \mu_\infty|(\mu(y))^{-1} dy < +\infty$$

by Assumption 4.2 and since $(\mu(y))^{-1}$ is bounded on $[0, +\infty[$. Similar reasoning can be used for $\bar{\mu}(\tau)$.

By (45), $\frac{\bar{\mu}(\tau(y))}{\bar{\rho}(\tau(y))} = \frac{\mu(y)}{\rho(y)}$ and hence if either (41) or (42) holds for $\rho(y), \mu(y)$ on $]\bar{y}, \infty[$, then the same property holds for $\bar{\mu}(\tau(y)), \bar{\rho}(\tau(y))$ on the interval $]\tau(\bar{y}), \infty[$. □

It is obvious that the statements of Lemma 4.1 are valid for $\bar{\gamma}_A(\tau)$.

Key information for our treatment is provided by the *limit-case equation*, which corresponds to the vectors of parameters $A_\infty = (K_\infty, \Omega_\infty) = (\beta\rho_\infty, \beta\mu_\infty)$, $\beta > 0$. For such choice of parameters

$$\gamma_{A_\infty}(y) = \Omega_\infty\rho(y) - K_\infty\mu(y) = \beta(\mu_\infty(\rho_\infty+\hat{\rho}(y))-\rho_\infty(\mu_\infty+\hat{\mu}(y))) = \beta\hat{\gamma}_\infty(y),$$

where

$$\hat{\gamma}_\infty(y) = \mu_\infty\hat{\rho}(y) - \rho_\infty\hat{\mu}(y) \tag{46}$$

satisfies

$$\lim_{y\to\infty}\hat{\gamma}_\infty(y) = 0.$$

We call

$$(\mu(y)u')' + \beta\hat{\gamma}_\infty(y)u = 0 \qquad (47)$$

the limit-case equation.

The following fact is easy to verify.

Remark 4.1 If (41) (respectively (42)) holds, then $\hat{\gamma}_\infty(y)$ is positive (respectively negative) on the corresponding interval $]\bar{y}, +\infty[$.

5 Results

We formulate here main results of the paper; the proofs are provided in the next Section. Our first result establishes non-existence of solutions whenever (42) holds for all $y \geq 0$.

Theorem 5.1 *Let assumptions 4.1–4.2 hold and (42) hold for all $y \geq 0$ (cf. the graph (a) in Fig. 2). Then there are no admissible values of parameters K, Ω for which solutions of (40)–(4)–(5) exist.*

If (42) does not hold on a subinterval of $[0, +\infty[$, then one can guarantee existence of solutions at least for sufficiently large K, Ω. We put $\frac{\check{\mu}}{\check{\rho}} = \min_{y\in[0,+\infty[} \frac{\mu(y)}{\rho(y)}$.

Theorem 5.2 *Let Assumptions 4.1–4.2–4.3 hold and moreover $\frac{\mu(y)}{\rho(y)} < \frac{\mu_\infty}{\rho_\infty}$ be fulfilled for y in a non-null subinterval of $[0, +\infty[$. Then for each $N = 1, 2, \ldots$ there exist $K_1 < K_2 < \cdots$ such that $\forall K > K_N$ there are at least N values $\Omega_j \in \left]\frac{\check{\mu}}{\check{\rho}}K, \frac{\mu_\infty}{\rho_\infty}K\right[, j = 1, \ldots, N$, such that the solution of (40)–(4)–(5) exists for each (K, Ω_j).*

Remark 5.1 The graphs (a) and (b) in Fig. 1 meet assumptions of the Theorem.

Finally, there is a case in which for each $K > 0$ one finds a numerable set of $\Omega_j \in \left]\frac{\check{\mu}}{\check{\rho}}K, \frac{\mu_\infty}{\rho_\infty}K\right[$ such that the solution exists for (K, Ω_j). It happens when the limit-case Eq. (47) is oscillatory (see Sect. 3.4).

Theorem 5.3 *Let assumptions 4.1–4.2 and (41) hold and the limit-case equation (47) be oscillatory.*

Then for each $K > 0$ there exists a numerable set of $\Omega_m \in \left]\frac{\check{\mu}}{\check{\rho}}K, \frac{\mu_\infty}{\rho_\infty}K\right[, m = 1, \ldots$, such that:

i) *for $A_m = (K, \Omega_m)$ the solution of (40)–(4)–(5) exists;*

ii) Ω_m *increase with m and accumulate (only) to* $\bar{\Omega} = \frac{\mu_\infty}{\rho_\infty} K$;
iii) *for the vector solutions* $Z(y; A_m)$ *there holds*

$$\text{Arg } Z(y; A_m) \in [(m - 1/2)\pi, m\pi] \text{ for } y \text{ sufficiently large.}$$

Remark 5.2 Assumptions 4.1–4.2 together with (41) hold for the graph (a) in Fig. 1, but the conditions which guarantee oscillatory property for the limit-case equation (such as (23) and (24)) can hardly be identified graphically.

5.1 Homogeneous Substrate Example

Consider a particular case in which the properties of the medium become depth-independent starting from some depth. For the model under discussion, this means such that $\mu(y)$ and $\rho(y)$ are constant: $\mu(y) \equiv \mu_s$, $\rho(y) \equiv \rho_s$ on $[y_s, +\infty[$ (as an example see the graph (b) in Fig. 2.

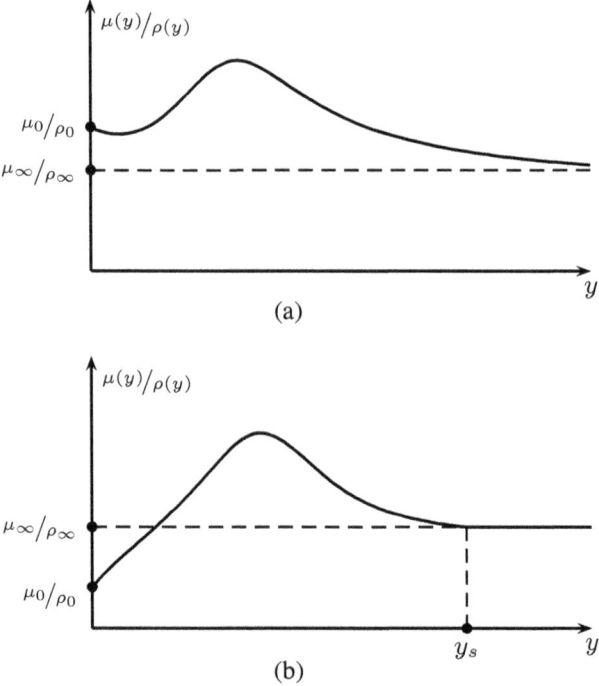

Fig. 2 (**a**) The graph illustrates Theorem 5.1. (**b**) The functions $\rho(y)$ and $\mu(y)$ in the homogeneous substrate example become constant when $y \geq y_s$

We denote $a_s = (\rho_s, \mu_s)$. Then, $a_\infty = \lim_{y\to\infty} a(y) = a_s$ and $\hat{a}(y) = a(y) - a_\infty$ vanishes on $[y_s, +\infty[$. If

$$\forall y \in [0, +\infty[: \frac{\mu(y)}{\rho(y)} \geq \frac{\mu_s}{\rho_s},$$

then the assumptions of Theorem 5.1 are met and solutions of (40)–(4)–(5) do not exist. If $\frac{\mu(y)}{\rho(y)} < \frac{\mu_s}{\rho_s}$ on some non-null subinterval of $[0, +\infty[$, then the assumptions of Theorem 5.2 are met and hence its claim holds.

6 Proofs

Since substitution (7) transforms parametric equation (40) into its standard form (43) and Assumptions 4.1–4.3 are maintained under (7), there is no loss of generality in taking $\mu(y) \equiv 1$ in (40).

The *proof of Theorem 5.1* is easy. Pick some $A = (K, \Omega)$ with positive components. There are two options: $\frac{\mu_\infty}{\rho_\infty} \leq \frac{\Omega}{K}$ or $\frac{\mu_\infty}{\rho_\infty} > \frac{\Omega}{K}$.

In the first case the coefficient $\gamma_A(y)$ in Eq. (40) is non-negative on some interval $[y_0, +\infty[$. Then, by Proposition 3.2, there exists a fundamental system of solutions of the form (20) and hence none of the solutions of (40) tend to the origin as $y \to +\infty$.

If otherwise $\frac{\Omega}{K} < \frac{\mu_\infty}{\rho_\infty} \leq \frac{\mu(y)}{\rho(y)}$ $\forall y \in [0, +\infty[$, then $\gamma_A(y) < 0$ on $[0, +\infty[$. By (28), for a solution $Z(y, A) = \begin{pmatrix} w \\ u \end{pmatrix}$ there holds $\frac{d}{dy}(u(y)w(y)) > 0$. This enters in contradiction with the boundary conditions (4)–(5), according to which $u(0)w(0) = 0$ and $\lim_{y\to +\infty}(u(y)w(y)) = 0$.

Proof (of Theorem 5.3) We start with a *sketch of the proof*.

Take a vector of parameters $A_\infty = (\bar{K}, \bar{\Omega})$ collinear to $a_\infty = (\rho_\infty, \mu_\infty)$ and consider its perturbation $A_{\infty,s} = (\bar{K}, \bar{\Omega} - s)$. It is immediate to see that for each $s > 0$ equation (40) with $A = A_{\infty,s}$ and the coefficient

$$\gamma_{A_{\infty,s}}(y) = \gamma_{A_\infty}(y) - s\rho(y) = -s\rho_\infty + (\bar{\Omega} - s)\hat{\rho}(y) - \bar{K}\hat{\mu}(y) \tag{48}$$

meets the assumptions of Proposition 3.1, and hence the equation

$$u'' + \gamma_{A_{\infty,\bar{s}}}(y)u = 0 \tag{49}$$

possesses a decaying solution $Z^+(y, A_{\infty,s})$.

Next we consider the solutions $Z^0(y, A_{\infty,s})$ of the same equation with the boundary condition (4). The goal is to detect the values $s > 0$, for which the solutions $Z^0(y, A_{\infty,s})$ and $Z^+(y, A_{\infty,s})$ meet at some intermediate point $\bar{y} \in [0, +\infty[$, i.e., admit at $\bar{y}$ the same value (mod π). In such a case they (or their opposites) can be concatenated into solutions of (40)–(4)–(5). The possibility of such *meeting* follows from Propositions 3.7 and 3.8, according to which the vectors $Z^0(\bar{y}, A_{\infty,s})$ and $Z^+(\bar{y}, A_{\infty,s})$ taken for a sufficiently large $\bar{y} \in [0, +\infty[$ rotate in opposite directions as s grows from some $\bar{s} > 0$.

Increasing $\bar{y}$ if necessary, one can assume that $\gamma_{A_{\infty,s}}(y) < 0$ on $(\bar{y}, +\infty[$ and Arg $Z^+(\bar{y}, A_{\infty,s}) \in]\pi/2, \pi[$ for $\forall s \geq \bar{s}$. On the other hand, for small $s > 0$, Arg $Z^0(\bar{y}, A_{\infty,s})$ is close to Arg $Z^0(\bar{y}, A_\infty)$, which, due to the oscillation property of the limit-case equation, tends to $+\infty$ as $\bar{y} \to +\infty$. Therefore, for each natural m one can find (again increasing $\bar{y}$ when necessary) small $\bar{s} > 0$ such that Arg $Z^0(\bar{y}, A_{\infty,\bar{s}}) > \pi m$. As s will grow from $\bar{s}$ to $\bar{\Omega}$, Arg $Z^0(\bar{y}, A_{\infty,\bar{s}})$ will decrease from the value greater than πm to the value less than π and during this evolution it becomes equal (mod π) to Arg $Z^+(\bar{y}, A_{\infty,s})$ for m distinct values of s.

Now we provide the detailed proofs of the statements i)–iii) of the Theorem.

i) Pick $\bar{K} > 0$ and take $\bar{\Omega} = \frac{\mu_\infty}{\rho_\infty} \bar{K}$, so that $A_\infty = (\bar{K}, \bar{\Omega})$ is collinear with a_∞. Consider the limit-case equation (47) with the parameter A_∞ and choose the solution $Z^0(\cdot; A_\infty)$ which satisfies the boundary condition (4). As long as Eq. (47) is oscillatory, Arg $Z^0(y; A_\infty)$ tends to infinity as $y \to +\infty$. Hence, for each natural m $\exists y_m \in [0, +\infty[$ such that Arg $Z^0(y_m; A_\infty) > \pi m$.

By the continuity of the trajectories of (40) with respect to the parameter A, one can find $\bar{s} > 0$ such that for any $s \in]0, \bar{s}]$ and for $A_{\infty,s} = (\bar{K}, \bar{\Omega} - s)$ there holds Arg $Z^0(y_m; A_{\infty,s}) > \pi m$. For the function $\gamma_{A_{\infty,\bar{s}}}(y)$ defined by (48) one can find $\bar{y} \geq y_m$ such that $\gamma_{A_{\infty,\bar{s}}}(y) < 0$ on $[\bar{y}, +\infty[$. It follows from Remark 3.3iii) that Arg $Z^0(\bar{y}; A_{\infty,\bar{s}}) > \pi m$.

The second-order equation (49) for $s = \bar{s}$ meets the assumptions of Proposition 3.1 and hence has the decaying solution $Z^+(y; A_{\infty,\bar{s}})$. By Proposition 3.6, there holds:

$$\forall y \geq \bar{y}: \text{Arg } Z^+(y; A_{\infty,\bar{s}}) \in]\pi/2, \pi[\ (\text{mod } \pi).$$

Letting s grow from $\bar{s}$ towards $\bar{\Omega}$, we note that the values of $\gamma_{A_{\infty,s}}(y) = \gamma_{A_\infty}(y) - s\rho(y)$ on $[0, +\infty[$ diminish; in particular, $\gamma_{A_{\infty,s}}(y) < 0$ for $y \in [\bar{y}, +\infty[$ for all $s \geq \bar{s}$. According to Proposition 3.7, the function $s \to$ Arg $Z^0(\bar{y}; A_{\infty,s})$ decreases monotonously from the value Arg $Z^0(\bar{y}; A_{\infty,\bar{s}}) > \pi m$ to the value Arg $Z^0(\bar{y}; A_{\infty,\bar{\Omega}}) \in]0, \pi[$.

Consider now the decaying solutions $Z^+(y; A_{\infty,s})$. Proposition 3.8 implies that for chosen $\bar{y}$ Arg $Z^+(\bar{y}; A_{\infty,s})$ grows with the growth of s while remaining (mod π) in the interval $]\pi/2, \pi[$. During this evolution there occur (at least) m values of s_j, $j = 1, \ldots, m$, for which

$$\text{Arg } Z^+(\bar{y}; A_{\infty,s_j}) = \text{Arg } Z^0(\bar{y}; A_{\infty,s_j}) - \pi n \ (n\text{-integer}).$$

Then the concatenations

$$Z(y; A_{\infty,s_j}) = \begin{cases} Z^0(y; A_{\infty,s_j}), & y \leq \tilde{y}, \\ (-1)^n Z^+(y; A_{\infty,s_j}), & y \geq \tilde{y}, \end{cases} \quad (50)$$

are the decaying solutions of the corresponding equations

$$u'' + \left((\bar{\Omega} - s_j)\rho(y) - \bar{K}\mu(y)\right)u = 0,$$

and (50) satisfies the boundary condition (4)–(5).

ii) Let $\tilde{\Omega} \in]0, \bar{\Omega})$ be a limit point of $\Omega_n = \bar{\Omega} - s_n$, $n = 1, \ldots$. Then $\tilde{\Omega} = \bar{\Omega} - \tilde{s} < \bar{\Omega}$.

Consider $\gamma_{A_{\infty,\tilde{s}}}$. There exists $\tilde{y}$, such that $\gamma_{A_{\infty,\tilde{s}}} < 0$ on $[\tilde{y}, +\infty[$. Pick the decaying solution $Z^+(y; A_{\infty,\tilde{s}})$. According to the aforesaid $\forall y \in [\tilde{y}, +\infty[$: Arg $Z^+(y; A_{\infty,\tilde{s}}) \in]\pi/2, \pi[$ (mod π).

Consider the solution $Z^0(y; A_{\infty,\tilde{s}})$ which meets the initial condition (4). If Arg $Z^+(\tilde{y}; A_{\infty,\tilde{s}}) \neq$ Arg $Z^0(\tilde{y}; A_{\infty,\tilde{s}})$ (mod π), then the inequality holds for values of s close to $\tilde{s}$, and in particular for all s_n but a finite number of them, and this results in a contradiction.

Let Arg $Z^0(\tilde{y}; A_{\infty,\tilde{s}}) -$ Arg $Z^+(\tilde{y}; A_{\infty,\tilde{s}}) = \pi m$. Since the function Arg $Z^0(\tilde{y}; A_{\infty,s}) -$ Arg $Z^+(\tilde{y}; A_{\infty,s})$ decreases with the growth of s, one concludes:

$$\text{Arg } Z^+(\tilde{y}; A_{\infty,s}) \neq \text{Arg } Z^0(\tilde{y}; A_{\infty,s}) \pmod{\pi}$$

for all $s \neq \tilde{s}$ from a sufficiently small neighborhood of $\tilde{s}$ and hence for all s_n but a finite number of them, which leads us to the same contradiction.

iii) By the construction provided in i), for each natural m, there exist $A_m = (\bar{K}, \Omega_m)$ and the decaying solution $Z(y, A_m)$ of (40)–(4)–(5), which converges to the origin in such a way that Arg $Z(y, A_m) \in [\pi(m - 1/2), \pi m]$ for sufficiently large y.

To prove its uniqueness, we assume on the contrary that there exists another $A' = (\bar{K}, \Omega')$ and a decaying solution of (40)–(4)–(5) such that for $y \in [y_0, +\infty[$ $\gamma_{A_m}(y) < 0$, $\gamma_{A'}(y) < 0$ and both Arg $Z(y, A')$ and Arg $Z(y, A_m)$ belong to $[\pi(m - 1/2), \pi m]$ for $y \in [y_0, +\infty[$.

Let, for example, $\Omega' > \Omega_m$. Then $\gamma_{A_m}(y) < \gamma_{A'}(y)$ and hence Arg $Z(y_0, A_m) <$ Arg $Z(y_0, A')$. This enters in contradiction with the result of Proposition 3.8. □

Proof (of Theorem 5.2) Let us pick $\bar{K} > 0$ and take $\bar{\Omega} = \tilde{\Omega} = \frac{\mu_\infty}{\rho_\infty}\bar{K}$, so that $A_\infty = (\bar{K}, \bar{\Omega})$ is collinear with a_∞. By assumptions of the Theorem, the function $\gamma_{A_\infty}(y)$ admits positive values on some non-null subinterval $]\underline{c}, c[\subset [0, +\infty[$. The same holds true for $\gamma_{\beta A_\infty}$ with $\beta A_\infty = (\beta \bar{K}, \beta \bar{\Omega})$, $\beta > 0$.

Our proof can be accomplished along the lines of the proof of Theorem 5.3 if one proves that for each N there exists $\beta_N > 0$, for which the solution $Z^0(y, \beta_N A_\infty)$ with initial condition (4) satisfies Arg $Z^0(c, \beta_N A_\infty) > \pi N$.

Consider the equation

$$u''(y) + \gamma_{\beta A_\infty}(y)u = u''(y) + \beta \gamma_{A_\infty}(y)u = 0$$

on the interval $[0, c]$. It is known [3, §A.3, §A.5] that the number of zeros of the solution $u(y, \gamma_{\beta A_\infty}(\cdot))$, or, the same, the increment of Prüfer's angle

$$\text{Arg } Z(y, \gamma_{\beta A_\infty}(\cdot)) - \text{Arg } Z(0, \gamma_{\beta A_\infty}(\cdot)),$$

grows as

$$\pi^{-1} \beta^{1/2} \int_0^y \left(\max(\gamma_{A_\infty}(\eta), 0) \right)^{1/2} d\eta + O(\beta^{1/3}) \tag{51}$$

as $\beta \to +\infty$. Hence choosing sufficiently large $\beta > 0$, we can get a solution $Z^0(y; \beta A_\infty)$ satisfying the boundary condition (4) with the property that Arg $Z^0(c; \beta A_\infty) > N\pi$. Proposition 3.3 yields $Z^0(y; \beta A_\infty) > N\pi, \forall y > c$.

The rest of the proof follows the proof of Theorem 5.3. One can also conclude from (51) that $N(k) \sim k$ as $k \to \infty$, where $N(k)$ is the number of surface-wave solutions with a given wave number $k = K^{1/2}$. □

7 Research Outlooks

The present study can be extended to various models which differ both mathematically and physically.

For one, it appears interesting to study the cases in which $\rho(y)$ and $\mu(y)$ have various types of asymptotic behavior at infinity, e.g., they may be seen as asymptotically periodic, that is, as perturbations of the periodic functions with the perturbing terms asymptotically vanishing at infinity.

Another modification of the model, which is of particular practical relevance, corresponds to a stack of perfectly welded layers, so that the variation of the material properties across the structure is discontinuous (more specifically, piecewise continuous; in particular, piecewise constant in the case of homogeneous layers). While extending Theorem 5.1 onto this case is straightforward, the generalization of the complete formulation of Theorem 5.2 and particularly of Theorem 5.3 is a more subtle matter which calls for an additional study. One of the relevant issues is an extension of the classical results on the oscillation and asymptotic integration of second-order differential equations, which are cited in Sects. 3.2–3.6 and employed later on, onto the case of the equations with discontinuous coefficients. This may require developing the averaging-type results for the discontinuously heterogeneous media.

A similar problem may be posed with respect to another form of the wave differential equation. For instance, the torsional acoustic wave in a radially heterogeneous (functionally graded or discretely layered) rod is described in cylindrical coordinates by the equation which is different from (3) in that it has a regular singular point at the axis $r = 0$. This point is tantamount to the infinite depth of the half-space in rectangular coordinates, so that the surface torsional wave is supposed to decay away from the rod's cylindrical surface towards the axis.

Coming back to the half-space case, one more option is to alter the boundary condition. Such a modification, being mathematically straightforward, is noteworthy for it enables considering a different physics, namely the electromagnetic waves of the so-called TE or TM polarization. The Maxwell equations admitting these wave solutions can be cast to the same form as Eq. (3), in which the mechanical displacement $u(y)$ is to be replaced with electric or magnetic field, $\rho(y)$ with inverse magnetic permeability or inverse dielectric permittivity, and $\mu(y)$ with dielectric permittivity or magnetic permeability, respectively, for the TE or TM waves. At the same time, the boundary condition at the half-space surface $y = 0$ can no longer be of a homogeneous type (cf. Eq. (4)) since, by contrast to sound, light does propagate in vacuum. Therefore, the surface electromagnetic wave must stay continuous at the boundary $y = 0$ and decay in the both directions away from it. The continuity condition may be suitably formulated in terms of the appropriately defined electromagnetic impedance (see [11]), in which terms the essence of the surface-wave existence considerations does not much differ from the present case of shear acoustic waves.

Further generalization of the problem of existence of surface acoustic waves in heterogeneous media can certainly be due to augmenting the number of unknowns and/or the number of variables. The former option implies passing from a scalar differential equation to a vector one, which may describe acoustic waves in anisotropic elastic media or else the waves of miscellaneous coupled-field nature (e.g., acoustoelectric, acoustomagnetic, thermoacoustic waves, etc.) The analytical results on the existence of vector-type surface waves in a spatially heterogeneous (aperiodic) half-space are basically confined to certain particular settings, such as the case of an arbitrarily heterogenous material of finite depth lying on a homogeneous substrate (see [17] for a background), the case of the Rayleigh dispersion branch in a functionally graded anisotropic half-space at high ω and k (validating the WKB asymptotics, see [16] and the bibliography therein), or else the special cases of spatial dependence of the material coefficients which allow for an explicit solution of the underlying wave equation [10, 12]. A different line of attack of the "vector wave" problem in its general formulation, can be extending the present paper's methods via engaging the results on the asymptotic integration of the linear systems of ordinary differential equations together with the multidimensional Sturmian theory (see [2]).

Finally, the most far-reaching variant of the surface-wave problem in a half-space is concerned with the material model which permits spatial variation of properties in more than one coordinate and hence leads to the wave equation in the PDE form. This form of the surface-wave problem has been amply dealt with in the canonical case of periodicity, which is broadly relevant to physical acoustics, and in the framework of the high-frequency asymptotic ray method, which is commonly used in seismic and fluid mechanics [4, 8, 9]. The treatment of the surface-wave problem in its general formulation remains to be a challenging task with plenty of room for further mathematical advances.

8 Conclusions

The present study has dealt with the bi-parametric Sturm-Liouville problem describing shear acoustic waves propagating with (non-zero) frequency ω and wave number k in a functionally graded half-space semi-bounded in the depth coordinate y and infinite in others. The density $\rho(y)$ and shear modulus $\mu(y)$, which appear in the coefficients of the homogeneous wave equation, are assumed to be arbitrary (aperiodic) continuous functions of y with a finite limit at $y \to \infty$. We provided criteria for non-existence/existence of surface-wave solutions in terms of the relation between the ratio $\frac{\rho(y)}{\mu(y)}$ and its limit value $\frac{\rho_\infty}{\mu_\infty}$ at $y \to \infty$. If $\frac{\rho(y)}{\mu(y)} \geq \frac{\rho_\infty}{\mu_\infty}$ for all y, then there are no surface waves. If, otherwise, there exists an interval $]y_1, y_2[$ where $\frac{\rho(y)}{\mu(y)} < \frac{\rho_\infty}{\mu_\infty}$, then the surface waves do exist for any k greater than some $k_0 > 0$, and the number $N(k)$ of solutions grows with the growth of k. An apparently most interesting result is a possibility of existence of an infinite number of solutions for any given k. For this to occur, the requirement $\frac{\mu(y)}{\rho(y)} < \frac{\mu_\infty}{\rho_\infty}$ for all sufficiently large y must be supplemented by a condition on the functions $\rho(y), \mu(y)$, which guarantees the so-called oscillatory property to the particular limit-case second-order equation (47).

A remark is in order that we have not elaborated on the properties of the dispersion spectrum $\omega(k)$ associated with the solutions of the problem. It may be expected that this spectrum consists of disjoint dispersion curves as it is in the case of periodic $\rho(y)$ and $\mu(y)$ (see [17]). However, this has to be confirmed by analysis to be carried out elsewhere.

References

1. Achenbach, J.D., Balogun, O.: Antiplane surface waves on a half-space with depth-dependent properties. Wave Motion **47**(1), 59–65 (2010)
2. Arnold, V.I.: The Sturm theorems and symplectic geometry. Funct. Anal. Appl. **19**, 251–259 (1985)
3. Atkinson, F.V., Mingarelli, A.B.: Multiparameter Eigenvalue Problems. Sturm-Liouville Theory. CRC Press, Boca Raton (2011)
4. Babich, V.M., Kiselev, A.P.: Elastic Waves. High Frequency Theory. CRC Press, Boca Raton (2018)

5. Bellman, R.: Stability Theory of Differential Equations. McGraw-Hill, New York (1953)
6. Birman, V., Byrd, L.W.: Modeling and analysis of functionally graded materials and structures. Appl. Mech. Rev. **60**(5), 195–216 (2007)
7. Biryukov, S.V., Gulyaev, Y.V., Krylov, V.V., Plessky, V.P.: Surface Acoustic Waves in Inhomogeneous Media. Springer, Berlin (1995)
8. Brekhovskikh, L.M., Godin, O.: Acoustics of Layered Media II. Springer Series on Wave Phenomena. Springer, Belin (1999)
9. Cerveny, V.: Seismic Ray Theory. Cambridge University Press, Cambridge (2010)
10. Collet, B., Destrade, M., Maugin, G.A.: Bleustein-Gulyaev waves in some functionally graded materials. Eur. J. Mech A/Solid **25**(5), 695–706 (2006)
11. Darinskii, A.N., Shuvalov, A.L.: Surface electromagnetic waves in anisotropic superlattices. Phys. Rev. A **102**, 033515 (2020)
12. Destrade, M.: Seismic Rayleigh waves on an exponentially graded, orthotropic half-space. Proc. Roy. Soc. A. **463**(2078), 495–502 (2007)
13. Fedoryuk, M.V.: Asymptotic Analysis. Springer, Berlin (1993)
14. Hartman, Ph.: Ordinary Differential Equations. SIAM, Philadelphia (2002)
15. Kennett, B.L.N.: Seismic Wave Propagation in Stratified Media. Cambridge University Press, Cambridge (1983)
16. Shuvalov, A.L.: The high-frequency dispersion coefficient for the Rayleigh velocity in a vertically inhomogeneous anisotropic halfspace. J. Acoust. Soc. Am. **123**(5), 2484–2487 (2008)
17. Shuvalov, A.L., Poncelet, O., Deschamps, M.: General formalism for plane guided waves in transversely inhomogeneous anisotropic plates. Wave Motion **40**(4), 413–426 (2004)
18. Shuvalov, A.L., Poncelet, O., Golkin, S.V.: Existence and spectral properties of shear horizontal surface acoustic waves in vertically periodic half-spaces. Proc. R. Soc. A **465**(2105), 1489–1511 (2009)
19. Ting, T.C.T.: Existence of anti-plane shear surface waves in anisotropic elastic half-space with depth-dependent material properties. Wave Motion **47**(6), 350–357 (2010)
20. Xiaoshan, C., Feng, J., Kishimoto, K.: Transverse shear surface wave in a functionally graded material infinite half space. Philos. Mag. Lett. **92**(5), 245–253 (2012)

Modeling of Microstructures in a Cosserat Continuum Using Relaxed Energies: Analytical and Numerical Aspects

Muhammad Sabeel Khan and Klaus Hackl

1 Introduction

This paper focuses on the treatment of a non-quasiconvex, and therefore ill-posed variational model for granular materials arises as a consequence of the particle counter rotations at the microscale. In continuum mechanics non-quasiconvex potentials may arise due to various reasons, e.g., in the case of strain-softening plasticity [34, 42] they can be caused by non-monotone constitutive behavior, in the case of single slip plasticity they can be due to single slip constraints on the deformation of crystal in association with cross-hardening [22, 23], for twinning induced plasticity they stem from multi-phase energy potentials corresponding to different martensitic variants [8, 12, 32, 35, 38].

So far, different approaches have been discussed in the literature to treat non-quasiconvex variational problems. One possibility is to use regularization techniques which are based on a gradient-type enhancement of the original non-quasiconvex energy function in (5). But the regularization method has its own limitations as far as the physical properties of the unrelaxed problems are concerned.

Contrary to this is the method of relaxation which is a more effective and natural way to deal with non-quasiconvex energies. There are two ways to relax the original non-quasiconvex energy minimization problem (5). Either to enlarge the space of admissible deformations $\left(W^{1,p\in(1,\infty)}\left(\Omega, \mathbb{R}^n\right)\right)$ to the space of parametrized measures [8, 53, 74] or to replace the original non-quasiconvex energy with its relaxed energy envelope. The methodology of constructing a relaxed minimization problem by using parametrized measures is discussed by Carstensen and Roubíček [13, 14], Nicolaides and Walkington [46, 47], Pedregal [51–53], and Roubíček [55–

M. Sabeel Khan · K. Hackl (✉)
Lehrstuhl für Mechanik-Materialtheorie, Ruhr-Universität Bochum, Bochum, Germany
e-mail: klaus.hackl@rub.de

57]. The references which suggest to replace the non-quasiconvex energy with its corresponding relaxed energy function are found in Carstensen et al. [15], Conti and Ortiz [22], Conti and Theil [23], Hackl and Heinen [35], Govindjee et al. [32], Miehe and Gürses [34]. Numerical schemes for calculating relaxed envelopes have been worked out by Aranda and Pedregal [4], Bartels [10], Carstensen et al. [16], Carstensen and Plechac [12], Carstensen and Roubíček [14], Chipot [19], Chipot and Collins [20], Collins et al. [21], Dolzmann and Walkington [30], Pedregal [52], and Roubíček [56]. For a detailed discussion on the methods of relaxation the reader is referred to the work by Dacorogna [25], Ball [7] and the references therein.

Exact analytical results for the relaxed energy are known only for few variational problems in the literature so far. For example, the work of DeSimone and Dolzmann [29] where they give an exact envelope of the relaxed energy potential for the free energy of the nematic elastomers undergoing a transition from isotropic to nematic-phase. Dret and Raoult [44] compute an exact quasiconvex envelope for the Saint Venant-Kirchhoff stored energy function expressed in terms of singular values. Some analytical examples of quasiconvex envelopes are also mentioned by Raoult in [54] for different models in nonlinear elasticity. Kohn and Strang [39, 40] gave an exact formula (see Theorem 1.1 in [39]) for the relaxed energy for a variational problem which has its emergence from the shape optimization problems for electrical conduction. Another exact relaxed result is given by Conti and Theil in [23] for the incremental variational problem for rate-independent single slip elastoplasticity. Conti and Ortiz [22] determine an exact analytical expression for the relaxed energy in single crystal plasticity with a non-convex constraint on the deformation of the crystal requiring all material points must deform via single slip. They extended their analytical expression in [24] to the case of crystal plasticity with arbitrary hardening features. Kohn and Vogelius studied the inverse problem of applied potential tomography and come up with an analytical formula [41] for the relaxed energy by using results from homogenization. In a similar manner but this time with the use of Fourier analysis Kohn presents in Theorem 3.1 of [38] an exact analytical expression for a two-well energy function with application to solid-solid phase transitions.

In this paper, we provide an exact relaxation for the non-quasiconvex energy which arises during our study on the rotational microstructures in granular materials. Due to a large number of industrial applications and their use in everyday life granular materials have been studied extensively throughout the past years. Numerous investigations have been performed in order to model the mechanical behavior of these materials [2, 3, 18, 31, 48–50, 59–61, 63, 65, 66]. In this work, the focus is to consider the counter rotations of granular particles at the microscale and to develop a mechanical model that can predict the formation of distinct deformation patterns that are related to the microstructures in these materials. For an overview on the experimental observations of such patterns the reader is referred to the book by Aranson and Tsimring [5]. For this purpose the continuum description of granular materials is used, specifically the theory of Cosserat continuum.

The present work is organized as follows. In Sect. 2, the intergranular kinematics is discussed and an interaction energy potential contributing to the strain energy of the material is proposed. In Sect. 3, a relaxed variational model for granular materials is presented where we state and prove a theorem on the explicit computation of the relaxed envelope. Employing this result, the exact relaxed energy is derived where all the material regimes are explicitly characterized. In Sect. 4, numerical results demonstrating on the properties of computed relaxed potential are presented. In Sect. 5, application to various continuum problems is shown. Finally in Sect. 6 conclusions are drawn.

2 Intergranular Interactions and Counter Rotations

Intergranular interactions and particle counter rotations in a granular medium subjected to deformation are intriguing and experimentally well recognized [48, 59] phenomenon that contribute in the development of material microstructures [9, 60, 64]. Because of intricate nature of particle rotations and complex behavior of granular materials under deformation it is therefore difficult to understand the intergranular cohesive interactions completely. In literature almost no comprehensive study appears which discuss the intergranular interactions and the arising phenomenon in detail that can truly justify the naturally observed microstructural patterns in deforming granular materials. Although the particle rotations at the microscale have been considered by a number of authors, see, e.g., [1, 17, 18, 49, 60, 64], the essence of their counter rotations especially their interactions in observing the formation of distinct deformation patterns is not well understood. It is therefore our aim to reconsider the intergranular kinematics of counter-rotating particles at microscale and to develop an interaction energy potential for a granular medium that arises as a consequence of these particle counter rotations.

Here, we develop an interaction energy potential that takes into account the intergranular kinematics at the continuum scale and define two new material parameters as a suitable measure for the observation of microstructural phases of granular materials. In this spirit, consider the granular material where two neighboring particles are in contact with each other as shown in Fig. 1. These particle interactions lead to two important modes of deformations called translational and microrotational

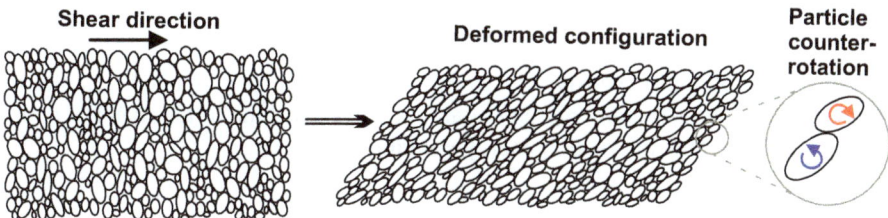

Fig. 1 Schematic of a granular medium subjected to shear with phenomenon of particle counter rotations

motions of the particles which can play a crucial role in the dissipation of the material energy [1, 49] at the continuum scale and therefore contribute to the material strain energy. These independent translational and rotational motions of the granules at the microscale are interlinked with a suitable deformation measure analogous to the concept used in the theory of generalized continuum. Consider now that at the continuum scale the translational motion of the two interacting particles is represented by the vector field $\{u_i \, \mathbf{e}_i\} : \mathbb{R}^d \mapsto \mathbb{R}^d$ and the rotational motion is represented by a field vector analogous to the microrotational vector $\{\varphi_i \, \mathbf{e}_i\} : \mathbb{R}^d \mapsto \mathbb{R}^d$ of the Cosserat continuum. Associated with these deformation field vectors are the strain measures. Corresponding to translational and microrotational vector field these measures are the deformation tensor $[u_{j,i} \, \mathbf{e}_i \otimes \mathbf{e}_j] : \mathbb{R}^d \mapsto \mathbb{R}^{d \times d}$ and $[\varphi_{j,i} \, \mathbf{e}_i \otimes \mathbf{e}_j] : \mathbb{R}^d \mapsto \mathbb{R}^{d \times d}$, respectively. The symmetric part of $u_{j,i} \, \mathbf{e}_i \otimes \mathbf{e}_j$ is the classical strain tensor $\varepsilon_{ij} \, \mathbf{e}_i \otimes \mathbf{e}_j$. An investigation of the rotating phenomenon of the interacting particles reveals that the macroscopic shear $\left(\varepsilon_{ij} - \frac{1}{d} \varepsilon_{kk} \delta_{ij} \right) \mathbf{e}_i \otimes \mathbf{e}_j$ influence the microrotational deformation $\varphi_{j,i} \, \mathbf{e}_i \otimes \mathbf{e}_j$ of the granular particles. This leads us to suggest a proportionality relation between the gradient of the microrotational vector field and the macroscopic shear strain which in mathematical terms is given by

$$\sqrt{\sum_{i,j=1}^{d} (\varphi_{j,i})^2} \propto \sqrt{\sum_{i,j=1}^{d} \left(\varepsilon_{ij} - \frac{1}{d} \varepsilon_{kk} \delta_{ij} \right)^2}, \tag{1}$$

where d is the dimension of the problem under consideration. This proportionality relation is solved with the introduction of the length scale parameter β with the dimension of the inverse of a length. Thus we can write

$$\sqrt{\sum_{i,j=1}^{d} (\varphi_{j,i})^2} = \beta \sqrt{\sum_{i,j=1}^{d} \left(\varepsilon_{ij} - \frac{1}{d} \varepsilon_{kk} \delta_{ij} \right)^2}. \tag{2}$$

Equation (2) is indeed the simplest possible assumption taking into account such an intergranular relationship. More complex forms can be envisioned, but we will demonstrate in the sequel that the present one already leads to a very intricate kinetics.

This brief but comprehensive discussion on intergranular kinematics enables us to propose an interaction energy potential that will contribute to the material strain energy function. This interaction energy potential is stated as

$$\mathscr{I} = \alpha \left(\sum_{i,j=1}^{d} (\varphi_{j,i})^2 - \beta^2 \sum_{i,j=1}^{d} \left(\varepsilon_{ij} - \frac{1}{d} \varepsilon_{kk} \delta_{ij} \right)^2 \right)^2, \tag{3}$$

where Einstein's summation convention is assumed. In tensorial notation it takes the following form

$$\mathscr{I} = \alpha \left(\|\nabla \boldsymbol{\varphi}\|^2 - \beta^2 \, \|\text{sym dev } \nabla \mathbf{u}\|^2 \right)^2, \tag{4}$$

where α and β are non-negative material constants, α is the interaction modulus having information regarding frictional effect in the interacting particles and β is related to the particle size having information regarding intrinsic length scale in Cosserat continuum. The proposed interaction energy potential not only bridges the gap between microstructural properties and the macroscopic behavior of the material but also enables us to characterize different microstructural regimes in granular materials.

3 A Relaxed Variational Model for Granular Materials

3.1 Variational Model

The mechanical response of granular materials can be computed from variational models defined within the context of Cosserat continuum theory. Let Ω be a bounded domain with Lipschitz boundary $\partial \Omega$ and $\mathbf{u} : \Omega \subset \mathbb{R}^d \mapsto \mathbb{R}^d$ be the displacement vector field where d being the dimension of the problem under consideration, $\boldsymbol{\Phi} : \Omega \subset \mathbb{R}^d \mapsto \mathfrak{so}(d) := \{R \in \mathbb{M}^{d \times d} \mid R^T = -R\}$ be the microrotations such that the micromotions of the particles are collected in the vector field $\boldsymbol{\varphi} = axl(\boldsymbol{\Phi}) : \Omega \subset \mathbb{R}^d \mapsto \mathbb{R}^d$, then the deformed configuration of these materials can be completely determined from the following minimization problem

$$\inf_{\mathbf{u}, \boldsymbol{\Phi}, \boldsymbol{\varphi}} \left\{ I(\mathbf{u}, \boldsymbol{\Phi}, \boldsymbol{\varphi}) \, ; \, (\mathbf{u}, \boldsymbol{\Phi}, \boldsymbol{\varphi}) \in W^{1,p}\left(\Omega, \mathbb{R}^d\right) \times W^{1,p}(\Omega, \mathfrak{so}(d)) \times W^{1,p}\left(\Omega, \mathbb{R}^d\right) \right\}. \tag{5}$$

along with the prescribed boundary conditions $\mathbf{u}|_{\partial \Omega_u} = \mathbf{u}_\circ$ and $\boldsymbol{\varphi}|_{\partial \Omega_\varphi} = \boldsymbol{\varphi}_\circ$. Here $W^{1,p}$ is the space of admissible deformations (also known as Sobolev space) with $p \in (1, \infty)$ related to the growth of the energy function W. The integral functional I is defined as

$$I(\mathbf{u}, \boldsymbol{\Phi}, \boldsymbol{\varphi}) = \int_\Omega W(\nabla \mathbf{u}, \boldsymbol{\Phi}, \nabla \boldsymbol{\varphi}) \, dV - \ell(\mathbf{u}, \boldsymbol{\varphi}), \tag{6}$$

where the potential ℓ takes the contribution of external forces $\mathbf{b}$, external couples $\mathbf{m}$, traction forces $\mathbf{t}_u$, and traction moments $\mathbf{t}_\varphi$ such that

$$\ell(\mathbf{u}, \boldsymbol{\varphi}) = \int_\Omega (\mathbf{b} \cdot \mathbf{u} + \mathbf{m} \cdot \boldsymbol{\varphi}) \, dV + \int_{\partial \Omega_u} \mathbf{t}_u \cdot \mathbf{u} \, dS + \int_{\partial \Omega_\varphi} \mathbf{t}_\varphi \cdot \boldsymbol{\varphi} \, dS. \tag{7}$$

In reality, the deformation of granular media is a dissipative process which should not be discussed in terms of energies and displacements. In this sense, our model only covers the initiation of material microstructures. For a full description of extended time-intervals, the variables $\mathbf{u}, \boldsymbol{\Phi}, \varphi$ would have to be replaced by their corresponding velocities and the energy W by a dissipation function. An exposition of this procedure in the case of rigid elasticity can be found in [67–69].

Within the framework of generalized elasticity the mechanical response of granular materials can be determined with the specification of an energy potential that depends, in an independent way, on the particle displacement and microrotations. It is therefore possible to replace the energy potential W in the integral functional (6) by the following Cosserat energy function

$$W^{csrt}(\nabla \mathbf{u}, \boldsymbol{\Phi}, \nabla \varphi) = \frac{1}{2} e(\mathbf{u}, \varphi) : \mathbb{C} : e(\mathbf{u}, \varphi) + \frac{1}{2} \kappa(\varphi) : \overline{\mathbb{C}} : \kappa(\varphi), \qquad (8)$$

which do not only depend on the gradients of the macro and micromotions of the particles but also on a relative macro-rotational deformation tensor $\boldsymbol{\Phi}$ that associates the macro-deformation with the micro-deformation of the particles. Here, $e = \nabla \mathbf{u} - \boldsymbol{\Phi}$ is the Cosserat deformation strain tensor, $\kappa = \nabla \varphi$ is the rotational deformation strain tensor, $\mathbb{C}$ and $\overline{\mathbb{C}}$ are the fourth order constitutive tensors of elastic constants.

The earlier discussion in Sect. 2 on the intergranular interactions and counter rotations of the particles leads us to introduce an enhanced energy potential for the granular materials. In this spirit, the interaction energy potential (4) is integrated with the Cosserat energy function (8) to model the microstructures of the granular materials. This enables us to define a new enhanced energy potential for the granular materials in a Cosserat medium which is given by

$$W(\nabla \mathbf{u}, \boldsymbol{\Phi}, \nabla \varphi) = \underbrace{W^{csrt}(\nabla \mathbf{u}, \boldsymbol{\Phi}, \nabla \varphi)}_{\text{Cosserat energy function}} + \underbrace{\alpha \left(\|\nabla \varphi\|^2 - \beta^2 \|\text{dev sym } \nabla \mathbf{u}\|^2 \right)^2}_{\text{Interaction energy potential}}.$$

(9)

In an isotropic elastic Cosserat medium the enhanced energy potential (9) takes the form

$$W(\nabla \mathbf{u}, \boldsymbol{\Phi}, \nabla \varphi) = \left(\frac{\lambda}{2} + \frac{\mu}{d}\right)(\text{tr } \varepsilon)^2 + \mu \|\text{dev } \varepsilon\|^2 + \mu_c \|\text{asy } \nabla \mathbf{u} - \boldsymbol{\Phi}\|^2 + \frac{\bar{\lambda}}{2}(\text{tr } \kappa)^2$$
$$+ \bar{\mu} \|\text{sym } \kappa\|^2 + \bar{\mu}_c \|\text{asy } \kappa\|^2 + \alpha \left(\|\text{sym } \kappa\|^2 + \|\text{asy } \kappa\|^2 - \beta^2 \|\text{dev } \varepsilon\|^2 \right)^2.$$

(10)

Here, $\lambda, \mu, \mu_c, \bar{\lambda}, \bar{\mu}, \bar{\mu}_c$ are the Cosserat material constants.

The non-convexity and hence the non-quasiconvexity of the energy potential (10) along some chosen strain paths can be seen from Fig. 2. Such non-quasiconvex energy potential when enters in (6) will lead to work with non-quasiconvex energy minimization problem whose general analytical solutions are always of interest. But,

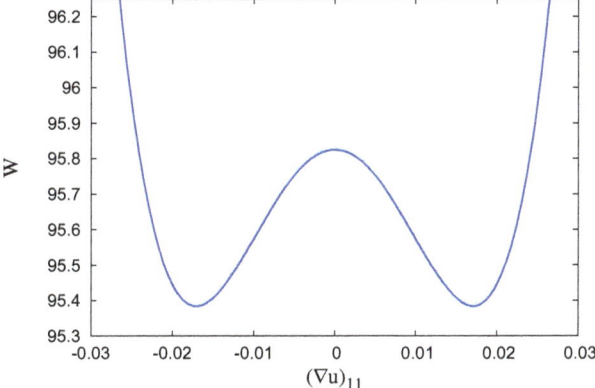

Fig. 2 Unrelaxed energy (10) curve for E $= 2.0 \times 10^2$ (MPa), $\nu = 0.3$, $\mu_c = 1.0 \times 10^{-2}$ (MPa), $\bar{\lambda} = 1.15 \times 10^2$ (N), $\bar{\mu} = 7.69 \times 10^1$ (N), $\bar{\mu}_c = 1.00 \times 10^1$ (N), $\alpha = 1.0 \times 10^1$ (N mm^2) and $\beta = 1.20 \times 10^2$ (mm^{-1})

the solutions to such non-quasiconvex energy minimization problems do not exist in general, which is highly due to fine scale oscillations of the gradients of minimizing deformations. Here, in this case, the non-existence of these solutions is due to the possible displacement and microrotation field fluctuations at fine scales. The fine scale oscillations of the minimizing displacement and microrotation field variables will lead to the development of internal structures in the material. Formation of such microstructures can be extended microstructures [6, 33] which is distributed through the material domain or the localized microstructures [11, 27] which appear in the form of narrow shearing bands. Moreover, the existence of the unique minimizing translational and microrotational deformations are not guaranteed in this situation.

Thus to avoid these problems and to resolve the internal structures of the materials in consideration it is therefore necessary to compute a quasiconvex (relaxed) energy potential W^{rel}. The relaxed potential when entered in the minimization problem (5) now assures the ellipticity of the resulting boundary value problem, since it satisfies the Legendre-Hadamard condition (see definition by Ball and Dacorogna [7, 25]). The study by Morrey [45], Dacorogna [25, 26] gives sufficient justification for the relation of Legendre-Hadamard (ellipticity) condition with the constitutive description of a related mechanical problem.

If possible to compute the exact relaxed envelope of the corresponding non-quasiconvex energy in the energy minimization problem (5) one does not only guarantee general solutions of the associated energy minimization problem but also can predict on the formation of both the extended and localized microstructures in the materials. It is worth mentioning that, in this case, we enable to compute an exact relaxed (quasiconvex) energy envelope corresponding to the non-quasiconvex energy potential in (10).

Since quasiconvex envelopes possess only degenerate ellipticity, only existence of minimizers can be guaranteed, no uniqueness. For numerical purposes it is

3.2 Computation of Relaxed Energy Envelope

In this section, we present our main result concerning the solutions of non-quasiconvex energy minimization problem in (5). In this respect, we compute an exact quasiconvex envelope of the energy function in (10). For other cases where it was possible to construct exact relaxed envelopes corresponding to energy minimization problems addressing different mechanical aspects the reader is referred to the work by Conti and Theil [23], Conti and Ortiz [22], Conti et al. [24], DeSimone and Dolzmann [29], Dret and Raoult [44], Kohn [38], Kohn and Strang [39, 40], Kohn and Vogelius [41], Raoult [54]. The quasiconvex envelope which here termed as the relaxed energy W^{rel} is thus stated as

Theorem 1 Assume $d = 3$, $\lambda, \mu, \mu_c, \bar{\lambda}, \bar{\mu}, \bar{\mu}_c, \alpha, \beta \geq 0$, $\mu_\circ = \min\{\bar{\mu}, \bar{\mu}_c\}$. Let

$$f = \mu_\circ s + \mu c + \alpha \left(s - \beta^2 c\right)^2, \qquad h = \begin{cases} (\bar{\mu} - \bar{\mu}_c) \, \|\text{sym } \kappa\|^2 & \text{if } \bar{\mu} \geq \bar{\mu}_c \\ (\bar{\mu}_c - \bar{\mu}) \, \|\text{asy } \kappa\|^2 & \text{otherwise} \end{cases}$$

and define g by

$$g = \min_{\substack{s,c;\; c \geq \|\text{dev } \varepsilon\|^2, \\ s \geq (\|\text{sym } \kappa\|^2 + \|\text{asy } \kappa\|^2)}} f(s, c). \tag{11}$$

Then, the quasiconvex envelope of the Cosserat strain energy defined in (10) is given by

$$W^{rel} = \left(\frac{\lambda}{2} + \frac{\mu}{d}\right)(\text{tr } \varepsilon)^2 + \mu_c \, \|\text{asy } \nabla \mathbf{u} - \boldsymbol{\Phi}\|^2 + \frac{\bar{\lambda}}{2}(\text{tr } \kappa)^2 + h$$

$$+ g\left(\|\text{sym } \kappa\|^2, \|\text{asy } \kappa\|^2, \|\text{dev } \varepsilon\|^2\right). \tag{12}$$

Proof Consider the rank-one line $\kappa_t = \kappa + t\, \mathbf{a} \otimes \mathbf{b}$; $\mathbf{a}, \mathbf{b} \in \mathbb{R}^d$, $t \in \mathbb{R}$, then

$$W(\varepsilon, \kappa_t) = \left(\frac{\lambda}{2} + \frac{\mu}{d}\right)(\text{tr } \varepsilon)^2 + \mu \, \|\text{dev } \varepsilon\|^2 + \mu_c \, \|\text{asy } \nabla \mathbf{u} - \boldsymbol{\Phi}\|^2 + \frac{\bar{\lambda}}{2}(\text{tr } \kappa)^2$$

$$+ \bar{\mu} \, \|\text{sym } \kappa_t\|^2 + \bar{\mu}_c \, \|\text{asy } \kappa_t\|^2 + \alpha \left(\|\text{sym } \kappa_t\|^2 + \|\text{asy } \kappa_t\|^2 - \beta^2 \|\text{dev } \varepsilon\|^2\right)^2. \tag{13}$$

Now, for any $s \geq \|\kappa\|^2$ we can select $t_- < t \leq 0$ such that $\|\kappa_t\|^2 = s$. A lamination in this direction gives

$$W^{rc} \leq \left(\frac{\lambda}{2} + \frac{\mu}{d}\right)(\operatorname{tr} \boldsymbol{\varepsilon})^2 + \mu_c \|\operatorname{asy} \nabla \mathbf{u} - \boldsymbol{\Phi}\|^2 + \frac{\bar\lambda}{2}(\operatorname{tr} \boldsymbol{\kappa})^2 + h$$
$$+ \min_{s \geq \|\operatorname{sym} \boldsymbol{\kappa}\|^2 + \|\operatorname{asy} \boldsymbol{\kappa}\|^2} \left\{\mu_\circ s + \mu\|\operatorname{dev} \boldsymbol{\varepsilon}\|^2 + \alpha\left(s - \beta^2\|\operatorname{dev} \boldsymbol{\varepsilon}\|^2\right)^2\right\}. \tag{14}$$

Here, rc in the superscript stands for rank-one convex envelope. Working along the rank-one line $\quad e_t = e + t\,\mathbf{c}\otimes\mathbf{d}; \quad \mathbf{c},\mathbf{d}\in\mathbb{R}^d$ and following the arguments above, we obtain

$$W^{rc} \leq \left(\frac{\lambda}{2} + \frac{\mu}{d}\right)(\operatorname{tr} \boldsymbol{\varepsilon})^2 + \mu_c \|\operatorname{asy} \nabla \mathbf{u} - \boldsymbol{\Phi}\|^2 + \frac{\bar\lambda}{2}(\operatorname{tr} \boldsymbol{\kappa})^2 + h$$
$$+ \min_{c \geq \|\operatorname{dev} \boldsymbol{\varepsilon}\|^2} \left\{\mu_\circ\left(\|\operatorname{sym} \boldsymbol{\kappa}\|^2 + \|\operatorname{asy} \boldsymbol{\kappa}\|^2\right) + \mu c \right. \tag{15}$$
$$\left. + \alpha\left(\|\operatorname{sym} \boldsymbol{\kappa}\|^2 + \|\operatorname{asy} \boldsymbol{\kappa}\|^2 - \beta^2 c\right)^2\right\}.$$

Hence the upper bound is proved. The lower bound is based on Lemma 1 below and on the fact that, for $h_1 : [0,\infty)^d \mapsto \mathbb{R}^d$ convex and non-decreasing in each variable and $h_2 : \mathbb{R}^{d\times d} \mapsto \mathbb{R}^d$ component-wise convex, the function $h_1 \circ h_2$ is convex. This completes the proof.

Lemma 1 *Let $f : [0,\infty)^2 \mapsto [0,\infty)$ be convex. Then the function g defined by*

$$g(x) = \inf_{s_1 \geq x_1, s_2 \geq x_2} f(s) \tag{16}$$

is convex and non-decreasing in each variable.

Proof Fix $x', x'', \lambda \in (0,1)$. For any $\epsilon > 0$ there are s', s'' such that $x' \leq s'$, $x'' \leq s''$, and

$$f(s') \leq g(x') + \epsilon, \quad f(s'') \leq g(x'') + \epsilon. \tag{17}$$

Then $\lambda s' + (1-\lambda)s'' \geq \lambda x' + (1-\lambda)x''$, and since f is convex we obtain

$$g(\lambda x' + (1-\lambda)x'') \leq f(\lambda s' + (1-\lambda)s'') \leq \lambda f(s') + (1-\lambda)f(s'')$$
$$\leq \lambda g(x') + (1-\lambda)g(x'') + \epsilon. \tag{18}$$

Therefore g is convex. Monotonicity is clear from the definition.

To compute the exact relaxed envelope in (12) one needs to solve the minimization problem (11). The stationarity conditions to this minimization problem are as follows

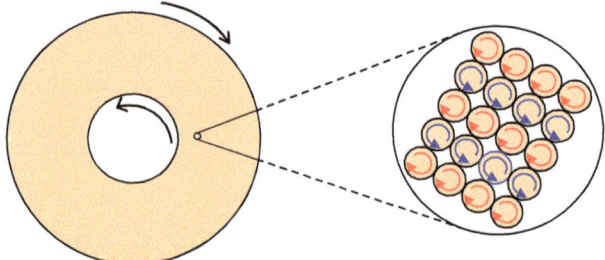

Fig. 3 A Couette shear cell where the two arrows indicate the shearing direction of the inner and outer boundaries of the annular domain. In the inset the microstructure pattern due to microrotational motions of the particles is shown

3.2.1 Stationarity Conditions

(1). for $s = \|\text{sym } \boldsymbol{\kappa}\|^2 + \|\text{asy } \boldsymbol{\kappa}\|^2$ and $c \geq \|\text{dev } \boldsymbol{\varepsilon}\|^2$: $\quad \dfrac{\partial g}{\partial c} = 0, \ \dfrac{\partial g}{\partial s} \geq 0,$ (19a)

(2). for $s = \|\text{sym } \boldsymbol{\kappa}\|^2 + \|\text{asy } \boldsymbol{\kappa}\|^2$ and $c = \|\text{dev } \boldsymbol{\varepsilon}\|^2$: $\quad \dfrac{\partial g}{\partial c} \geq 0, \ \dfrac{\partial g}{\partial s} \geq 0,$ (19b)

(3). for $c = \|\text{dev } \boldsymbol{\varepsilon}\|^2$ and $s \geq \|\text{sym } \boldsymbol{\kappa}\|^2 + \|\text{asy } \boldsymbol{\kappa}\|^2$: $\quad \dfrac{\partial g}{\partial s} = 0, \ \dfrac{\partial g}{\partial c} \geq 0.$ (19c)

On the basis of these three stationarity conditions the material energy can be characterized into the following three phases

3.2.2 Material Regime with Rotational Microstructure (Phase 1)

This phase is corresponding to the material regime where there are microstructures due to the microrotations (which are in fact the rotational degrees of freedom assembled in the microrotational vector field $\boldsymbol{\varphi}$) of the continuum particles. A schematic representation of such microstructure is given in Fig. 3. The enhanced energy potential (10) is non-convex in this microstructural phase. It is observed that whenever the norm of the curvature strain tensor is dominating over the norm of the macroscopic shear strain tensor for some specific choice of the material parameters μ, α, and β, the material experiences a microstructure in microrotations. This microstructural material phase is characterized by the following inequality

$$\|\boldsymbol{\kappa}\|^2 \geq \beta^2 \|\text{dev } \boldsymbol{\varepsilon}\|^2 + \frac{\mu}{2\alpha\beta^2}. \tag{20}$$

It is important to note the effect of shear modulus μ, internal length scale (e.g., the diameter of particles) β and the coherency interaction modulus or frictional modulus α in conjunction with the curvature and macroscopic shear strains which plays very crucial role in the observation of this internal structural phase of the material. Using the first stationarity condition (19a) the minimizers of the problem in (11) are obtained as

$$s = \|\text{sym } \kappa\|^2 + \|\text{asy } \kappa\|^2, \qquad c = \frac{1}{\beta^2}\left(\|\text{sym } \kappa\|^2 + \|\text{asy } \kappa\|^2\right) - \frac{\mu}{2\alpha\beta^4}. \tag{21}$$

Thus, the scalar convex function g is given by

$$g = \begin{cases} \left(\bar{\mu} - \bar{\mu}_c + \mu_\circ + \dfrac{\mu}{\beta^2}\right)\|\text{sym } \kappa\|^2 + \left(\mu_\circ + \dfrac{\mu}{\beta^2}\right)\|\text{asy } \kappa\|^2 - \dfrac{\mu^2}{4\alpha\beta^4} & \text{if } \bar{\mu} \geq \bar{\mu}_c \\[2mm] \left(\mu_\circ + \dfrac{\mu}{\beta^2}\right)\|\text{sym } \kappa\|^2 + \left(\bar{\mu}_c - \bar{\mu} + \mu_\circ + \dfrac{\mu}{\beta^2}\right)\|\text{asy } \kappa\|^2 - \dfrac{\mu^2}{4\alpha\beta^4} & \text{if } \bar{\mu} < \bar{\mu}_c. \end{cases} \tag{22}$$

The relaxed energy of the material in this phase is obtained as

$$W_1^{rel} = \begin{cases} \begin{cases} \left(\dfrac{\lambda}{2} + \dfrac{\mu}{d}\right)(\text{tr } \boldsymbol{\varepsilon})^2 + \mu_c\|\text{asy }\nabla\mathbf{u} - \mathscr{E}\cdot\boldsymbol{\varphi}\|^2 - \dfrac{\mu^2}{4\alpha\beta^4} \\[2mm] + \dfrac{\bar{\lambda}}{2}(\text{tr }\kappa)^2 + (\bar{\mu} - \bar{\mu}_c)\|\text{sym }\kappa\|^2 + \left(\mu_\circ + \dfrac{\mu}{\beta^2}\right)\|\kappa\|^2 \end{cases} & \text{if } \bar{\mu} \geq \bar{\mu}_c, \\[8mm] \begin{cases} \left(\dfrac{\lambda}{2} + \dfrac{\mu}{d}\right)(\text{tr } \boldsymbol{\varepsilon})^2 + \mu_c\|\text{asy }\nabla\mathbf{u} - \mathscr{E}\cdot\boldsymbol{\varphi}\|^2 - \dfrac{\mu^2}{4\alpha\beta^4} \\[2mm] + \dfrac{\bar{\lambda}}{2}(\text{tr }\kappa)^2 - (\bar{\mu} - \bar{\mu}_c)\|\text{asy }\kappa\|^2 + \left(\mu_\circ + \dfrac{\mu}{\beta^2}\right)\|\kappa\|^2. \end{cases} & \text{if } \bar{\mu} < \bar{\mu}_c \end{cases} \tag{23}$$

3.2.3 Material Regime with No Microstructure (Phase 2)

This phase is connected with the material regime where there is no internal structure in the material. The second stationarity condition (19b) clearly shows that the minimizers of the functional in (11) are $\left(\|\text{sym }\kappa\|^2 + \|\text{asy }\kappa\|^2\right)$ and $\|\text{dev }\boldsymbol{\varepsilon}\|^2$, respectively. This indicates that the original energy potential in (10) is convex in this material phase. The criteria for the recognition of this material phase are given by the following inequality relation

$$\beta^2\|\text{dev }\boldsymbol{\varepsilon}\|^2 - \frac{\mu_\circ}{2\alpha} \leq \|\kappa\|^2 \leq \beta^2\|\text{dev }\boldsymbol{\varepsilon}\|^2 + \frac{\mu}{2\alpha\beta^2}. \tag{24}$$

The function g in this phase is given by

$$g = \bar{\mu} \|\text{sym } \kappa\|^2 + \bar{\mu}_c \|\text{asy } \kappa\|^2 + \mu \|\text{dev } \varepsilon\|^2$$
$$+ \alpha \left(\|\text{sym } \kappa\|^2 + \|\text{asy } \kappa\|^2 - \beta^2 \|\text{dev } \varepsilon\|^2 \right)^2. \quad (25)$$

The relaxed energy potential in this phase is thus the original energy potential (10) itself and we write

$$W_2^{rel} = \left(\frac{\lambda}{2} + \frac{\mu}{d}\right) \left(\text{tr } \varepsilon\right)^2 + \mu \|\text{dev } \varepsilon\|^2 + \mu_c \|\text{asy } \nabla \mathbf{u} - \mathcal{E} \cdot \boldsymbol{\varphi}\|^2 + \frac{\bar{\lambda}}{2}\left(\text{tr } \kappa\right)^2$$
$$+ \bar{\mu} \|\text{sym } \kappa\|^2 + \bar{\mu}_c \|\text{asy } \kappa\|^2 + \alpha \left(\|\text{sym } \kappa\|^2 + \|\text{asy } \kappa\|^2 - \beta^2 \|\text{dev } \varepsilon\|^2 \right)^2. \quad (26)$$

3.2.4 Material Regime with Translational Microstructure (Phase 3)

This phase constitutes an unexpected outcome of the theory presented. It consists of laminates formed by alternating displacements as, for example, formed by phase-transforming materials. It would be interesting to see whether such structures can be observed experimentally.

The phase is related to the material regime where there is a microstructure in translational motions (which are in fact the displacement degrees of freedom of the continuum particles and are assembled in the displacement vector field $\mathbf{u}$) of the continuum particles. A schematic representation of such microstructure formation is shown in Fig. 4. The enhanced energy potential (10) thus becomes non-convex in this phase. Using the third stationarity condition (19c) it is observed that the norm of the macroscopic shear strain tensor is dominating over the norm of the rotational strain tensor. The material is said to be in this phase whenever the following criteria is satisfied

$$\beta^2 \|\text{dev } \varepsilon\|^2 - \frac{\mu_o}{2\alpha} \geq \|\kappa\|^2. \quad (27)$$

It is important to note the effect the coherency modulus α and the Cosserat material modulus μ_o in the characterization of this microstructural phase. The minimizers of the functional in (11) are obtained after solving the third stationarity condition (19c) which are given as

$$c = \|\text{dev } \varepsilon\|^2 \quad \text{and} \quad s = \beta^2 \|\text{dev } \varepsilon\|^2 - \frac{\mu_o}{2\alpha}. \quad (28)$$

Thus minimum potential g in (11) takes the following form

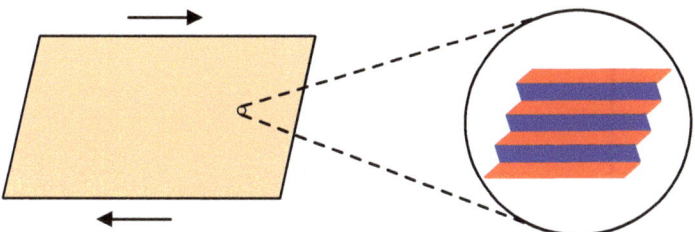

Fig. 4 A rectangular specimen under shear. In the inset the microstructure pattern formed due to the translational motions of the continuum particles is shown

$$g = \begin{cases} (\bar{\mu} - \bar{\mu}_c)\|\text{sym}\,\kappa\|^2 + (\mu_\circ\beta^2 + \mu)\,\|\text{dev}\,\varepsilon\|^2 - \dfrac{\mu_\circ^2}{4\alpha} & \text{if} \quad \bar{\mu} \geq \bar{\mu}_c \\[2mm] (\bar{\mu}_c - \bar{\mu})\|\text{asy}\,\kappa\|^2 + (\mu_\circ\beta^2 + \mu)\,\|\text{dev}\,\varepsilon\|^2 - \dfrac{\mu_\circ^2}{4\alpha} & \text{if} \quad \bar{\mu} < \bar{\mu}_c. \end{cases} \qquad (29)$$

Hence the relaxed energy potential in this phase is obtained as

$$W_3^{rel} = \begin{cases} \begin{cases} \left(\dfrac{\lambda}{2} + \dfrac{\mu}{d}\right)(\text{tr}\,\varepsilon)^2 + \mu_c\,\|\text{asy}\,\nabla\mathbf{u} - \mathscr{E}\cdot\boldsymbol{\varphi}\|^2 + \dfrac{\bar{\lambda}}{2}(\text{tr}\,\kappa)^2 \\[2mm] +(\bar{\mu} - \bar{\mu}_c)\|\text{sym}\,\kappa\|^2 + \left(\mu_\circ\beta^2 + \mu\right)\|\text{dev}\,\varepsilon\|^2 - \dfrac{\mu_\circ^2}{4\alpha} \end{cases} & \text{if} \quad \bar{\mu} \geq \bar{\mu}_c \\[8mm] \begin{cases} \left(\dfrac{\lambda}{2} + \dfrac{\mu}{d}\right)(\text{tr}\,\varepsilon)^2 + \mu_c\,\|\text{asy}\,\nabla\mathbf{u} - \mathscr{E}\cdot\boldsymbol{\varphi}\|^2 + \dfrac{\bar{\lambda}}{2}(\text{tr}\,\kappa)^2 \\[2mm] -(\bar{\mu} - \bar{\mu}_c)\|\text{asy}\,\kappa\|^2 + (\mu_\circ\beta^2 + \mu)\,\|\text{dev}\,\varepsilon\|^2 - \dfrac{\mu_\circ^2}{4\alpha}. \end{cases} & \text{if} \quad \bar{\mu} < \bar{\mu}_c \end{cases} \qquad (30)$$

3.2.5 Relaxed Energy

The total relaxed energy thus comprises all the three energies in each of the phase and it acquires finally the following form

$$W^{rel} = \begin{cases} W_1^{rel} & \text{if } \|\kappa\|^2 \geq \beta^2 \|\text{dev }\varepsilon\|^2 + \dfrac{\mu}{2\alpha\beta^2} \\ \\ W_2^{rel} & \text{if } -\dfrac{\mu_\circ}{2\alpha} \leq \|\kappa\|^2 - \beta^2 \|\text{dev }\varepsilon\|^2 \leq \dfrac{\mu}{2\alpha\beta^2} \\ \\ W_3^{rel} & \text{if } \|\kappa\|^2 \leq \beta^2 \|\text{dev }\varepsilon\|^2 - \dfrac{\mu_\circ}{2\alpha} \\ \\ \end{cases} \quad (31)$$

where W_1^{rel}, W_2^{rel}, and W_3^{rel} are explicitly given as in (23), (26), and (30), respectively. The computation of this analytical expression for the relaxed energy corresponding to non-quasiconvex energy function in (10) thus enable us to predict all microstructural features of the material which are carried safely from the microscopic to macroscopic computational scale. Hence we have extracted all possible information regarding the development of microstructural regimes in the granular materials pertinent to observing its macro-mechanical behavior. For practical applications it is now more efficient and effective to reformulate the original non-quasiconvex problem in (5) to a relaxed energy minimization problem using this relaxed potential.

3.2.6 Nonlinear Constitutive Relations

The proposed granular material model is completed with the formulation of constitutive relations between stress and strain tensors in a Cosserat medium. The constitutive structure of the proposed theory thus comprises of three phases (as discussed above) where in each phase the force-stress are explicitly related to the Cosserat strain tensors according to the following formulas:

$$\sigma = \begin{cases} 2\left(\dfrac{\lambda}{2} + \dfrac{\mu}{d}\right)(\text{tr }\varepsilon)\mathbf{I} + 2\mu_c\left(\text{asy }\nabla\mathbf{u} - \boldsymbol{\Phi}\right), & \text{(Phase 1)} \\ \\ \begin{aligned} & \lambda(\text{tr }\varepsilon)\mathbf{I} + 2\mu\varepsilon + 2\mu_c\left(\text{asy }\nabla\mathbf{u} - \boldsymbol{\Phi}\right) \\ & - 4\alpha\beta^2\left(\|\kappa\|^2 - \beta^2\|\text{dev }\varepsilon\|^2\right)(\text{dev }\varepsilon), \end{aligned} & \text{(Phase 2)} \\ \\ \lambda(\text{tr }\varepsilon)\mathbf{I} + 2\mu\varepsilon + 2\mu_\circ\beta^2(\text{dev }\varepsilon) + 2\mu_c\left(\text{asy }\nabla\mathbf{u} - \boldsymbol{\Phi}\right). & \text{(Phase 3)} \end{cases}$$
$$(32)$$

The couple stress tensor is related to the curvature strain tensors by the following formulas:

$$\boldsymbol{\mu} = \begin{cases} \begin{cases} \bar{\lambda}\left(\operatorname{tr}\boldsymbol{\kappa}\right)\mathbf{I} + 2\left(\bar{\mu}-\bar{\mu}_c\right)\left(\operatorname{sym}\boldsymbol{\kappa}\right) + 2\left(\mu_\circ + \dfrac{\mu}{\beta^2}\right)\boldsymbol{\kappa} & \text{if } \bar{\mu} \geq \bar{\mu}_c, \\ \bar{\lambda}\left(\operatorname{tr}\boldsymbol{\kappa}\right)\mathbf{I} - 2\left(\bar{\mu}-\bar{\mu}_c\right)\left(\operatorname{asy}\boldsymbol{\kappa}\right) + 2\left(\mu_\circ + \dfrac{\mu}{\beta^2}\right)\boldsymbol{\kappa} & \text{if } \bar{\mu} < \bar{\mu}_c. \end{cases} & \text{(Phase 1)} \\[2ex] \begin{cases} \bar{\lambda}\left(\operatorname{tr}\boldsymbol{\kappa}\right)\mathbf{I} + 2\bar{\mu}\left(\operatorname{sym}\boldsymbol{\kappa}\right) + 2\bar{\mu}_c\left(\operatorname{asy}\boldsymbol{\kappa}\right) \\ + 4\alpha\left(\|\operatorname{sym}\boldsymbol{\kappa}\|^2 + \|\operatorname{asy}\boldsymbol{\kappa}\|^2 - \beta^2\|\operatorname{dev}\boldsymbol{\varepsilon}\|^2\right)\boldsymbol{\kappa} \end{cases} & \text{(Phase 2)} \\[2ex] \begin{cases} \bar{\lambda}\left(\operatorname{tr}\boldsymbol{\kappa}\right)\mathbf{I} + 2\left(\bar{\mu}-\bar{\mu}_c\right)\left(\operatorname{sym}\boldsymbol{\kappa}\right) & \text{if } \bar{\mu} \geq \bar{\mu}_c, \\ \bar{\lambda}\left(\operatorname{tr}\boldsymbol{\kappa}\right)\mathbf{I} - 2\left(\bar{\mu}-\bar{\mu}_c\right)\left(\operatorname{asy}\boldsymbol{\kappa}\right) & \text{if } \bar{\mu} < \bar{\mu}_c. \end{cases} & \text{(Phase 3)} \end{cases}$$

(33)

4 Numerical Results

Based on one-dimensional numerical computations the mechanical response of the material is analyzed along some chosen macroscopic strain paths. A simple shear and a tension-compression tests are briefly presented to observe the development of microstructures which is characterized by the activation of different material regimes as discussed in the Sect. 3.2.

4.1 A Simple Shear Test

Consider a two-dimensional domain $\Omega = (0, X_1) \times (0, X_2)$ where $(X_1, X_2) \in \mathbb{R}^2$. We choose the macroscopic strain paths as follows

$$\begin{aligned} \boldsymbol{\varepsilon} &= \frac{\gamma}{2}\left(\mathbf{e}_1 \otimes \mathbf{e}_2 + \mathbf{e}_2 \otimes \mathbf{e}_1\right), \\ \mathbf{e} &= \gamma\, \mathbf{e}_2 \otimes \mathbf{e}_1 + \varphi_3\left(\mathbf{e}_2 \otimes \mathbf{e}_1 - \mathbf{e}_1 \otimes \mathbf{e}_2\right), \\ \boldsymbol{\omega}_e &= \left(\frac{\gamma}{2} + \varphi_3\right)\left(\mathbf{e}_2 \otimes \mathbf{e}_1 - \mathbf{e}_1 \otimes \mathbf{e}_2\right), \\ \boldsymbol{\kappa} &= b\left(\mathbf{e}_1 \otimes \mathbf{e}_3 + \mathbf{e}_2 \otimes \mathbf{e}_3\right). \end{aligned} \qquad (34)$$

Here, γ is the macroscopic shear, φ_3 is the material microrotational degree of freedom, and b is some fixed curvature. We assume that φ_3 linearly depends on both of the material coordinates X_1 and X_2 such that $\varphi_3 = b(X_1 + X_2)$. In this analysis we take $b = \dfrac{\pi}{6}$ and calculate φ_3 for all those material points which lies on the line $X_1 + X_2 = 1$. Other than Lame's constants $\lambda = \dfrac{\nu E}{(1+\nu)(1-2\nu)}$

Table 1 Material parameters for the analytical computations in a simple shear test

Parameter	Numerical value	Units	Parameter	Numerical value	Units
E	2.0×10^2	(MPa)	$\bar{\lambda}$	λ	(N)
μ_c	1.0×10^{-1}	(MPa)	$\bar{\mu}$	μ	(N)
ν	0.3	(—)	$\bar{\mu}_c$	μ_c	(N)
α	5.0×10^{-1}	(N mm^2)	β	1.0×10^1	(mm^{-1})

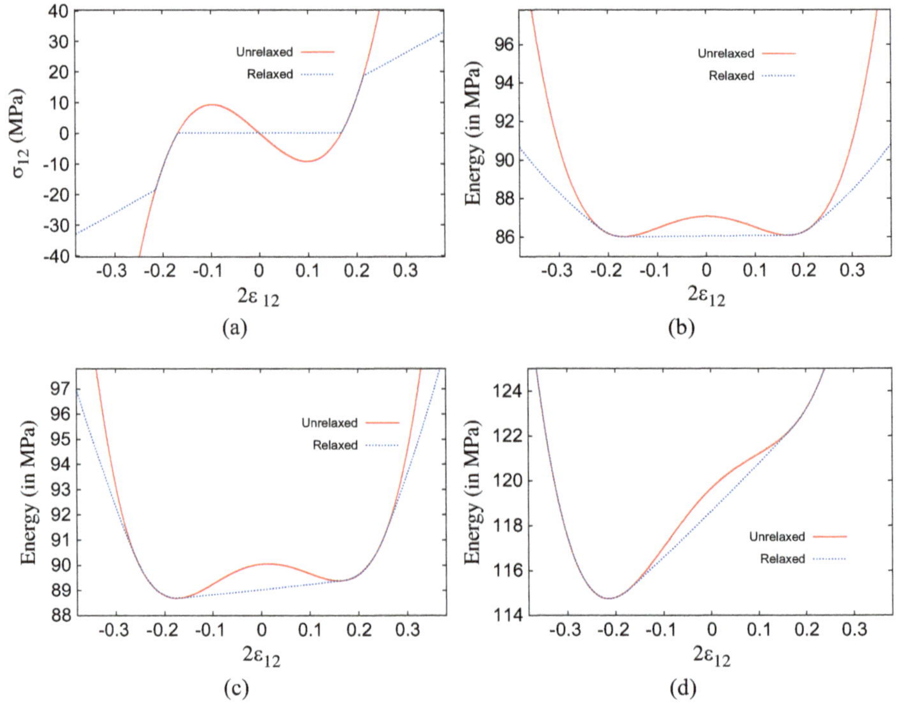

Fig. 5 (a) Relaxed and unrelaxed stress-strain curve in different material regimes; (b) Relaxed and unrelaxed curve for the Cosserat coupled modulus $\mu_c = 0.1$; (c) Relaxed and unrelaxed curve for the Cosserat coupled modulus $\mu_c = 1.0$; and (d) Relaxed and unrelaxed curve for the Cosserat coupled modulus $\mu_c = 10.0$

and $\mu = \dfrac{E}{2(1+\nu)}$ there are eight additional material parameters that are pertinent to the material microstructures and are described in Table 1. Initially the material experiences a rotational microstructure. Upon further loading it enters a regime without microstructure. Further, upon increasing the load it changes its state to a translational microstructure. It is observed that all three phases of the material structure coexist. In Fig. 5a, the constitutive response of the material is shown, where it is observed that the non-monotone stress-strain curve is replaced by its energetically equivalent Maxwell line corresponding to a uniformly vanishing stress. This regime corresponds to a rotational microstructure. In the microstructure free material regime a nonlinear constitutive response is seen. Whereas, in the material regime with translational microstructure, we observe a linear constitutive response in this one-dimensional analysis. The corresponding non-convex and relaxed energy plots are shown in Fig. 5b. In Fig. 5c and d, the relaxed and unrelaxed energy is plotted for two different values of the Cosserat modulus $\mu_c = 1.0$ and $\mu_c = 10.0$, respectively. These figures demonstrate that not only the particle size in granular material effects the development of microstructures but also the Cosserat shear modulus does have influence in the development of material microstructures in granular materials.

4.2 A Tension-Compression Test

In this example the material behavior in a plane strain tension-compression test is investigated. The macroscopic strain tensors for this analysis take the following form

$$\boldsymbol{\varepsilon} = \delta \, \mathbf{e}_1 \otimes \mathbf{e}_1,$$
$$\boldsymbol{e} = \delta \, \mathbf{e}_1 \otimes \mathbf{e}_1 + \varphi_3 \left(\mathbf{e}_2 \otimes \mathbf{e}_1 - \mathbf{e}_1 \otimes \mathbf{e}_2 \right), \quad (35)$$
$$\boldsymbol{\omega}_e = \varphi_3 \left(\mathbf{e}_2 \otimes \mathbf{e}_1 - \mathbf{e}_1 \otimes \mathbf{e}_2 \right).$$

Here δ is the macroscopic stretch. The Cosserat rotational strain tensor $\boldsymbol{\kappa}$ is taken to be the same as mentioned in the previous test. Moreover, the microrotational degree of freedom, φ_3 at each material point is calculated according to a similar assumption as in the case of the simple shear test. The material parameters are chosen as described in Table 2. It is observed that all the three phases of material structure coexist in this case. The constitutive behavior in the material microstructural and non-microstructural regimes is shown in Fig. 6a where contrary to the case of the shear test it is observed that the stress does not vanish in the regime where the material displays a rotational microstructure. Here, the non-monotone stress-strain curve is replaced by a monotone one. This is due to the non-constant slope of the relaxed energy envelope in the globally non-convex range of the unrelaxed energy potential, as seen in the magnified inset in Fig. 6b. Moreover, the properties

Table 2 Material parameters for the analytical computations in a tension-compression test

Parameter	Numerical value	Units	Parameter	Numerical value	Units
E	2.0×10^2	(MPa)	$\bar{\lambda}$	1.15×10^2	(N)
μ_c	1.0×10^{-2}	(MPa)	$\bar{\mu}$	7.69×10^1	(N)
ν	0.3	(—)	$\bar{\mu}_c$	1.00×10^1	(N)
α	1.0×10^{-1}	(N mm^2)	β	1.20×10^2	(mm^{-1})

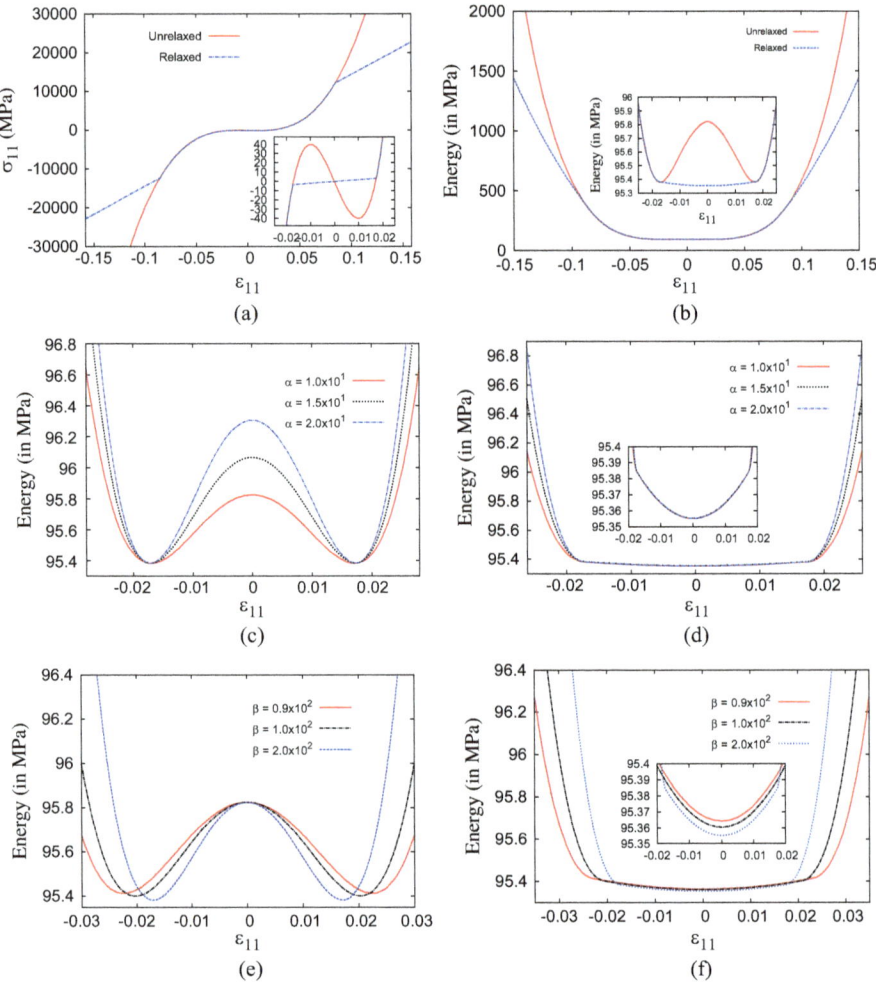

Fig. 6 (**a**) Relaxed and unrelaxed stress-strain curve in different regimes of the material; (**b**) Relaxed and unrelaxed energy curve in different material regimes; (**c**) Unrelaxed energy curves for varying values of the material parameter α; (**d**) Relaxed energy curves for varying values of the material parameter α; (**e**) Unrelaxed energy curves for varying values of the material parameter β; and, (**f**) Relaxed energy curves for varying values of the material parameter β

of unrelaxed and relaxed energy envelope are studied for different values of the interaction modulus α and the material parameter β related to particle size. A two-well energy structure is seen in Fig. 6c for three different values of the interaction modulus. Both wells have the same local minimum. In Fig. 6c, it is observed that by varying the interaction modulus the local minima of the energy envelope do not change. The computed relaxed energy is plotted in Fig. 6d where it is seen that by varying the interaction modulus the global minima of all the three energy curves do not change. The influence of the particle size on the material strain energy is observed in Fig. 6e and f. It is seen that the particle size does not only influence the range of local non-convexity of the energy potential but also its global non-convexity range. It is important to note that the local maxima of the energy potential do not change with the varying particle size. This is contrary to the case seen in Fig. 6c. Moreover, the local and global minima of the potential are shifted and get lower values with increased value of the material parameter β, as seen in Fig. 6e and f.

5 Finite Element Results

In the absence of body force **b** and body couple **m** the system of linear and angular momentum weak-balance equations are solved numerically using a finite element discretization of the material domain. Here we present numerical experiments to demonstrate on the important features of the exact quasiconvex energy envelopes. The computations are performed to simulate granular material behavior in a Couette shear cell, under compression and in indentation. The geometry of the model is reduced to two dimensions thereby allowing to compute three degrees of freedom at each point of the material domain. Two of them are the displacements u_1 and u_2 and third is the microrotation φ_3. In the computations these degrees of freedom are approximated on each node of an element Ω^e using biquadratic interpolation functions. A plain strain assumption is used in all the three cases in consideration.

5.1 Extended Microstructure in a Couette Shear Cell

Couette shear cells have been used to analyze the shear flows in granular materials in a number of numerical and experimental studies. For an overview on the comparison between numerical and experimental results obtained in a granular Couette shear the reader is referred to the paper by Lätzel et al. [43]. These investigations provide a clear evidence on the formation of different microstructural patterns in granular materials under shear deformation. These appear as a result of localized deformations near the rotating cylinder in a Couette annular geometry [28, 36, 58, 71, 72]. These investigations have shown that under intense shearing different deformation patterns develop. The formation depends upon the interactions of the granular particles at microscale. The kinematic of these particle rotations is shown in Fig. 7c where two possible kind of particle rotations, namely counter rotations and identical rotations can be seen.

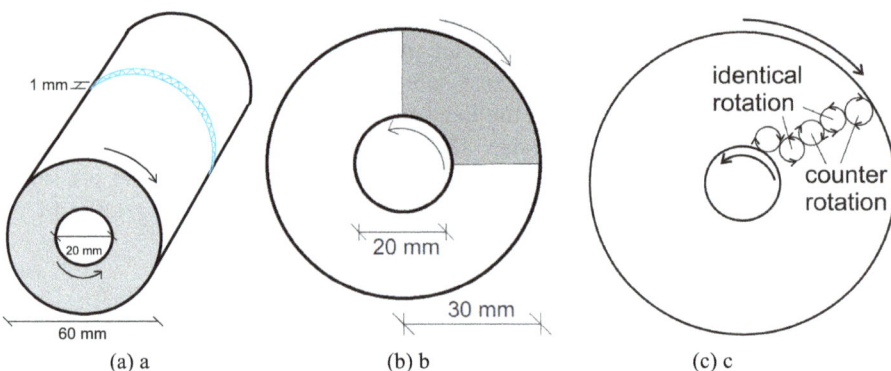

Fig. 7 (**a**) Geometry of two circular rotating cylinders, (**b**) Reduced Couette geometry and boundary conditions, (**c**) Kinematics of particle rotations: a schematic of a rotating particle chain exhibiting different sense of rotation

Here, we are able to perform a numerical experiment to show that all material regimes introduced above may occur simultaneously. For this purpose, a Couette annular geometry is taken into consideration to observe the formation of microstructure using the proposed theory. The granular material is confined between two concentric rigid circular cylinders as shown in Fig. 7a. The cylinders are subjected to rotations in opposite direction. Due to symmetry we model only the first quadrant. The width of the annulus is taken to be 20 mm. The inner circular boundary is at a radius of 10 mm from the origin of the annulus. The circular boundaries are supposed to rotate in opposite direction. The boundary conditions for the numerical simulation using the proposed model uses fixed displacement along the circular boundaries whereas a small microrotation is prescribed at the boundaries.

Our intension with this study is to observe the development of microstructural phases within the annular domain subjected to rotational deformation. The both rotational and translational microstructures develop as shown in Fig. 8. In Fig. 8a the material exhibits a rotational microstructure. In Fig. 8b–d, it is shown that for particular parameters as listed in Table 3, all the material phases coexist. Moreover, the deformed configurations in Fig. 8 indicate that decreasing the value of β causes the material to behave more softly. Also with decrease in particle size, rotational microstructures are more pronounced.

5.2 *Localized Deformations in Granular Materials*

Localization of deformation has been observed both numerically [27, 70] and experimentally [37, 62] in a number of physical situations. Our emphasis with this study is to show that it is possible to observe this phenomenon with the application of exact relaxed potentials. To illustrate on the formation of these localized

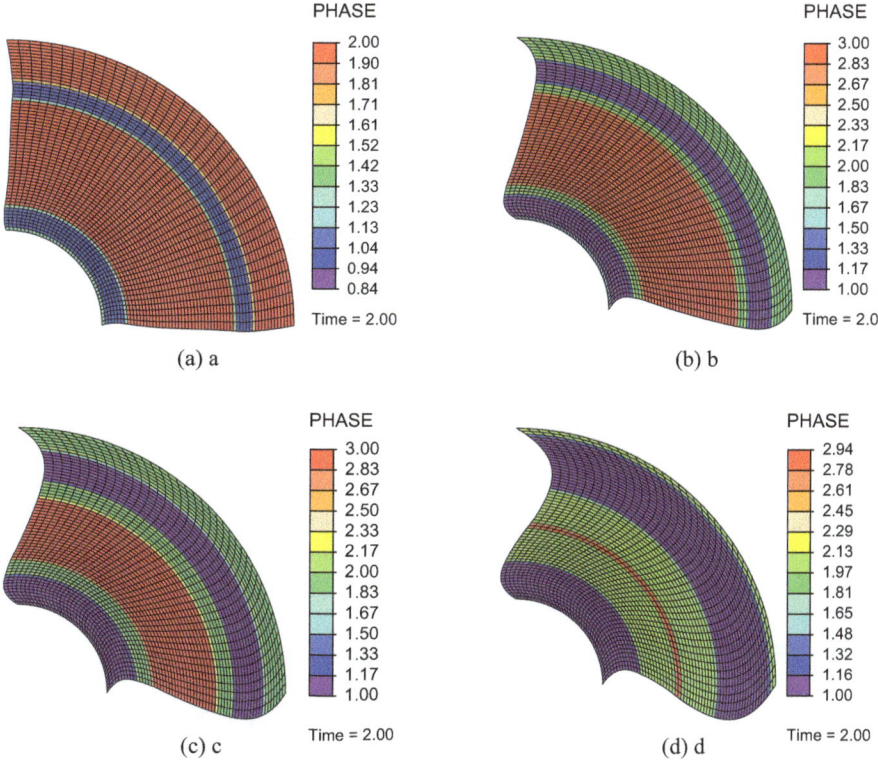

Fig. 8 Couette shear cell: (**a**) coexistence of areas with no microstructure (red) and rotational microstructure (blue), (**b–d**) coexistence of areas with translational microstructure (red), no microstructure (green) and rotational microstructure (purple)

Table 3 Material parameters for the shear test in a Couette geometry

Figure	E	ν	μ_c	α	β	$\bar{\mu}$	$\bar{\mu}_c$
–	(MPa)	–	(MPa)	(N mm^2)	(mm^{-1})	(N)	(N)
8a	2.0×10^2	0.3	2.0×10^0	2.0×10^5	5.8×10^{-1}	8.0×10^1	5.0×10^1
8b	2.0×10^2	0.3	2.0×10^2	2.0×10^5	5.8×10^{-1}	8.0×10^1	5.0×10^1
8c	2.0×10^2	0.3	2.0×10^2	2.0×10^5	4.0×10^{-1}	8.0×10^1	5.0×10^1
8d	2.0×10^2	0.3	2.0×10^2	2.0×10^5	2.0×10^{-1}	8.0×10^1	5.0×10^1

deformation bands a tension-compression test performed on a rectangular specimen is presented. The formation of microstructural zones in the specimen clearly predicts the localized deformation mechanism observed by Kaus and Podladchikov [37].

5.2.1 A Rectangular Specimen in Compression

In this example, a rectangular specimen of a granular material is considered with a small imperfection in the form of a weak element at the center of the specimen as shown in Fig. 9. The material parameters used for the simulation are given in Table 4. The geometry and boundary conditions are shown in Fig. 9, where the vertical displacements on both top and bottom of the specimen are constrained. The material points are allowed to move horizontally at both the top and bottom boundary of the specimen except the point at the left lower corner of the specimen which is fixed in both the horizontal and vertical direction. Additionally a frictional boundary condition is used where a microrotation of the continuum points is allowed at both the top and bottom boundary of the specimen. A maximum displacement of -34.8 cm is applied at the top boundary in vertical direction. We consider 1000 loading steps with a load step size of 4.35×10^{-3}.

Two different mesh sizes for discretizing the specimen into finite elements are used in the analysis, see Fig. 10. It is observed that microstructure in the material develops in zones where material failure may possibly occur. The red color zone is corresponding to the material phase exhibiting translational microstructure. The width of the microstructural band is not affected by mesh size. This is highly due

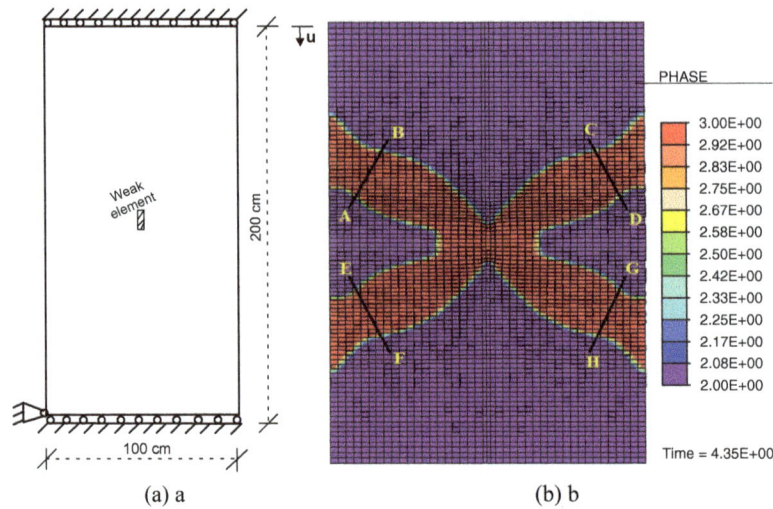

(a) a (b) b

Fig. 9 (a) Geometry and boundary conditions of the rectangular specimen with weak element. (b) Selected lines along the width of the microstructural zone

Table 4 Material parameters for the specimen with introduced imperfection in compression

–	E	ν	μ_c	α	β	$\overline{\mu}$	$\overline{\mu}_c$
–	(MPa)	–	(MPa)	(N mm^2)	(mm^{-1})	(N)	(N)
Mesh	2.0×10^5	0.3	2.0×10^1	5.0×10^1	1.5	7.0	2.0×10^1
Weak element	2.0×10^3	0.3	2.0×10^1	1.0×10^3	1.5	3.0×10^2	4.0

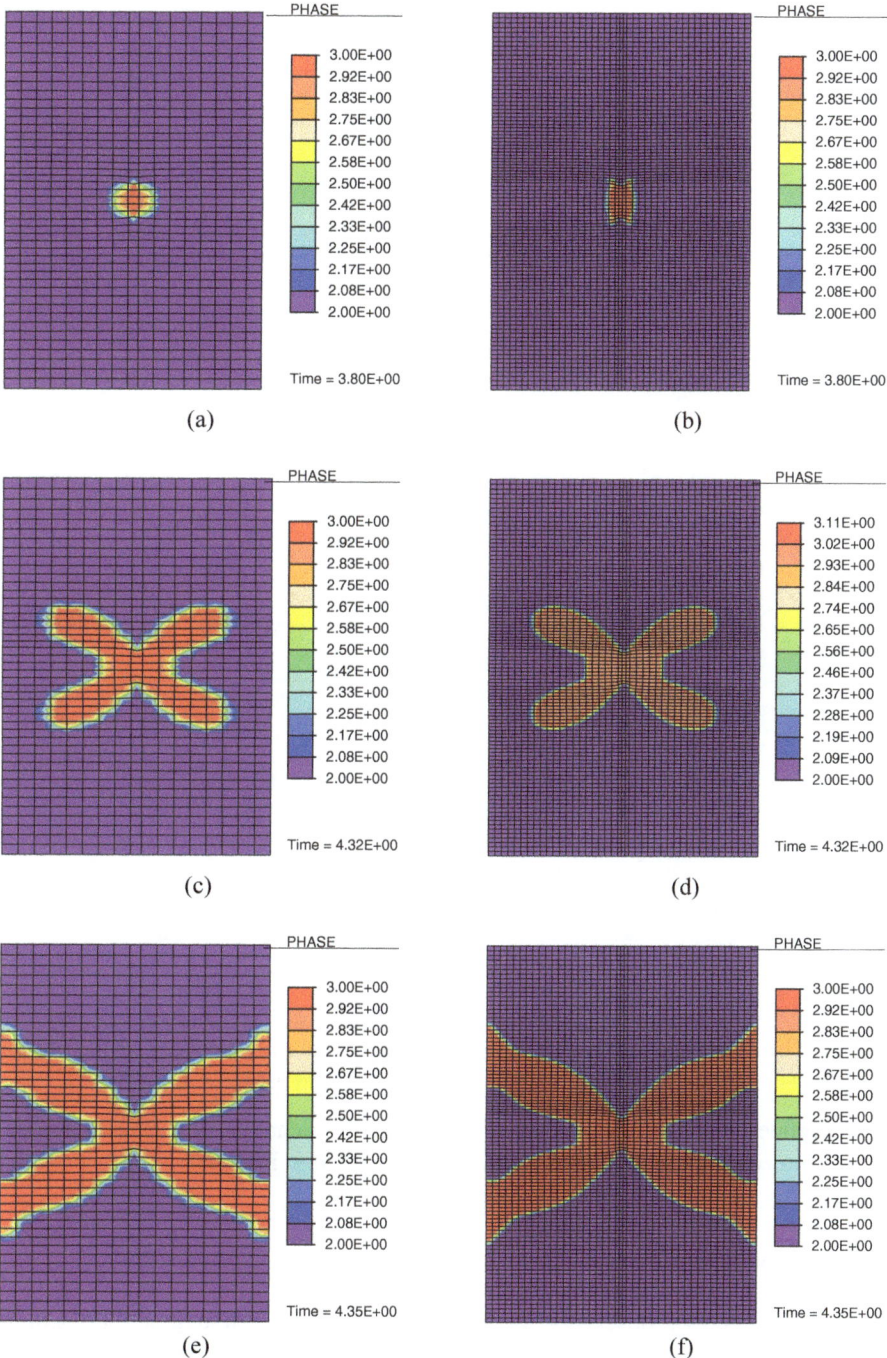

Fig. 10 Rectangular specimen in compression, deformed configuration and material phases. First column: coarse mesh consisting of 765 elements. Second column: fine mesh consisting of 4214 elements

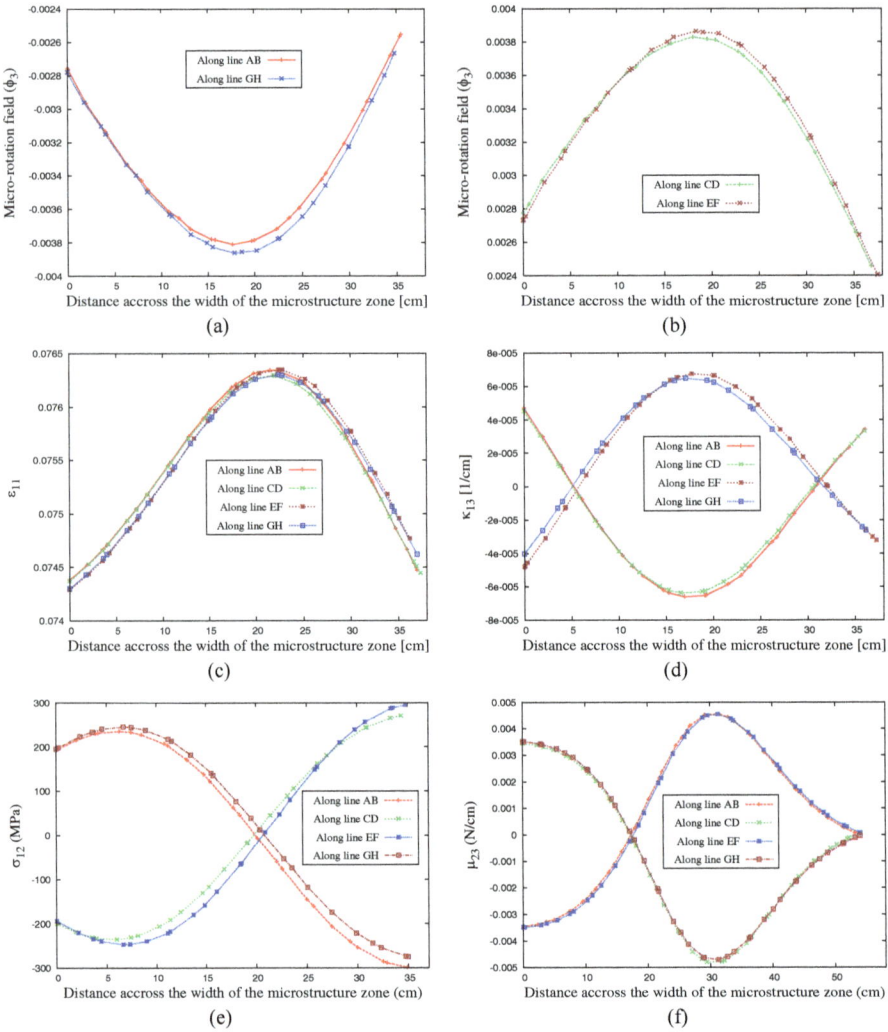

Fig. 11 Distribution of Cosserat rotation φ_3 (**a**) and (**b**), strains ε_{11} (**c**), torsion κ_{13} (**d**), stress σ_{12} (**e**) and couple stress μ_{23} (**f**) along the width of microstructural zones

to the properties of the exact relaxed potentials. The behavior of selected variables across the zones exhibiting microstructure is depicted in Fig. 11

5.2.2 Indentation Test

The significance of observing the mechanical response of a granular medium subjected to indentation is evident from the load bearing capacity problems in geotechnical engineering. Particle rotations have always played an important role in load bearing capacity problems and are therefore of keen interest to many researchers [5, 9, 64] where Cosserat continuum was used to describe the behavior of granular medium subjected to indentation.

In this example we study the microstructure formation in a granular medium under indentation with the application of relaxation theory within the framework of Cosserat continuum. A plain strain assumption is used in this analysis. A body of dimensions 200×100 cm^2 is subjected to indentation by a flat rigid indenter with a dimension of 50×5 cm^2 as shown in Fig. 12. The geometry of the granular medium is discretized into 2560 finite elements whereas the geometry of the indenter is discretized into 250 finite elements. The indenter can only move in vertical direction and this constraint is applied by fixing the horizontal degrees of freedom of all the nodal points of indenter. Both the horizontal and vertical degrees of freedom on the right and left boundary of the granular medium are fixed. The continuum points can move only in horizontal direction at the base of the granular medium which is ensured by fixing the vertical degrees of freedom. The punching of the indenter is controlled by the applied vertical displacements, where a maximum displacement of 3.76 cm is applied at the top nodes of the indenter mesh in 1390 loading steps with a step size of 1.4×10^{-3}. The material parameters used for the indenter and the granular medium are shown in Table 5. A large number of experiments have been performed on the granular foundations subjected to indentation revealing similar bands of localized deformations as shown in this investigation. Also numerical simulation using a finite element scheme for the Cosserat continuum by Walsh and Tordesillas [73] has shown such kind of microstructure formation. We show a numerical solution where the development of microstructure has been predicted in the localized zones around the indenter. The nucleation and the evolution of the microstructural zone can be observed as the indenter moves downward into the material domain. The zone developed beneath the indenter is corresponding to translational microstructure. The results in Fig. 13 are in accordance to the generalized Prandtl's solution of a rigid flat punch problem and is in agreement with the experimental investigation [73].

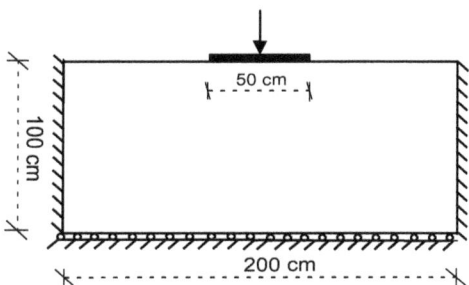

Fig. 12 Geometry of the granular medium under indentation along with the prescribed boundary conditions

Table 5 Material parameters for the indentation test on a granular medium

–	E $(\frac{N}{cm^2})$	ν	μ_c $(\frac{N}{cm^2})$	α (N cm^2)	β (cm^{-1})	$\bar{\mu}$ (N)	$\bar{\mu}_c$ (N)
Granular medium	2.0×10^4	0.3	2.0	5.0×10^4	0.5	7.0×10^3	2.0×10^2
Indenter	2.0×10^{12}	0.3	2.0	5.0×10^3	0.5	7.0×10^3	2.0×10^2

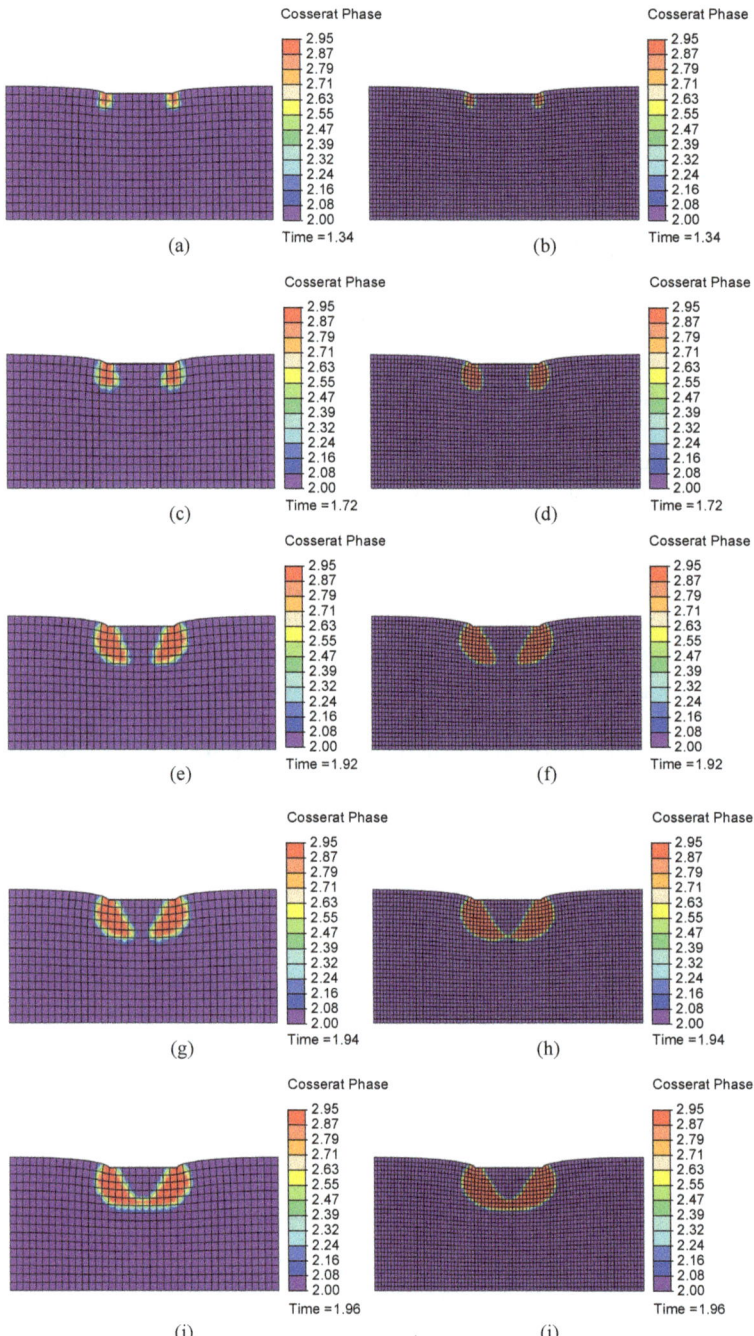

Fig. 13 Microstructure development beneath indenter in a granular foundation

6 Conclusion

In nature granular materials exhibit distinct patterns under deformation. The formation of these patterns is strongly influenced by counter rotations of the interacting particle at the microscale. In this article, we study the counter rotations of the particles and the formation of rotational microstructures in granular materials.

By employing the direct methods in the calculus of variations it turns out to be possible to derive an exact quasiconvex envelope of the energy potential. It is worth mentioning that there are no further assumptions necessary to derive this quasiconvex envelope. The computed relaxed potential yields all the possible displacement and microrotation field fluctuations as minimizers. Hence, by doing so we do not only resolve the issues concerning related non-quasiconvex variational problem but also guarantee the existence and uniqueness of energy minimizers. Moreover, the independence of these minimizers on the discretization of the spatial domain is ensured. We conclude with the result that the granular material behavior can be divided into three different regimes. Two of the material regimes are exhibiting microstructures in rotational and translational motions of the particles, respectively, and the third one is corresponding to the case where there is no internal structure of the deformation field.

The proposed model is analyzed numerically employing material point and finite element calculations. Moreover, it has been shown that all possible phases can coexist in a specific material body at the same instant in time.

Acknowledgments The first author gratefully acknowledges the funding by Higher Education Commission (HEC) of Pakistan and highly appreciate the support by Deutscher Akademischer Austausch Dienst (DAAD) for this research work. The second author gratefully acknowledges the funding by the German Research Foundation (DFG) in the framework of Project C4 of the Collaborative Research Center "Interaction Modeling in Mechanized Tunneling" (SFB 837).

References

1. Alonso-Marroquín, F., Vardoulakis, I., Herrmann, H.J., Weatherley, D., Mora, P.: Effect of rolling on dissipation in fault gouges. Phys. Rev. E **74**, 301–306 (2006). https://doi.org/10.1103/PhysRevE.74.031306
2. Alsaleh, M.I., Voyiadjis, G.Z., Alshibli, K.A.: Modeling strain localization in granular materials using micropolar theory: mathematical formulations. Int. J. Numer. Anal. Meth. Goemech. **30**, 1501–1524 (2006). https://doi.org/10.1002/nag.533
3. Alshibli, K.A., Alsaleh, M.I., Voyiadjis, G.Z.: Modelling strain localization in granular materials using micropolar theory: numerical implementation and verification. Int. J. Numer. Anal. Meth. Goemech. **30**, 1525–1544 (2006). https://doi.org/10.1002/nag.534
4. Aranda, E., Pedregal, P.: Numerical approximation of non-homogeneous, non-convex vector variational problems. Numer. Math. **89**, 425–444 (2001). https://doi.org/10.1007/s002110100294
5. Aranson, I., Tsimring, L.: Granular Patterns. Oxford University Press, Oxford (2009)
6. Bagnold, R.A.: The Physics of Blown Sand and Desert Dunes. Methuen, London (1941)

7. Ball, J.M.: Convexity conditions and existence theorems in nonlinear elasticity. Arch. Ration. Mech. Anal. **63**, 337–403 (1976). https://doi.org/10.1007/BF00279992
8. Ball, J.M., James, R.D.: Fine phase mixtures as minimizers of energy. Arch. Ration. Mech. Anal. **100**, 13–52 (1987). https://doi.org/10.1007/bf00281246
9. Bardet, J.P.: Observation on the effects of particle rotations on the failure of idealized granular materials. Mech. Mater. **8**, 159–182 (1994). https://doi.org/10.1016/0167-6636(94)00006-9
10. Bartels, S.: Numerical analysis of some non-convex variational problems. PhD thesis, Christian-Alberechts-Universität, Kiel (2001)
11. Bauer, E., Huang, W.: Numerical investigation of strain localization in a hypoplastic Cosserat material under shearing. In: Desai (ed.) Proceedings of the 10th International Conference on Computer Methods and Advances in Geomechanics, pp. 525–528. Taylor & Francis, Routledge (2001)
12. Carstensen, C., Plecháč, P.: Numerical solution of the scalar double-well problem allowing microstructure. Math. Comp. **66**, 997–1026 (1997). https://doi.org/10.1090/S0025-5718-97-00849-1
13. Carstensen, C., Roubíček, T.: Numerical approximation of young measures in non-convex variational problems. Tech. Rep., 97–18. Universität Kiel (1997)
14. Carstensen, C., Roubíček, T.: Numerical approximation of young measures in non-convex variational problems. Numer. Math. **84**, 395–415 (2000). https://doi.org/10.1007/s002119900122
15. Carstensen, C., Hackl, K., Mielke, A.: Nonconvex potentials and microstructures in finite-strain plasticity. Proc. R. Soc. Lond. A **458**, 299–317 (2002). https://doi.org/10.1098/rspa.2001.0864
16. Carstensen, C., Conti, S., Orlando, A.: Mixed analytical-numerical relaxation in finite single-slip crystal plastictiy. Contin. Mech. Thermodyn. **20**, 275–301 (2008). https://doi.org/10.1007/s00161-008-0082-0
17. Chang, C.S., Hicher, P.Y.: An elasto-plastic model for granular materials with microstructural consideration. Int. J. Solids Struct. **42**, 4258–4277 (2005). https://doi.org/10.1016/j.ijsolstr.2004.09.021
18. Chang, C.S., Ma, L.: Elastic material constants for isotropic granular solids with particle rotation. Int. J. Solids Struct. **29**, 1001–1018 (1992). https://doi.org/10.1016/0020-7683(92)90071-Z
19. Chipot, M.: The appearance of microstructures in problems with incompatible wells and their numerical approach. Numer. Math. **83**, 325–352 (1999). https://doi.org/10.1007/s002110050452
20. Chipot, M., Collins, C.: Numerical approximation in variational problems with potential wells. SIAM J. Numer. Anal. **29**, 1002–1019 (1992). https://doi.org/10.1137/0729061
21. Collins, C., Kinderlehrer, D., Luskin, M.: Numerical approximation of the solution of a variational problem with a double well potential. SIAM J. Numer. Anal. **28**, 321–332 (1991). https://doi.org/10.1137/0728018
22. Conti, S., Ortiz, M.: Dislocation microstructures and the effective behavior of single crystals. Arch. Ration. Mech. Anal. **176**, 103–147 (2005). https://doi.org/10.1007/s00205-004-0353-2
23. Conti, S., Theil, F.: Single-slip elastoplastic microstructures. Arch. Ration. Mech. Anal. **178**, 125–148 (2005). https://doi.org/10.1007/s00205-005-0371-8
24. Conti, S., Hauret, P., Ortiz, M.: Concurrent multiscale computing of deformation microstructure by relaxation and local enrichment with application to single-crystal plasticity. Multiscale Model. Simul. **6**, 135–157 (2007). https://doi.org/10.1137/060662332
25. Dacorogna, B.: Direct Methods in the Calculus of Variations. Springer, Berlin (1989)
26. Dacorogna, B.: Necessary and sufficient conditions for strong ellipticity of isotropic functions in any dimension. Discrete Contin. Dynam. Syst. Ser. B. **1**, 257–263 (2001). https://doi.org/10.3934/dcdsb.2001.1.257
27. de Borst, R.: Simulation of strain localization: a reappraisal of the Cosserat-continuum. Eng. Comput. **8**, 317–332 (1991). https://doi.org/10.1108/eb023842
28. Debrégeas, G., Tabuteau, H., di Meglio, J.-M.: Deformation and flow of a two-dimensional foam under continuous shear. Phys. Rev. Lett. **87**, 178305 (2001). https://doi.org/10.1103/PhysRevLett.87.178305

29. DeSimone, A., Dolzmann, G.: Macroscopic response of nematic elastomers via relaxation of a class of SO(3)-invariant energies. Arch. Ration. Mech. Anal. **161**, 181–204 (2002). https://doi.org/10.107/s002050100174
30. Dolzmann, G., Walkington, N.J.: Estimates for numerical approximations of rank one convex envelopes. Numer. Math. **85**, 647–663 (2000). https://doi.org/10.1007/PL00005395
31. Ehlers, W., Volk, W.: On shear band localization phenomena of liquid-saturated granular elastoplastic porous solid materials accounting for fluid viscosity and micropolar solid rotations. Mech. Cohes.-Frict. Mat. **2**, 301–320 (1997). https://doi.org/10.1002/(SICI)1099-1484(199710)2:4<301::AID-CFM34>3.0.CO;2-D
32. Govindjee, S., Hackl, K., Heinen, R.: An upper bound to the free energy of mixing by twin-compatible lamination for n-variant martensitic phase transformations. Contin. Mech. Therm. **18**, 443–453 (2007). https://doi.org/10.1007/s00161-006-0038-1
33. Gudehus, G., Nübel, K.: Evolution of shear bands in sand. Géotechnique **54**, 187–201 (2004). https://doi.org/10.1680/geot.2004.54.3.187
34. Gürses, E., Miehe, C.: On evolving deformation microstructures in non-convex partially damaged solids. J. Mech. Phys. Solids **59**, 1268–1290 (2011). https://doi.org/10.1016/j.jmps.2011.01.002
35. Hackl, K., Heinen, R.: An upper bound to the free energy of n-variant polycrystalline shape memory alloys. J. Mech. Phys. Solids **56**, 2832–2843 (2008). https://doi.org/10.1016/j.jmps.2008.04.005
36. Howell, D., Behringer, R.P., Veje, C.: Stress fluctuations in a 2D granular Couette experiment: a continuous transition. Phy. Rev. Lett. **82**(26), 5241–5244 (1999). https://doi.org/10.1103/PhysRevLett.82.5241
37. Kaus, J.P.K., Podladchikov, Y.Y.: Initiation of localized shear zones in viscoelastic rocks. J. Geophys. Res. **111**, **B04412**, 1–18 (2006). https://doi.org/10.1029/2005JB003652
38. Kohn, R.V.: The relaxation of a double-well energy. Contin. Mech. Thermodyn. **3**, 193–236 (1991). https://doi.org/10.1007/BF01135336
39. Kohn, R.V., Strang, G.: Optimal design and relaxation of variational problems I. Commun. Pure Appl. Math. **39**, 113–137 (1986). https://doi.org/10.1002/cpa.3160390107
40. Kohn, R.V., Strang, G.: Optimal design and relaxation of variational problems II. Commun. Pure Appl. Math. **39**, 139–182 (1986). https://doi.org/10.1002/cpa.3160390202
41. Kohn, R.V., Vogelius, M.: Relaxation of a variational method for impedance computed tomography. Commun. Pure Appl. Math. **40**, 745–777 (1987). https://doi.org/10.1002/cpa.3160400605
42. Lambrecht, M., Miehe, C., Dettmar, J.: Energy relaxation of non-convex incremental stress potentials in a strain-softening elastic-plastic bar. Int. J. Solids Struct. **40**, 1369–1391 (2003). https://doi.org/10.1016/S0020-7683(02)00658-3
43. Lätzel et al.: Comparing simulation and experiment of a 2D granular Couette shear device. Eur. Phys. J. E **11**, 325–333 (2003). https://doi.org/10.1140/epje/i2002-10160-7
44. Le Dret, H., Raoult, A.: The quasiconvex envelope of the Saint Venant-Kirchhoff stored energy function. Proc. Roy. Soc. Edinb. **125A**, 1179–1192 (1995). https://doi.org/10.1017/S0308210500030456
45. Morrey, C.B.: Quasi-convexity and the lower semicontinuity of multiple integrals. Pac. J. Math. **2**, 25–53 (1952). See http://projecteuclid.org/euclid.pjm/1103051941
46. Nicolaides, R.A., Walkington, N.J.: Computation of microstructure utilizing Young measures representations. In: Rogers, C.A., Rogers, R.A. (eds.) Recent Advances in Adaptive and Sensory Materials and Their Applications, pp. 131–141. Technomic Publ., Lancaster (1992)
47. Nicolaides, R.A., Walkington, N.J.: Strong convergence of numerical solutions to degenrate variational problems. Math. Comp. **64**, 117–127 (1992). http://www.jstor.org/stable/2153325
48. Oda, M., Kazama, H.: Microstructure of shear bands and its relation to the mechanisms of dilatancy and failure of dense granular soils. Géotechnique **48**, 465–481 (1998). https://doi.org/10.1680/geot.1998.48.4.465
49. Papanicolopulos, S.A., Veveakis, E.: Sliding and rolling dissipation in Cosserat plasticity. Granul. Matter **13**, 197–204 (2011). https://doi.org/10.1007/s10035-011-0253-8

50. Pasternak, E., Mühlhaus, H.B.: Cosserat continuum modelling of granulate materials. In: Valliappan, S., Khalili, N. (eds.) Computational Mechanics - New Frontiers for New Millennium, pp. 1189–1194. Elsevier Science, Amsterdam (2001)
51. Pedregal, P.: Numerical approximation of parametrized measures. Numer. Funct. Anal. Optim. **16**, 1049–1066 (1995). https://doi.org/10.1080/01630569508816659
52. Pedregal, P.: On numerical analysis of non-convex variational problems. Numer. Math. **74**, 325–336 (1996). https://doi.org/10.1007/s002110050219
53. Pedregal, P.: Parametrized Measures and Variational Principles. Birkhäuser, Basel (1997)
54. Raoult, A.: Quasiconvex envelopes in nonlinear elasticity. In: Schröder, J., Neff, P. (eds.) Poly-, Quasi- and Rank-one Convexity in Applied Mechanics, pp. 17–51. Springer, Vienna (2010). https://doi.org/10.1007/978-3-7091-0174-2
55. Roubíček, T.: Finite element approximation of a microstructure evolution. Math. Methods Appl. Sci. **17**, 377–393 (1994). https://doi.org/10.1002/mma.1670170505
56. Roubíček, T.: Numerical approximation of relaxed variational problems. J. Convex Anal. **3**, 329–347 (1996). http://eudml.org/doc/233027
57. Roubíček, T.: Relaxation in Optimization theory and Variational Calculus. Valter de Gruyter, Berlin (1997)
58. Savage, S.B., Sayed, M.: Stresses developed by dry cohesionless granular materials sheared in an annular shear cell. J. Fluid Mech. **142**, 391–430 (1984). https://doi.org/10.1017/S0022112084001166
59. Sawada, K., Zhang, F., Yashima, A.: Rotation of granular material in laboratory tests and its numerical simulation using TIJ-Cosserat continuum theory. Comput. Methods 1701–1706 (2006). https://doi.org/10.1007/978-1-4020-3953-9_104
60. Suiker, A.S.J., de Borst, R., Chang, C.S.: Micro-mechanical modelling of granular material. Part 1: derivation of a second-gradient micro-polar constitutive theory. Acta Mech. **149**, 161–180 (2001). https://doi.org/10.1007/BF01261670
61. Suiker, A.S.J., de Borst, R., Chang, C.S.: Micro-mechanical modelling of granular material. Part 2: Plane wave propagation in infinite media. Acta Mech. **149**, 181–200 (2001). https://doi.org/10.1007/BF01261671
62. Tejchman, T., Gudehus, G.: Silo-music and silo-quake experiments and a numerical Cosserat approach. Powder Technol. **76**, 201–212 (1993). https://doi.org/10.1016/S0032-5910(05)80028-2
63. Tejchman, J., Niemunis, A. FE-studies on shear localization in an anisotropic micro-polar hypoplastic granular material. Granul. Matter. **8**, 205–220 (2006). https://doi.org/10.1007/s10035-006-0009-z
64. Tordesillas, A., Peters, J.F., Muthuswamy, M.: Role of particle rotations and rolling resistance in a semi-infinite particulate solid indented by a rigid flat punch. ANZIAM J. **46**, C260-275 (2005)
65. Tordesillas, A., Walsh, S.D.C.: Incorporating rolling resistance and contact anisotropy in micromechanical models of granular media. Powder Technol. **124**, 106–111 (2002). https://doi.org/10.1016/S0032-5910(01)00490-9
66. Tordesillas, A., Walsh, S.D.C., Gardiner, B.: Bridging the length scales: micromechanics of granular media. BIT Numer. Maths. **44**, 539–556 (2004). https://doi.org/10.1023/B:BITN.0000046817.60322.ed
67. Trinh, B.T., Hackl, K.: Performance of mixed and enhanced finite elements for strain localization in hypoplasticity. Int. J. Num. Anal. Methods Geomech. **35**, 1125–1150 (2012). https://doi.org/10.1002/nag.1042
68. Trinh, B.T., Hackl, K.: Modelling of shear localization in solids by means of energy relaxation. Asia Pac. J. Comput. Eng. **1**, 1–21 (2014)
69. Trinh, B.T., Hackl, K.: A model for high temperature creep of single crystal superalloys based on nonlocal damage and viscoplastic material behavior. Contin. Mech. Thermodyn. **26**, 551–562 (2014)
70. Trinh, T.B., Hackl, K.: Modelling of shear localization in solids by means of energy relaxation. Asia Pac. J. Comput. Eng. **1** (2014). https://doi.org/10.1186/s40540-014-0009-0

71. Utter, B., Behringer, R.P.: Multiscale motion in the shear band of granular Couette flow. AIP Conf. Proc. **1145**, 339–342 (2009). https://doi.org/10.10631/1.3179928
72. Veje, C.T., Daniel, W., Howell, W.: Kinematics of a two-dimensional granular Couette experiment at the transition to shearing. Phys. Rev. E **59**, 739–745 (1999). https://doi.org/10.1103/PhysRevE.59.739
73. Walsh, S.D.C., Tordesillas, A.: Finite element methods for micropolar models of granular materials. Appl. Math. Modell. **30**(10), 1043–1055 (2006). https://doi.org/10.1016/j.apm.2005.05.016
74. Young, L.C.: Generalized curves and the existence of an attained absolute minimum in the calculus of variations. C. R. Sci. Lettres Varsovie, C III **30**, 212–234 (1937)

The Polar-Isogeometric Method for the Simultaneous Optimization of Shape and Material Properties of Anisotropic Shell Structures

Christian Fourcade, Paolo Vannucci, Dosso Felix Kpadonou, and Paul de Nazelle

1 Introduction

It is well known that structural optimization problems can be of different types. For instance, shape optimization is concerned with the best shape of a structure, e.g. of a beam[1]; topological optimization considers the optimal distribution of the matter inside a given volume, for prescribed boundary conditions and applied forces [2, 3]. To these two very classical structural optimization problems, modern materials and technologies have recently added some new problems, e.g. the one concerning the optimization of the material properties [4, 5], which is typical of composite materials. In such problems, the best elastic and/or strength properties are to be found, [6–11]. An even more recent problem is the search for the best distribution of the local material properties for a structure whose mechanical characteristics can vary pointwise, [12–15].

C. Fourcade
Renaut SAS, Guyancourt, France
e-mail: christian.fourcade@renault.com

P. Vannucci (✉)
LMV - UMR8100 CNRS and University of Versailles Saint Quentin, Versailles, France
e-mail: paolo.vannucci@uvsq.fr

D. F. Kpadonou
Laboratoire de Mathématiques de Versailles, UMR 8100, Université de Versailles, Versailles, France
e-mail: felix.kpadonou@ens.uvsq.fr

P. de Nazelle
Institut de Recherche Technologique SystemX, Palaiseau, France
e-mail: paul.denazelle@irt-systemx.fr

We consider in the present study a new kind of structural optimization problem: the simultaneous optimization of the shape and material properties distribution of an anisotropic shell.

The problem of concern, specifically, is the following one: we consider a connected planar domain, bounded by one or more non-intersecting closed curves. We imagine that such a domain defines a plate, with specified boundary conditions on the border and that this plate is realized using an orthotropic material, whose elastic properties (moduli and direction of the symmetry axes) can vary pointwise. The plate is submitted to given forces, on the boundaries and on the field, and the mass of the plate is known.

We imagine to let evolve the plate into a shell, preserving at least one of the curves defining the original plate, in such a way to transform it into a shell of constant thickness with the global constraint that the mass of such a shell can exceed the mass of the original plate by an a priori fixed amount at most. We want to determine the shell that ensures the highest stiffness. This means to determine, at the same time, the optimal shape of the final shell and the optimal distribution of the elastic properties. In fact, we consider that the elastic properties, defined by the elastic moduli and by the orientation of the orthotropy axes, can vary pointwise through the shell. Actually, to be as much as possible close to real-world problems, we consider the anisotropic shell to be a laminated shell composed by orthotropic layers whose elastic moduli and orthotropy axes direction can change locally. This is particularly motivated by the possibility, given by modern technologies of additive manufacturing, of placing into an isotropic matrix reinforcing fibers varying locally in quantity and direction.

This new kind of structural optimization problems touches hence at an almost unexplored aspect of structural mechanics: the relation between geometry and material distribution in obtaining an optimal structural behavior, topic of very few studies, see e.g. [16].

The approach that we have used in this paper is an *isogeometric-like* one. The word *isogeometric* refers, usually, to numerical techniques in which the solution, or a quantity of interest, of a given problem is discretized using the basis functions describing the exact geometry in an iso-parametric sense. Introduced by T.R. Hugues [17], these methods were first implemented in the frame of structural and computational fluid dynamics [18–20].

Extensive research has recently been devoted to the isogeometric method, whose principle is based on a direct integration of numerical analysis, optimization, and design process in the same environment. The design variables are the control points associated with the B-splines or NURBS (Non-Uniform Rational Basis Splines) functions used to parameterize the shape of the structure and sometimes their weights [21, 22]. The present research follows previous works done in the same direction, [23, 24]; the most important innovation presented here is the design of the anisotropy properties fields jointly to the design of the optimal shell shape, that is done using the same isogeometric-like technique used for the parameterization of the shell shape. This technique, developed during the PhD thesis [25], is contemporary to the first two published studies using jointly polar parameters and spline parametrization of the design variables, [26, 27]. However, unlike these papers, where just the elastic properties are optimized for a fixed geometry, a plate, in our study the design concerns at the same time the elastic properties and the shell shape.

The word isogeometric is normally reserved to approaches where the interpolation functions for the elements representing the structure in a finite element approach are also B-splines or NURBS. This is not the case in our study, where a standard finite element formulation has been used. However, we precise that in this paper the structure behavior is defined by the classical Nagdhi's model (deep shell model written in curvilinear coordinates), so that the state equations of the optimization problem are set up on the domains of charts which define the geometry. Since the three-dimensional structure is parametrized by standard CAD functions, the Naghdi's equations are posed on a square or on a triangle and constitute (through the first and second fundamental forms) an isogeometric-like mechanical interpolation of the structure. Note moreover that, compared with the current industrial standards, this approach of the shape optimization is naturally interfaced with CAD software and allows to simplify the classical optimization process by eliminating the re-meshing steps and the phases of conversion of geometries (from FEM models to CAD models) which deteriorate the optimization results and require dedicated software. For these reasons, we have used the term *isogeometric-like* to denote our approach, sometimes, for the sake of shortness, simply reduced to *isogeometric*, in the sense specified above, and, in the end, we have called *polar-isogeometric approach* the technique presented in this paper, as based, on one side, on the polar formalism and, on the other side, on an isogeometric-like method.

The paper is organized as follows: Sect. 2 introduces the general mathematical statement of the optimization problem considered in this study, while Sect. 3 focuses on the description of the shell model equation governing the behavior of the structure. Section 4 describes the parameterization used for the geometry in the standard isogeometric framework. In Sect. 5 we describe the polar formalism technique used to represent the elastic tensor. The section ends with the parametrization used to represent these polar parameters; moreover, we introduce some sufficient conditions to be satisfied by the control points of such parameters in order to ensure the pointwise satisfaction of the admissibility constraints on the elastic tensor. In Sect. 6 we give the formulation of the optimization problem in the polar-isogeometric framework. The design variables are the control points driving the geometry and the polar parameters. Section 7 presents some numerical examples concerning the optimal design of anisotropic shell structures. The paper ends with a conclusion and an outlook on possible future developments.

2 Definition of the Optimization Problem

A shell is a three-dimensional elastic body with one dimension (the thickness) small in regard to its other characteristic dimensions. As such, it can be identified to a finite surface, generally chosen to be its middle-surface, denoted in the following by Ω. We consider here anisotropic shells, like those constituted of composite materials, and in particular, referring to the recent additive manufacturing technologies of fiber placement, we focus on shells whose anisotropic elastic properties can vary pointwise.

Assuming that the shell thickness is constant, we are interested in the optimal design of both shell geometry and material distribution. Hence, a design is expressed as a couple $(\Omega, \mathbb{E})$, where Ω is a parametric surface embedded in the Euclidean space $\mathcal{E}^3$, and $\mathbb{E}$ is the elastic tensor of its constitutive materials which can vary pointwise on Ω. We consider as optimal design problem the maximization of the structure stiffness; as well known, [28], this is equivalent to the minimization of the compliance (the work done by the applied forces) which, in linear elasticity, is exactly twice the strain energy stored in the structure (Clapeyron's theorem). So that, defining the displacement field $U_{\Omega, \mathbb{E}}$ associated with a given applicant design $(\Omega, \mathbb{E})$ as the solution of the state equation:

$$\text{find } U_{\Omega, \mathbb{E}} \in \mathbf{W} \text{ such that}$$
$$a_{\Omega, \mathbb{E}}(U_{\Omega, \mathbb{E}}, V) = l_\Omega(V), \ \forall \ V \in \mathbf{W}, \tag{1}$$

where:

- $a_{\Omega, \mathbb{E}}$ and l_Ω (defined in Eq. (25)) are, respectively, the bilinear form associated with the strain energy and the linear form associated with the work of the applied loads; these functions are parametrized by the shape Ω and parametrized by the elastic tensor $\mathbb{E}$,
- and $\mathbf{W}$ is the space of virtual displacements which are independent of Ω and $\mathbb{E}$,

the design criteria j takes the form:

$$j(\Omega, \mathbb{E}) = a_{\Omega, \mathbb{E}}(U_{\Omega, \mathbb{E}}, U_{\Omega, \mathbb{E}}). \tag{2}$$

We can thus formalize the maximization of the structure stiffness as the following optimization problem:

$$\text{Find } (\Omega^*, \mathbb{E}^*) \in \mathcal{E}_S \times \mathcal{E}_M, \text{ such that}$$
$$j(\Omega^*, \mathbb{E}^*) \leq j(\Omega, \mathbb{E}) \quad \forall (\Omega, \mathbb{E}) \in \mathcal{E}_S \times \mathcal{E}_M, \tag{3}$$

where $\mathcal{E}_S$ is a shape design space of admissible geometries which takes into account the regularity constraint, the boundary conditions specified on the geometry, while $\mathcal{E}_M$ is a "material design space" which takes into account the admissibility constraints on the elastic tensor.

Within the isogeometric framework, we will assume that Ω is defined as the image of a rectangular domain $\omega \subset \mathbb{R}^2$ throughout a mapping Φ, which is moreover assumed to be a linear combination

$$\Phi(\xi^1,\xi^2) := \sum_{i,j} p_{ij} B_{ij,d}(\xi^1,\xi^2), \quad \text{where } (\xi^1,\xi^2) \in \omega, \qquad (4)$$

of bivariate B-spline function $B_{ij,d}$ (defined in Sect. 4). We can thus interpret the control points $p_{ij} \in \mathbb{R}^3$ of the spline surface Ω as shape optimization variables and suppose that $\mathcal{E}_S$ is a subset of a finite dimensional space. On the other hand we will see in Sect. 5 that the polar formalism allows to parametrize an anisotropic elastic tensor $\mathbb{E}$ with the help of three independent variables which will in turn be interpolated by a formula of the form

$$\sum_{k,l} q_{kl} B_{kl,d}(\xi^1,\xi^2), \quad \text{where } (\xi^1,\xi^2) \in \omega, \qquad (5)$$

whose control points $q_{kl} \in \mathbb{R}^3$ will be considered as material optimization variables. To summarize we can say that this framework enables us to formulate the problem (3) over a finite dimensional space adapted to designers needs and to handle it with the help of standard optimization methods, such as steepest descent methods, etc.

There are different shell models in the literature, among others: the Love's classical theory [29], generalizing to shells the classical Kirchhoff's theory for plates, the Koiter's model [30] and the Naghdi's one [31] which is the corresponding for shells of the Reissner-Mindlin's [32] model of plates. We have used in our study the Naghdi's model, briefly introduced hereafter.

3 Naghdi's Shell Equations

In the following, unless otherwise specified, Greek indexes range in the set $\{1, 2\}$ while Latin indexes in $\{1, 2, 3\}$. The Einstein's summation convention is systematically used over repeated subscript and superscript. Let a bi-dimensional domain $\omega \subset \mathbb{R}^2$ and an injective mapping $\Phi : \omega \to \mathcal{E}^3$ of class C^2 be given, we denote $\Omega := \Phi(\omega)$ the image of ω by Φ and we assume that the Jacobian matrix of Φ is of rank 2 at any point $\xi := (\xi^1, \xi^2) \in \omega$. This means that the vectors

$$\mathbf{a}_\alpha := \frac{\partial \Phi}{\partial \xi^\alpha} = \Phi_{,\alpha} \qquad (6)$$

are linearly independent, span the tangent space to the surface Ω at $\Phi(\xi)$, and that the unit normal vector

$$\mathbf{a}_3 := \frac{\mathbf{a}_1 \wedge \mathbf{a}_2}{\|\mathbf{a}_1 \wedge \mathbf{a}_2\|} \qquad (7)$$

is well defined everywhere. In this context, the Naghdi's equations are mechanical equations intending to approach the elastic behavior of the three-dimensional body

$$\Omega_t = \left\{ X \in \mathcal{E}^3 \ : \ X = \Phi(\xi) + \xi^3 \mathbf{a}_3, \ (\xi, \xi^3) \in \omega \times [-\frac{t}{2}, \frac{t}{2}] \right\}, \tag{8}$$

by a system of partial differential equations defined on ω. Before writing down an expression of their weak form (i.e. virtual work principle), let us recall some usual notations of differential geometry on surfaces.

3.1 Geometric Preliminaries

The geometric elements presented below (first fundamental forms and element of differential calculous) are of common use in differential geometry on surfaces, they allow us to formulate the Nagdhi's equations, which (as a deep shell model) closely depend on the geometry of the shell's reference configuration.

3.1.1 First Fundamental Form

The first Fundamental form on Ω is the second-order, symmetric tensor field defined by the restriction of the scalar product on $\mathcal{E}^3$ to the tangent spaces to Ω. Its components in the so-called covariant basis $\mathbf{a}_\alpha$ are the entries

$$a_{\alpha\beta} := \mathbf{a}_\alpha \cdot \mathbf{a}_\beta, \tag{9}$$

of a symmetric definite positive matrix $[a]$. Denoting $a := \det [a]$ we can define an infinitesimal area element dA on Ω as

$$dA := \sqrt{a}\, d\xi, \tag{10}$$

where $d\xi := d\xi^1 d\xi^2$ is the area element on ω. So that, for instance, the surface surf(Ω_0) of a set $\Omega_0 \subset \Omega$ is defined as:

$$\text{surf}(\Omega_0) = \int_{\Omega_0} dA = \int_{\Phi^{-1}(\Omega_0)(\subset \omega)} \sqrt{a}\, d\xi. \tag{11}$$

Denoting $a^{\alpha\beta}$ the entries of the matrix $[a]^{-1}$, the dual basis $(\mathbf{a}^\alpha)_{\alpha=1:2}$ (called contravariant basis) of the covariant basis $(\mathbf{a}_\beta)_{\beta=1:2}$ is defined by

$$\mathbf{a}^\alpha := a^{\alpha\beta} \mathbf{a}_\beta. \tag{12}$$

This means, for instance, that a vector **u** having the components u^α in the covariant basis will have the components u_β in the contravariant one and that the following "change of basis" relationships take place:

$$u_\beta = a_{\beta\alpha} u^\alpha \quad \text{or} \quad u^\alpha = a^{\alpha\beta} u_\beta. \tag{13}$$

We conclude this paragraph by noticing that if the parametrization of the surface Ω is normal, the vectors $\mathbf{a}_\alpha$ are orthogonal and the first fundamental form is diagonal. For instance, the mapping

$$(\xi^1, \xi^2) \in]0, 2\pi[\times]0, 1[:= \omega \mapsto (\sin\xi^1, \cos\xi^1, \xi^2) \in \mathcal{E}^3 \tag{14}$$

is a parameterization of a unit radius cylinder, whose first fundamental form reduces to identity; this means that the paper sheet ω can be rolled over the cylinder without changing the surfaces, lengths, or angles calculated on the flat sheet. This example also underlines that the first fundamental form itself cannot characterize a surface.

3.1.2 Second Fundamental Form

The second fundamental form is a symmetric second-order tensor filed on Ω, whose covariant and mixed components are, respectively, defined by

$$b_{\alpha\beta} := \mathbf{a}_{\alpha,\beta} \cdot \mathbf{a}_3 \quad \text{and} \quad b_\alpha^\beta = a^{\beta\sigma} b_{\sigma\alpha}. \tag{15}$$

A geometric interpretation of the second fundamental form in terms of curvatures of the surface Ω is proposed in [33] where it is proved that the curvature of a curve drawn on the intersection between the surface Ω and a plane containing $\mathbf{a}_3$ can be computed with the help of the covariant components $b_{\alpha\beta}$ of the second fundamental form. We can see that $b_{\alpha\beta} = 0$ if Ω is a plane surface and that $b_{11} = 1$ while $b_{12} = b_{21} = b_{22} = 0$ for the cylinder defined in the example (14).

3.1.3 Covariant Derivatives

Let be given a vector field **w**, we denote by w_i (resp. w^j) its covariant (resp. contravariant) coordinates

$$\mathbf{w} = w_j \mathbf{a}^j = w^k \mathbf{a}_k \text{ where we have adopted the convention } \mathbf{a}^3 := \mathbf{a}_3. \tag{16}$$

Letting $\Gamma_{\alpha\beta}^\lambda = \mathbf{a}_{\alpha,\beta} \cdot \mathbf{a}^\lambda$ be the Christoffel's symbols, the derivative $\mathbf{w}_{,\alpha}$ is defined by

$$\mathbf{w}_{,\alpha} = (w_{\alpha|\beta} - b_{\alpha\beta} w_3) \mathbf{a}^\alpha + (w_{3|\beta} + b_\beta^\sigma w_\sigma) \mathbf{a}_3, \tag{17}$$

where the symbol $_{|\beta}$ denotes the covariant derivative operation with respect to ξ^β defined below:

$$w_{\alpha|\beta} = w_{\alpha,\beta} - w_\sigma \Gamma^\sigma_{\alpha,\beta},$$
$$w_{3|\beta} = v_{3,\beta}. \tag{18}$$

We can check that the Christoffel symbols $\Gamma^\lambda_{\alpha\beta}$ are 0 for the cylinder given in example (14); in this case, the covariant derivatives $w_{\alpha|\beta}$ reduce to the classical partial derivatives $w_{\alpha,\beta}$.

It should be noticed to conclude that the geometric entities we have defined so far depend on the chosen parameterization for the surface Ω and that their formal computation can be very complicated (i.e. only numerically performed); this is particularly true for spline or Nurbs surfaces.

3.2 Weak Formulation of the Nagdhi's Equations

The Naghdi's shell theory, [31] applies to three-dimensional bodies Ω_t such as defined in (8), submitted to the volume forces $p^i \mathbf{a}_i$. It is obtained from the three-dimensional elasticity theory with the help of two a priori hypotheses:

- the first, of mechanical nature, consists to assume that if the thickness t is small enough, the stresses are approximately two-dimensional and vary linearly along the thickness (ie. along the $\mathbf{a}_3$ axis);
- the second one is geometric and makes the hypothesis that the normal fibers to the mid-surface Ω are rigid, but they do not have to remain orthogonal to the deformed configuration of Ω.

The Naghdi's model accounts for membrane, bending and shearing deformations of the shell. The unknowns of its linear version are the displacements $u_i \mathbf{a}^i$ of the points located on the mid-surface Ω and the rotation field $s_\alpha \mathbf{a}^\alpha$ of the normals $\mathbf{a}_3$, this means that we assume that the displacement $\mathbf{U}$ of a generic point $\Phi(\xi) + \xi^3 s_\alpha \mathbf{a}_\alpha$ in Ω_t is defined as:

$$\mathbf{U} = u_i \mathbf{a}^i + \xi^3 s_\alpha \mathbf{a}^\alpha. \tag{19}$$

Denoting, respectively, by $u := (u_1, u_2, u_3)$ and $s := (s_1, s_2)$ the covariant components of displacements and rotations, the Nagdhi's equations take the following weak form:

find a couple $[u, s]$ in an appropriate vector space W such that:

$$\int_\omega Q^{\alpha\beta\sigma\tau} \gamma_{\sigma\tau}(u)\gamma_{\alpha\beta}(v) + E^{\alpha 3 \beta 3} \gamma_{\beta 3}([u,s])\gamma_{\alpha 3}([v,r]) \sqrt{a}\, d\xi$$
$$+ \frac{t^2}{12} \int_\omega Q^{\alpha\beta\sigma\tau} \chi_{\sigma\tau}([u,s])\chi_{\alpha\beta}([v,r]) \sqrt{a}\, d\xi \qquad (20)$$
$$= \frac{1}{t} \int_\omega f^i v_i \sqrt{a}\, d\xi \quad \text{for all } [v, r] \in W,$$

where:

- denoting E^{ijkl} the contravariant components of the elastic tensor

$$\mathbb{E} = E^{ijkl} \mathbf{a}_i \otimes \mathbf{a}_j \otimes \mathbf{a}_k \otimes \mathbf{a}_l, \qquad (21)$$

we set:

$$Q^{\alpha\beta\sigma\tau} := E^{\alpha\beta\sigma\tau} - \frac{E^{\alpha\beta 33} E^{33\sigma\tau}}{E^{3333}}; \qquad (22)$$

note that when the shell is made of an isotropic material of Lamés coefficients λ and μ we have:

$$E^{\alpha 3 \beta 3} = \mu a^{\alpha\beta} \quad \text{and}$$
$$Q^{\alpha\beta\sigma\tau} = \frac{4\lambda\mu}{\lambda + 2\mu} a^{\alpha\beta} a^{\sigma\tau} + 2\mu \left(a^{\alpha\sigma} a^{\beta\tau} + a^{\alpha\tau} a^{\beta\sigma} \right), \qquad (23)$$

as usually stated in the literature;
- the functions $\gamma_{\alpha\beta}(u) = \frac{1}{2}(u_{\alpha|\beta} + u_{\beta|\alpha}) - b_{\alpha\beta} u_3$ are the covariant components of the change of metric tensor associated with the displacement field $u_i \mathbf{a}^i$,
- while $\gamma_{\alpha 3}([u, s]) := \frac{1}{2}\left(s_{\alpha|\beta} + s_{\beta|\alpha} - b_\alpha^\sigma \right)$ and

$$\chi_{\alpha\beta}([u, s]) := \frac{1}{2}\left(s_{\alpha|\beta} + s_{\beta|\alpha} - b_\alpha^\sigma d_{\sigma\beta}(u) - b_\beta^\sigma d_{\sigma\alpha}(u) \right) \quad \text{where:} \qquad (24)$$
$$d_{\lambda\mu}(u) := u_{\lambda|\mu} - b_{\lambda\mu} u_3$$

are, respectively, the covariant components of the transverse shear and the Naghdi's change of curvature tensors associated with the displacements $u_i \mathbf{a}^i$ of the mid-surface and rotations $s_\alpha \mathbf{a}^\alpha$ of the fibers of the shell,
- and at last, $f^i := \int_{-t}^{t} p^i d\xi^3$, are the contravariant components of the resultant on the mid-surface of volume forces applied to Ω_t; note that the resulting moments $\int_{-t}^{t} \xi^3 p^i d\xi^3$ generated on the mid-surface by the volume forces are not taken into account in this shell model; this limits its mechanical domain of validity to

shells of moderate curvatures. We can, however, load the free edges by given linear torques and forces.

The admissible displacements domain is a vector space W made up of test functions $\xi \in \omega \mapsto [v(\xi), r(\xi)] \in \mathbb{R}^5$ satisfying given boundary conditions defined on some given parts $\gamma_b \subset \partial \omega$ of the boundary of ω. For instance, the shell is said to be clamped (resp. simply supported) on the boundary $\Phi(\gamma_b) \subset \partial \Omega$ if $[v(\xi), r(\xi)] = [0, 0]$ (resp. $v(\xi) = 0$) for any $\xi \in \gamma_b$, while the boundary $\Phi(\gamma_b)$ is said to be a free if the functions v and s may take arbitrary values on γ_b.

Denoting by $a([u, s], [v, r])$ (resp. $l(v)$) the left-hand member (resp. the right-hand member) of Eq. (20), this equation rewrites as:

$$a([u, s], [v, r]) = l([v, r]) \quad \text{for all} \quad [v, r] \in W, \tag{25}$$

where the mappings

$$\begin{aligned} ([u, s], [v, r]) \in W \times W &\mapsto a([u, s], [v, r]) \in \mathbb{R} \\ \text{and} \quad [v, r] \in W &\mapsto l(v) \in \mathbb{R} \end{aligned} \tag{26}$$

are, respectively, bilinear and linear on W, they can even be assumed to be continuous if W is a (closed) subspace of $[H^1(\omega)]^5$. M. Bernadou [34] proved that if (i) the shell is assumed to be clamped on a non-zero part of its perimeter, (ii) the parametrization Φ of Ω is of class C^3 and (iii) the elastics tensor is a continuous function of $\xi \in \omega$, the bilinear form $a(.,.)$ is moreover coercive on W. So that we can use the Lax-Milgram's Lemma to ensure well-posedness of the variational equation (25).

At this stage, we have defined a well-posed state equation for the optimization problem (introduced in Sect. 2), which is parametrized by the first and the second fundamental form of a shape $\Phi(\omega)$. To complete its definition we have now to precise, in the two following Sections, the space $\mathcal{E}_S$ (resp. $\mathcal{E}_M$) of admissible shapes (resp. of admissible materials).

4 Parametrization by an "Isogeometric-Like" Approach

In its usual meaning, the isogeometric approach consists to interpolate the state variables of a Partial differential equation by the functions which are used to define the geometry on which it is posed in a computer-assisted design environment (specifying so, to the CAD framework, the "classical" iso-parametric Finite Elements interpolation methods). The practical value of this approach being to facilitate the interfacing between geometric modeler and simulation software. In the previous section we saw that the Nagdhi equations allowed us to parameterize the shells equations by their shapes, which led us to pose the shape optimization problem on a shape-independent functional space. The question we need to answer

Fig. 1 Example of middle-surface with the associated control points

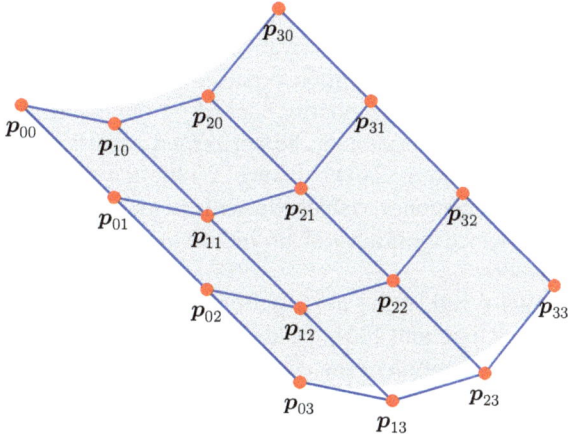

now is to describe the space $\mathcal{E}_S$ of the admissible shapes and to ensure that the geometric entities, such as first, second fundament form, Christoffel's symbols, etc., are efficiently calculable. The purpose of this section is to verify that the design space provided by the CAD's shape functions, such as Bézier [35], B-spline or NURBS, is suited to the needs, and that it, moreover, enables the three-dimensional shell equations to be laid down on the domain of the CAD charts whose geometry is extremely simple (thus very easy to mesh). So doing, the method proposed in this paper will be referred to as "Isogeometric-Like" approach tailored to shape optimization. We are thus interested to three-dimensional surfaces which are represented by linear combinations of numerical functions, defined on a rectangular domain $\omega := I_1 \times I_2$

$$\Phi(\xi^1, \xi^2) := \sum_{i,j} p_{ij} B_{i,d_1}(\xi^1) B_{j,d_2}(\xi^2) \tag{27}$$

where, see Fig. 1, the coefficients p_{ij} are given points in $\mathbb{R}^3$ which control the surface and the functions $t \mapsto B_{i,k}(t)$ are B-Splines (piecewise polynomial functions) defined on an interval I of the real line. The shape optimization problem is formulated in terms of control points and we can assume that the design space $\mathcal{E}_S$ of admissible shapes is a subset of $\mathbb{R}^{3*N}$, accounting for given CAD design constraints. The Nagdhi's equations being moreover posed on a fixed rectangular sub-domain of $\mathbb{R}^2$, there is no inconvenience to solve them by a classical F.E. method. In other words, we solve the discretized equations

$$a([u_h, s_h], [v_h, r_h]) = l([v_h, r_h]) \quad \forall [v_h, r_h] \in W_h, \tag{28}$$

posed on a space W_h, made up of locally polynomial functions u_h (resp. s_h) which interpolate the admissible déplacements (resp. admissible rotations) on the triangles of a given triangulation T_h of ω. According to the results obtained in the previous section, the bilinear form $a(.,.)$ (resp. the linear form $l(.)$) closely depends on the first and second derivatives of the shell's chart Φ, which are evaluated on the numerical integration points of the triangles of T_h. Formula (27) shows that these derivatives explicitly depend on the control points of the surface Ω and that their computations may be performed with the help of the algorithms classically implemented in CAD software. Within the shape optimization framework, a status of optimization variables we will be given to points p_{ij} and we will see that the compliance minimization problem can be handled with the help of a steepest descent method.

We recall in the next section the definition and the essential properties of the splines functions [36], these will be furthermore used in Sect. 5 to define a relevant parameterization of the elasticity tensor.

4.1 Univariate Splines Functions

A univariate spline is a piecewise polynomial function which is a linear combination of B-splines $B_{i,d}$, defined as follows:

- we consider a non-decreasing sequence

$$\Sigma = \{t_0, t_1, \cdots, t_m\}, \quad \text{with} \quad t_k \leq t_{k+1}, \tag{29}$$

of real numbers, referred to as knots vector. If they are r knots t_i equals τ, we say that τ is a node of multiplicity r; we adopt moreover the notation:

$$w_{i,j}(t) = \begin{cases} \frac{t - t_i}{t_{i+j} - t_i} & \text{if } t_{i+j} < t_i \\ 0 & \text{otherwise} \end{cases} \tag{30}$$

- and we define the B-spline $B_{i,d}$ of degree d associated with the given knot vector Σ by the Cox-de-Boor recurrence formula:

$$B_{i,0}(t) = \begin{cases} 1 & \text{if } t_i \leq t < t_{i+1} \\ 0 & \text{otherwise} \end{cases}$$
$$B_{i,d}(t) = w_{i,d}(t) B_{i,d-1}(t) + (1 - w_{i+1,d}(t)) B_{i+1,d-1}(t) \quad \text{for } d \geq 1. \tag{31}$$

The main properties of a B-spline are listed in the following Proposition, proved in [37], for instance.

Proposition 4.1 *With the above notations, the B-splines* $\left(B_{i,d}\right)_{i=0:m-d-1}$ *are defined on the real line and satisfy the properties listed below:*

1. the function $B_{i,d}(.)$ is a is a polynomial of degree d over each interval $[t_i, t_{i+d}[$;
2. it is zero outside the interval $[t_i, t_{i+d}[$;

3. we also have $B_{i,d}(t_i) = 0$ unless $t_i = t_{i+1} = \cdots = t_{i+d} < t_{i+d+1}$, in which case $B_{i,d}(t_i) = 1$;
4. let $t \in]t_i, t_{i+d+1}[$ be given, we have $B_{i,d}(t) = 1$ if and only if $x = t_{i+1} = \cdots = t_{i+d}$;
5. we have $0 < B_{i,d}(t) \leq 1$ for all $t \in]t_i, t_{i+d}]$;
6. the family $\left(B_{i,d}(.)\right)_{i=0}^{m-d-1}$ satisfies the unit partition property: let $[a, b]$ be an interval such that $t_d \leq a$ and $b \leq t_{m-d}$, then

$$\sum_{i=0}^{m-d-1} B_{i,d}(t) = 1, \quad \forall t \in [a, b]; \tag{32}$$

7. the function $B_{i,d}(.)$ is right-infinitely differentiable on $\mathbb{R}$, its derivative at a point t being defined by the formula

$$B'_{i,d}(t) = d \left[\frac{B_{i,d-1}(t)}{t_{i+d} - t_i} - \frac{B_{i+1,d-1}(t)}{t_{i+d+1} - t_{i+1}} \right], \tag{33}$$

with the convention $\frac{1}{0} := 0$, inherited from the formula (30);
8. the function $B_{i,d}$ is of class C^{d-k} in the neighborhood of a knot of multiplicity k. □

In the sequel, when we will talk about a spline of degree d, we will implicitly set $a := t_0$, $b = t_m$ and assume that the knot vector Σ is defined in such a way that $t_0 = \cdots = t_d = a$ and $t_{m-d} = \cdots = d_m = b$. Such a spline will be referred to as spline clamped at a and b in the meaning that $B_{0,d}(a) = B_{m-d-1,d}(b) = 1$.

Remark 4.1 When $a = 0$, $b = 1$ and the knot vector Σ is made up of numbers 0 and 1 with the multiplicities $d + 1$, the recurrence Eq. (31) reduces to

$$B_{i,d}(t) = t B_{i,d-1}(t) + (1 - t) B_{i+1,d-1}(t), \quad \forall t \in [0, 1], \tag{34}$$

so that the B-splines thus defined are the Bernstein polynomials, which make up a basis of the space of d-degree polynomials (defined on $[0, 1]$). □

4.1.1 Spline Curves

This algebraic frame being specified, we are interested in geometric curves whose parametric representations are defined, from a given family $(p_i)_{i=1}^n$ of points in $\mathbb{R}^3$, by formulas of the form

$$C(u) := \sum_{i=0}^{n} p_i B_{i,d}(u), \quad u \in [a, b] \text{ and } n := m - d - 1. \tag{35}$$

Such a curve is a Bézier curve if the knot vector satisfies the properties introduced in Remark 4.1, while it is called spline in the other cases. The practical value of this parameterization of geometric curves is, see the NURBS book [38], to produce very effective computational algorithms. For instance, the derivative $C'(u)$ of the spline (35) is itself a $d-1$ degree spline defined by:

$$C'(u) := d \sum_{i=0}^{n-1} \frac{p_{i+1} - p_i}{t_{i+p+1} - t_i} B_{i,d-1}(u). \tag{36}$$

This formula offers an easy access (see, for instance, the algorithm A3.3 in [38]) to geometric entities such as tangents, principal normals, bi-normals, etc. and, as well, to their derivatives with respect to the curve's control points. We conclude this paragraph in noticing that a Bézier representation of a curve is particularly simple to implement: for instance, writing a third-degree Béziers curve in the following form:

$$C(u) = [1, u, u^2, u^3] \begin{bmatrix} 1 & 0 & 0 & 0 \\ -3 & 3 & 0 & 0 \\ 3 & -6 & 3 & 0 \\ -1 & 3 & -3 & 1 \end{bmatrix} \begin{bmatrix} p_0 \\ p_1 \\ p_2 \\ p_3 \end{bmatrix}, \tag{37}$$

we see that all the geometric computations become almost analytic.

4.2 Bivariate Spline Functions

Let be given two knot vectors $\Sigma_\alpha = \{t_i^\alpha\}_{i=0:m_\alpha}$, ($\alpha = 1, 2$) partitioning the intervals I_α, we set $\omega := I_1 \times I_2$ and the following functions

$$\begin{aligned} (u, v) \in \omega \mapsto B_{ij,\boldsymbol{d}}(u, v)) &:= B_{i,d_1}(u) B_{j,d_2}(v) \in \mathbb{R}_+ \\ \text{where } \boldsymbol{d} \text{ is defined as } \boldsymbol{d} &:= (d_1, d_2) \end{aligned} \tag{38}$$

will be referred to as bivariate B-splines. They are piecewise polynomial functions verifying properties analogous to properties 1 to 6 of Proposition 4.1. We specifically use the unit partition property

$$\sum_{i,j} B_{ij,\boldsymbol{d}}(u, v) = 1 \quad \forall (u, v) \in \omega \tag{39}$$

in the proofs of Propositions 5.1 and 5.2. The partial derivatives of $B_{ij,\boldsymbol{d}}$ are defined with the help of formula (33) as

The Polar-Isogeometric Method for Shape and Material Optimization of. . . 103

$$\partial_u B_{ij,\boldsymbol{d}}(u,v) = B'_{i,d_1}(u) B_{j,d_2}(v) \quad \text{and}$$
$$\partial_v B_{ij,\boldsymbol{d}}(u,v) = B_{i,d_1}(u) B'_{j,d_2}(v) \tag{40}$$

and (see algorithms A3.7, A3.8 of NURBS book) computed by recycling the one-dimensional algorithms. Property 8) of Proposition 4.1 shows that $B_{ij,\boldsymbol{d}}$ is of class C^p with $p = \min(d_1 - k_1, d_2 - k_2)$ in a neighborhood of a couple of knots (t_i^1, t_j^2) of multiplicities k_1 and k_2. In the sequel, we will assume that $p \geq 3$.

4.2.1 Spline Surfaces

Given a control polygon $(p_{i,j}) \in \mathbb{R}^3$, the mapping

$$(\xi^1, \xi^2) \in \omega := I_1 \times I_2 \mapsto \Phi(\xi^1, \xi^2) := \sum_{i=0}^{m_1-d_1-1} \sum_{j=0}^{m_2-d_2-1} p_{i,j} B_{ij,\boldsymbol{d}}(\xi^1, \xi^2) \tag{41}$$

define a surface in $\mathcal{E}^3$, which is parametrized by the points p_{ij}. We will say that it is a Bézier patch when elementary functions $B_{ij,\boldsymbol{d}}$ are products of Bernstein polynomials; we show in Fig. 1 an example of a third order Bézier surface (i.e. a spline surface defined by knot vectors of the form $\Sigma_\alpha = \{0,0,0,0,1,1,1,1\}$ and $d = 3$).

Remark 4.2

1°) In so far as the splines are assumed to be clamped at the ends of the intervals I_1 and I_2, the edges of the surface are spline curves whose control points are the boundary points of the control polygon $(p_{i,j})$.
2°) It is an immediate consequence of the unit partition property (39) that spline surfaces satisfy the affine invariance property, in the meaning that the image $A_f \circ \Phi$ of a spline surface Φ by an affine mapping A_f remains a spline whose control points are the image by A_f of the control points of the initial spline.
3°) To make Nagdhi's equations meaningful, we will always assume that the mapping Φ defined in formula (41) is injective (in this respect, we represent in Fig. 2 the spline parameterization of the curve which has a double point). This requirement will be translated into appropriate additional constraints on the control points in the formulation of the shape optimization problem. □

5 Anisotropy Representation

An anisotropic material has elastic properties changing with the direction. Such properties are expressed by the elastic tensor. The main goal when designing such materials is to set up the optimal distribution of the elastic properties. In the case of

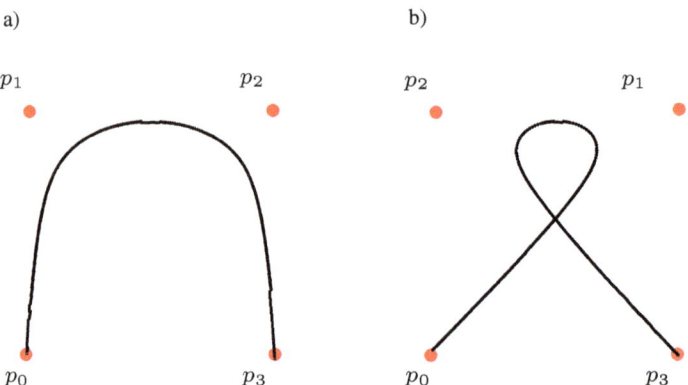

Fig. 2 Example of spline parameterization of a curve having a double point. The curves shown in (**a**) and (**b**) are splines controlled by the points $(p_i)_{i=0:3}$. We see that the curve in (**b**), which is obtained from the curve shown in (**a**) by switching the roles of p_1 and p_2, has a double point. This means that the spline parametrization of the curve in (**b**) is not injective

optimal design of planar anisotropic structures, it is suitable to make use of the polar formalism, introduced by Verchery [39] in 1979, to represent a planar elastic tensor using just invariants and angles. This formalism allows to easily represent rotations and the constraints on the design variables (the polar variables).

Moreover, the polar formalism allows to split the elastic tensor into its isotropic and anisotropic parts; hence it offers the possibility to target and explicitly tune the anisotropy. More details on the polar formalism can be found in [40–42]. For a complete presentation of anisotropic elasticity and of the polar method and its applications, the reader can refer to [5].

The polar formalism has successfully been applied to several different optimization problems concerning laminated structures, [6–16, 43–49] as well as to some theoretical problems, [41, 42, 50–58].

As said above, the mechanical properties of concern in this study are condensed in the elasticity tensor $\mathbb{E}$; this is a fourth-rank tensor whose components satisfy the minor and major symmetries of the indexes:

$$E^{ijkl} = E^{jikl} = E^{ijlk} = E^{klij}, \quad i, j, k, l = 1, 2, 3. \tag{42}$$

For the case of a plane tensor, which is of interest in our study, as detailed below, the independent elastic components reduce then to only 6: E^{1111}, E^{1112}, E^{1122}, E^{1212}, E^{1222} and E^{2222}, in the following generically indicated as $E^{\alpha\beta\lambda\mu}$.

In the polar formalism, the components $E^{\alpha\beta\lambda\mu}$ of the plane elastic tensor are expressed as

$$E^{1111}=T_0+2T_1+R_0\cos 4\Phi_0+4R_1\cos 2\Phi_1,$$
$$E^{1112}=R_0\sin 4\Phi_0+2R_1\sin 2\Phi_1,$$
$$E^{1122}=-T_0+2T_1-R_0\cos 4\Phi_0,$$
$$E^{1212}=T_0-R_0\cos 4\Phi_0, \tag{43}$$
$$E^{1222}=-R_0\sin 4\Phi_0+2R_1\sin 2\Phi_1,$$
$$E^{2222}=T_0+2T_1+R_0\cos 4\Phi_0-4R_1\cos 2\Phi_1.$$

In the above equation, T_0, T_1, R_0, and R_1 are elastic moduli, while Φ_0 and Φ_1 are (the polar) angles. In particular, it can be shown that T_0 and T_1 are the isotropy invariants while R_0, R_1, $\Phi_0 - \Phi_1$ are the anisotropy invariants. The elasticity is then represented through intrinsic quantities, tensor invariants, and angles, which is particularly suitable when working with orientation depending properties, like in anisotropy. The choice of one of the two polar angles fixes the frame: as $\Phi_0 - \Phi_1$ is an invariant, choosing Φ_0 or Φ_1 corresponds to fix a frame and also the value of the other angle.

It can be shown, see [40], that the polar invariants are linked to the elastic symmetries. In particular, ordinary orthotropy corresponds to the condition

$$\Phi_0 - \Phi_1 = K\frac{\pi}{4}; \quad K = 0, 1. \tag{44}$$

The value of K is very important in optimization problems; in fact, it has been seen in several cases that changing K from 0 to 1 or vice versa transforms an optimal solution into an anti-optimal one (i.e. the best to the worst), [45]. Taking into account for (44) in (43), we obtain for an orthotropic layer:

$$\begin{aligned}
E^{1111}(\Phi_1) &= T_0 + 2T_1 + R_0^K \cos 4\Phi_1 + 4R_1 \cos 2\Phi_1, \\
E^{1112}(\Phi_1) &= - R_0^K \sin 4\Phi_1 - 2R_1 \sin 2\Phi_1, \\
E^{1122}(\Phi_1) &= - T_0 + 2T_1 - R_0^K \cos 4\Phi_1, \\
E^{1212}(\Phi_1) &= T_0 - R_0^K \cos 4\Phi_1, \\
E^{1222}(\Phi_1) &= R_0^K \sin 4\Phi_1 - 2R_1 \sin 2\Phi_1, \\
E^{2222}(\Phi_1) &= T_0 + 2T_1 + R_0^K \cos 4\Phi_1 - 4R_1 \cos 2\Phi_1,
\end{aligned} \tag{45}$$

with

$$R_0^K = (-1)^K R_0. \tag{46}$$

Two other special orthotropies exist: square symmetry (i.e. with elastic properties periodic of $\frac{\pi}{2}$), corresponding to the condition $R_1 = 0$ and R_0-orthotropy, corresponding to $R_0 = 0$. For more details on this subject, the reader is referred to [40, 59].

To summarize, in the polar formalism, the following six parameters define the elastic tensor in any frame:

- two isotropic invariants T_0, T_1;
- three anisotropic invariants R_0, R_1, $\Phi_0 - \Phi_1$. For ordinarily orthotropic layers, these can be replaced by the two quantities R_0^K and R_1, still representing the three invariants (indeed $K \in \{0, 1\}$)
- the angle Φ_1, fixing the frame.

We finally remark that isotropy corresponds to $R_0 = R_1 = 0$.

5.1 Elastic Assumptions

In this paper, we consider the optimal design of a shell under the following assumptions: the shell is locally orthotropic everywhere and homogeneous through the thickness. The design concerns exclusively the anisotropic part of e, for both the direction and the elastic moduli.

This is a simplified setting, but it corresponds to a real situation, that of a shell composed by a quasi-homogeneous orthotropic laminate of identical orthotropic plies, see e.g. [60]. In such a case, the elastic behavior of the laminate is completely determined by a unique elastic tensor, describing at the same time the extension and the bending response of the shell, and there is no coupling between extension and bending. In addition, because the plies are identical, the isotropic part, i.e. the polar invariants T_0 and T_1, is everywhere equal to those of the basic layer, so they cannot be affected by the design process, once the material chosen. We precise, however, that here we simply address the problem specified hereon, just as a mathematical problem, regardless of whether or not it corresponds to the above laminate; that is why we still use the symbol e to denote the stiffness tensor.

We also assume that the through-the-thickness properties $E^{\alpha 3 \beta 3}$ are much less important for the process at hand than the in-plane ones, so they are simply considered as constant throughout the design process. This fact is justified for thin shells, as we assume the shells at hand to be; in addition, this approximation is consistent with the fact, always confirmed by the numerical results, that the optimal shell is the one working in a membrane regime, where shear and bending energy tends to zero; as a consequence, the transversal shear moduli $E^{\alpha 3 \beta 3}$ are inessential in this context.

Finally, the main consequence of such assumptions for the optimum design of the shell is that the number of elastic design variables is reduced to only three: R_0^K, R_1, and Φ_1: two elastic moduli and the orthotropy direction.

5.2 Constraints on the Polar Parameters

The polar elastic moduli cannot take arbitrary values, they are submitted to some constraints that can be of two types, depending upon whether the shell is homogeneous (i.e. composed by a unique layer) or not (i.e. it is a laminated shell):

- Elastic bounds, see [56], resulting from the positive definiteness of e, [5]:

$$T_1[T_0 + R_0^K] > 2R_1^2,$$
$$T_0 > |R_0^K|, \qquad (47)$$
$$R_1 \geq 0.$$

Such constraints must be satisfied locally by any elastic homogeneous sheet in a planar elastic state.
- Geometric bounds: it can be shown, see [41], that laminates composed by identical layers cannot realize all the possible combinations of the values of the elastic moduli. We could say, in some words, that laminates form a more restricted elastic class. Mathematically speaking, this corresponds to the fact that the bounds on e are not (47) but some other more restrictive ones, called *geometric bounds*, because linked to the stacking sequence. For the case of an orthotropic laminate composed of orthotropic layers, such geometric bounds are (here, K_L, $R_0^{K_L} = (-1)^{K_L} R_0^L$ and R_1^L are polar variables of the basic layer)

$$2\left(\frac{R_1}{R_1^L}\right)^2 - 1 \leq \frac{R_0^K}{R_0^{K_L}},$$
$$R_0^K \leq R_0^L, \qquad (48)$$
$$0 \leq R_1 \leq R_1^L.$$

Since Eq. (48) is more restrictive than (47), when the problem concerns the design of a laminated structure, Eq. (47) must be replaced by Eq. (48), otherwise, one could obtain some values for the components of $\mathbb{E}$ that cannot be realized in practice through a laminate composed by identical plies.

5.3 Parameterization of the Polar Parameters

We have to recall now that in the problem at hand, the elastic properties can vary pointwise throughout the shell. This means that the design variables R_0^K, R_1, and Φ_1 are actually variable fields defined on Ω (or via the chart Φ, on ω), which have to satisfy constraints (47) or (48) everywhere. In other words, the problem has an infinite dimension and carries an infinite number of constraints. A classical finite element discretization, like in [12, 14, 15], would yield to a huge number of design variables and constraints. Following an idea originally introduced in [26], we propose below a new approach (based on a spline parameterization of the polar variables) similar to the way in which the shell geometry is parametrized. Just as in [26], but unlike in [27] (where only the polar angle is described in terms of spline) we define here a parametrization of the invariants R_0^K and R_1, described, like the polar angle, by a spline surface each one, so obtaining a complete spline parametrization of the elastic tensor e.

As already said, we assume the polar parameters fields to be designed under the form of Bézier, B-spline or NURBS functions. We concentrate on the parameterizations of the polar moduli R_0^K are R_1 which are mainly involved in the elastic and geometric constraints. Indeed, our goal is to define a set of constraints on the control points ensuring pointwise satisfaction of inequalities (47). Two parameterizations for R_1 are relevant (here, we refer to the elastic bounds (47), but a similar procedure can be used for the geometric bounds (48)):

- a conformal parameterization for which R_1 is parametrized as the square root of a positive spline (which is not a spline); the interest of this change of variable being to simplify the constraints (47) which turn to linear. The disadvantage being that the elastic coefficients and the constraints are no longer differentiable at $R_1 = 0$;
- a direct parametrization, where R_1 is parameterized by a spline but, see Proposition 5.2, the constraints on the control points depend on the number of control points, this parametrization allows, however, make the elasticity tensor differentiable with respect to the control points of the polar variables.

The main results of this section, enounced in Propositions 5.1 and 5.2 are consequences of the Proposition 4.1.

5.3.1 Constraints for Conformal Parameterization

In this case, the polar parameters R_0^K and R_1^2 are parametrized by splines and we show in Proposition 5.1 that this allows to reduce the non-linear constraints (47) to linear ones, related to the control points of the square of the polar variable R_1.

Proposition 5.1 *Let $d = (d^1, d^2)$ and $\Sigma = (\Sigma_1, \Sigma_2)$ be two pairs of integers and knot vectors. Assume that the mappings $\xi \mapsto R_1^2(\xi) \in \mathbb{R}_+$ and $\xi \mapsto R_0^K(\xi)$ are splines functions, written according to their control points $(r_1^2)_{ij}$ and $(r_0^K)_{ij}$ as*

$$R_1^2(\xi) = \sum_{(i,j)\in I} (r_1^2)_{ij} B_{ij,d}(\xi), \quad R_0^K(\xi) = \sum_{(i,j)\in I} (r_0^K)_{ij} B_{ij,d}(\xi), \quad (49)$$

where, setting $m_\alpha := \mathrm{Card}\,(\Sigma_\alpha) - 1$, the indexes (i, j) range in the set

$$I := \{0, \cdots, m_1 - d_1\} \times \{0, \cdots, m_2 - d_2\}. \quad (50)$$

If the following inequalities

$$T_1[T_0 + (r_0^K)_{ij}] > 2(r_1^2)_{ij},$$
$$-T_0 < (r_0^K)_{ij} < T_0, \tag{51}$$
$$(r_1^2)_{ij} \geq 0$$

are satisfied for all $(i, j) \in I$, *the inequalities (47) take place for any* $\xi \in \omega$.

In an analogous way, for the case of laminate made of identical orthotropic layers, the constraints for the geometric bounds (48) take the following form:

$$2\frac{(r_1^2)_{ij}}{(R_1^L)^2} - 1 \leq \frac{(r_0^K)_{ij}}{R_0^{KL}},$$
$$(r_0^K)_{ij} \leq R_0^L, \qquad \forall (i, j) \in I \tag{52}$$
$$0 \leq (r_1^2)_{ij} \leq (R_1^L)^2;$$

with R_0^{KL} and R_1^L are the polar variables of the basic orthotropic layer, defined in Sect. 5.2. □

Proof As T_0 and T_1 do not depend on ξ, positiveness of B-splines $B_{ij,d}(.)$ and inequalities (51) entail that

$$T_1 \sum_{(i,j) \in I} [T_0 + (r_0^K)_{ij}] B_{ij,d}(\xi) > 2 \sum_{(i,j) \in I} (r_1^2)_{ij} B_{ij,d}(\xi) \quad \forall \xi \in \omega. \tag{53}$$

Property (39) shows that:

$$\sum_{(i,j) \in I} T_0 B_{ij,d}(\xi) = T_0, \quad \forall \xi \in \omega. \tag{54}$$

Reporting (54) into (53), we see that inequality (47) is satisfied for any $\xi \in \omega$. The other inequalities set out in the Proposition are proved in the same manner. □

It is worth noting, as the following example shows, that the proposed approach yields to sufficient but not necessary constraints.

Example 5.1 Let us consider the knot vector $\Sigma = \{0, 0, 0, 0, 1, 1, 1, 1\}$ and assume $d = 3$. The polynomial basis functions are the classical Bernstein functions

$$B_{0,3}(t) = (1-t)^3, \quad B_{1,3}(t) = 3(1-t)^2 t, \quad B_{2,3}(t) = 3(1-t)t^2, \quad b_{3,3}(t) = t^3. \tag{55}$$

In this case 0 and 1 are knots of multiplicities 4 and the curve $t \in [0, 1] \mapsto C_p(t)$ interpolates these endpoints which are below the horizontal line representing the

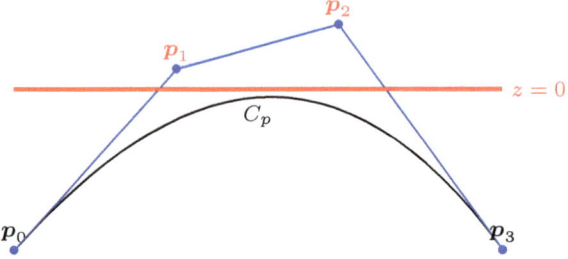

Fig. 3 Example of clamped B-spline. In this particular case, the control points: $p_0 = -2.0$, $p_1 = 0.25$, $p_2 = 0.8$ and $p_3 = -2.0$ are not all negative while $C_p(t) < 0$ for all $t \in [0, 1]$

zero ordinate line. However, as can be noticed on Fig. 3, the control points p_1 and p_2 are positives while the value of the spline polynomial is always negative.

5.3.2 Constraints for Direct Parameterization

We now assume that the polar variable R_1 is parameterized by a spline, considering the case of elastic bounds, this new parameterization only affects the inequality $(47)_1$ and we have the following result, proofed in [25].

Proposition 5.2 *Using the notations introduced in Proposition 5.1 and assuming that the polar moduli $\xi \mapsto R_1(\xi) \in \mathbb{R}$ is parameterized by the following spline*

$$R_1(\xi) = \sum_{(i,j) \in I} (r_1)_{ij} B_{ij,d}(\xi). \tag{56}$$

The inequality $(47)_1$ holds for any $\xi \in \omega$ if the following inequalities :

$$2[(r_1)_{ij}]^2 - \frac{T_1(T_0 + (r_0^K)_{ij})}{N} < 0, \quad \text{where} \quad N := \text{Card}(I), \tag{57}$$

are satisfied for all $(i, j) \in I$. □

The same treatment applied to the geometric bound constraint $(48)_1$ leads to the following constraints on the control points $(r_1)_{ij}$ (resp. $(r_0^K)_{ij}$) of the R_1 (resp. R_0^K) polar variable:

$$2(\frac{(r_1)_{ij}}{R_1^L})^2 - \frac{1}{N}(1 + \frac{(r_0^K)_{ij}}{R_0^{K^L}}) \leq 0 \quad , \quad \forall (i,j) \in I. \tag{58}$$

$$0 \leq (r_1)_{ij} \leq R_1^L$$

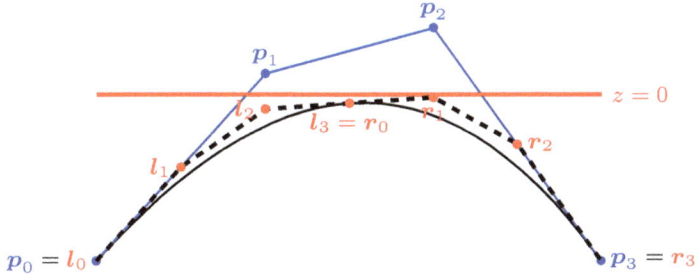

Fig. 4 Illustration of subdivision on the B-spline curve C_p. The *new* calculated control points $\{l_i, r_i\}$, $i \in \{0, \cdots, 3\}$ are both negative

5.3.3 Comments

The above-mentioned polar variables parameterization allows to reduce the number of design variables, which are now control points of splines. However, Fig. 3 of the Example 5.1 shows that defining the constraints on the control points can lead to narrow the design space. Nevertheless, it is possible to use spline modelling flexibility to introduce the constraints on additional control points to enlarge the exploration field. More specifically, once the parametrizations of the polar parameters are set, one can use subdivision (through *Casteljau algorithm*) see [61] or knot insertion algorithms to add new more interpolating control points, [36, 62]. This reduces to introduce new control parameters, which are linear combinations of originals and better adapted to define more accurate constraints. The strategy is illustrated in the following Example.

Example 5.2 (B-spline Flexibility for the Constraints) Let us return to the Example 5.1. Figure 4 shows the spline curve with its initial control points p_i and some news ones l_i and r_j, which are associated with the two splines obtained by inserting an interpolant knot point at the middle ($\xi = \frac{1}{2}$). These new control points satisfy the negativity constraint and then will help to check and better explore the admissible design.

Subdivision operation is a well known flexibility given by B-spline function. The idea was first defined for Bézier curves and surfaces with the *Casteljau algorithm*; see [61]. This algorithm allows to evaluate a Bezier function at some given parametric coordinates and also, at the same time, to split or subdivide the Bézier curve at that specific parametric coordinate. The subdivision technique has been generalized for B-spline functions by Cox-De-Boor [36, 62].

Hence, one can choose a certain level of subdivision a priori at which the sufficient constraints will be checked on the new computed control points and then obtain a better exploration of the design space. □

6 Formulating the Shape and Anisotropy Optimization Algorithm

The resolution of the optimization problem defined in Sect. 2 can now be formalized with the help of the meta-algorithm 6.1, where:

Algorithm 6.1 Main steps of the optimization algorithm

Input : Initial geometry Φ_0 (defined by its control points p_{ij}^0),
Initial elastic tensor e_0 (defined by its control points q_{kl}^0),
Geometric and material constraints.
Output : Optimal shape and optimal elastic tensor defined by they control points.
begin
   Initialisation: Meshing T_h of the (rectangular) domain ω and definition of the interpolation space W_h. Concerning the Nagdhi's equations, displacements and rotations of the mid-surface are interpolated by Lagrange P_2 triangular finite elements with 6 numerical integration points on each triangle. Enumerating in a vector $U \in \mathbb{R}^{N_{dof}}$ the degrees of freedom associated with this interpolation space, we write the bilinear form $a(.,.)$ (resp. the linear form $l(.)$) introduced in (25) as $a(u_h, v_h) = \langle [K_{\Phi,e}]U, V \rangle$ (resp. $l(v_h) = \langle F_\Phi, V \rangle$), where $\langle .,. \rangle$ denote the scalar product on $\mathbb{R}^{N_{dof}}$.
   Initialize $(\Phi, e) \longleftarrow (\Phi_0, e_0)$.
   while $\|\nabla j(\Phi, e)\| \geq Tol$ **do**
      Assembly (on the triangulation T_h) of the stiffness matrix $[K_{\Phi,e}]$ and the right hand member F_Φ of the shell's equations;
      Computaion and assembly of partial derivatives $[\partial_{p_{ij}} K_{\Phi,e}]$, $\partial_{p_{ij}} F_\Phi$ and $[\partial_{q_{kl}} K_{\Phi,e}]$;
      Resolution of the state equation $[K_{\Phi,e}]U = F_\Phi$;
      Computation of the derivatives of the compliance j with respect to the control points, p_{ij} of the shape $\Phi(\omega)$ and, q_{kl} of the elastic tensor e:

$$\partial_{p_{ij}} j(\Phi, e) = \langle 2\partial_{p_{ij}} F_\Phi - [\partial_{p_{ij}} K_{\Phi,e}]U, U \rangle \quad \text{and} \qquad (59)$$
$$\partial_{q_{kl}} j(\Phi, e) = -\langle [\partial_{q_{kl}} K_{\Phi,e}]U, U \rangle$$

      Update the geometry Φ and the elastic tensor e by modifying their control points in accordance with the gradient calculated above.
   end
end

- the assembly of the stiffness matrix and its derivatives is carried out with the help of the conventional finite element assembly procedures, where the geometric entries $\mathbf{a}_\alpha$, $a_{\alpha\beta}$ ..., defined in Sect. 3.1, are calculated on the numerical integration points of each triangle of T_h by using the B-spline differentiation algorithms; for instance, the tangent vectors (6) which write as

$$\mathbf{a}_\alpha(\xi^1, \xi^2) = \sum_{(i,j)} p_{i,j} \partial_\alpha B_{ij,\mathbf{d}}(\xi^1, \xi^2) \qquad (60)$$

are computed with the help of formulas (40) and (33);
- computation of the derivatives of the stiffness matrix with respect to the control points p_{ij} is a little more complicated and requires the calculation of the derivative of the bilinear form $a(.,.)$, defined in formulas (20) and (25), with respect to p_{ij}; it is a symmetric bilinear form which depends on the derivatives of Φ with respect to the control points; for instance, the derivative of the tangent vectors $\mathbf{a}_\alpha$ is the vector whose all components are $\partial_\alpha B_{ij,\mathbf{d}}$; we refer to [25] for the literal expression of the derivatives of the other geometric entities;
- the formula (59) defining the derivatives of the compliance j with respect to the control points (p_{ij}) (resp. q_{kl}) results from the fact that j is written as $j := \langle [K_{\Phi,e}]U, U \rangle$, so that we have:

$$\partial_{p_{ij}} j = \langle [\partial_{p_{ij}} K_{\Phi,e}]U, U \rangle + 2 \langle [K_{\Phi,e}]\partial_{p_{ij}} U, U \rangle, \tag{61}$$

for instance; differentiating the state equation $[K_{\Phi,e}]U = F_\Phi$, we obtain on the other hand:

$$[K_{\Phi,e}]\partial_{p_{ij}} U = \partial_{p_{ij}} F_\Phi - [\partial_{p_{ij}} K_{\Phi,e}]U. \tag{62}$$

The formula (59) is obtained by carrying forward (62) to (61).

Algorithm 6.1 is complemented by the following constraints, which are derived from design needs and mechanical well-posedness requirements:

- box constraints on shape control points, which reflect the designer's need for a shape contained in a given control polygon;
- material constraints are both box and inequality constraints, they are defined, according to the type of material taken into account, by formulae (51) or (52) for a conformal representation of the polar variables;
- in the case of the examples described in Sect. 7, we impose box constraints on the surface A of the shell; this amounts to minimize the compliance of the shell under mass constraints; note that this surface, which is calculated with the help of the formula (11), is a differentiable function of the control points p_{ij} of the shape;

The following additional constraints are technical and intend to ensure the convergence of the optimization algorithm:

- to eliminate the singular cases identified in Remark 4.2-3, we assume that the shape control points vary on lines orthogonal to the initial surface; this condition is restrictive and must be adapted on a case-by-case basis, it arises from the fact that the splines describe algebraic surfaces rather than differentiable manifolds;
- as the compliance j is invariant by some affine transformations A_f applied to shape $\Omega := \Phi(\omega)$, in the meaning that:

$$j(A_f \circ \Phi, e) = j(\Phi, e), \tag{63}$$

we will assume that the points controlling the boundary of Ω remain fixed (i.e. are excluded from the scope of optimization).

After a brief description of the practical formulation and steps of the optimal design problem in the framework of the polar-isogeometric approach, we give in the next section some numerical examples.

7 Some Examples

For the numerical resolution of the examples considered in this section, we have used NLOPT, a free/open-source library for NonLinear OPTimization. It includes the implementation of numerous optimization algorithms adapted for global and local optimizations. The library involves different types of algorithms such as, among others, Moving Asymptote Method (MMA) or COBYLA (Constrained Optimization by Linear Approximation), which can be gradient-based or derivative-free. We have used the COBYLA algorithm which appears to yield the best optimization results among the different algorithms.

As specified in the Introduction, in all the examples, we always start from a shell in the form of a flat domain with a given uniform distribution of the anisotropic parameters (e.g. an isotropic distribution). Hence, we begin all the iterations starting from a plate; during the computation, the plate is more and more transformed into a shell and at the same time anisotropy changes pointwise. During this process, the elastic energy, at the beginning entirely stored as bending energy in the plate, transforms continuously more and more to membrane energy. The ideal situation, corresponding to the optimal shell, is the one where all the strain energy of the shell is in the form of membrane energy, i.e. when the bending energy of the shell vanishes everywhere.

This is the process driving the initial plate to the final stiffest shell, acting simultaneously upon shape, i.e. geometry, and anisotropy, i.e. elasticity. So, this process can help to investigate the mutual influence of these two aspects on the morphogenesis of optimal anisotropic shells.

Material Properties
In the forthcoming examples, the elastic coefficients of the basic material (i.e. of the basic layer, for the case of a laminate) are:

$$E_2 = E = 9000 \times 10^6 \text{ Pa}$$
$$E_1 = E_3 = 161 \times 10^6 \text{ Pa}$$
$$\nu_{12} = \nu_{23} = \nu = 0.26$$
$$\nu_{13} = 0.26 \text{ and } G_{13} = \frac{E}{1+\nu}$$
$$G_{12} = G_{23} = 61 \times 10^6 \text{ Pa}.$$

The polar parameters corresponding to the plane reduced elastic tensor are:

$$T_0 = 1.17 \times 10^9 \text{ Pa}, \quad T_1 = 1.16 \times 10^9 \text{ Pa}$$
$$R_0^K = 1.11 \times 10^9 \text{ Pa}, \quad R_1 = 1.11 \times 10^9 \text{ Pa}, \text{ and } \Phi_1 = 0.$$

We consider up to three different types of optimal design problems, in order to analyze the incidence of the anisotropy on the shape:

- optimal shape design with an isotropic material (by setting $R_0^K = R_1 = 0$);
- optimal shape design with the specified anisotropic material and a fixed material orientation throughout the shell;
- joint optimal design of the shape and anisotropy, included the material orientation.

In all the cases, geometric bounds (48) are used in the calculations. The shape is parameterized by cubic B-spline polynomial of clamped knot vectors and 4 control points in each coordinate direction. The polar parameters are defined through B-spline polynomial of degrees d of clamped knot vectors of the form

$$\Sigma_\alpha = \{\underbrace{0, \cdots, 0}_{d+1}, \underbrace{1, \cdots, 1}_{d+1}\}, \; \alpha \in \{1, 2\}.$$

The geometry is subjected to box constraints on the control points and to a bounded area constraint of the form

$$l_0 = (1 - \epsilon_{tol})A_0 \leq A \leq u_0 = (1 + \epsilon_{tol})A_0,$$

that is, the relative variation of the design area with respect to the initial shell area A_0 is ϵ_{tol}.

In order to tailor efficiently the locally variable elastic properties, it can be relevant to use an assembling of patches for the parameterization of the polar parameters. This does not present any additional difficulty since, provided that the material frame of two adjacent patches is consistent, the continuity on the polar parameters (if necessary) is easily obtained by equating the control parameters at the interfaces of the adjacent patches.

7.1 Optimal Design of a Circular Dome

The first case concerns the design of a circular dome submitted to its own weight and simply supported at its boundary. The geometry of the initial structure, a circular plate, and the problem data are represented in Fig. 5 and Table 1. For symmetry reasons, the optimization is performed on a quarter of the structure and symmetry conditions are imposed on the elastic displacement.

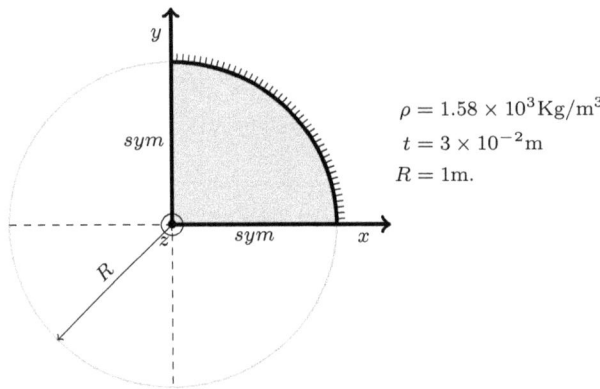

$\rho = 1.58 \times 10^3 \text{Kg/m}^3$
$t = 3 \times 10^{-2} \text{m}$
$R = 1\text{m}.$

Fig. 5 Circular plate geometry and boundary conditions

Table 1 Design problem relative to the circular plate

Geometry & loading:	Circular plate of radius $R = 1$ m subjected to its own weight	
Constraints:	• Symmetries with C^1 regularity throughout the planes of symmetry $x - z$ and $y - z$ • Fix place constraint on the simply supported boundary • Box constraints • Bounded area constraints $\epsilon_{tol} = 0.3$	
	Anisotropy	
	R_0^K, R_1	Φ_1
Knots	$\Sigma^\alpha = \{0, 0, 0, 0, 1, 1, 1, 1\}$	$\Sigma_{\Phi_1}^\alpha = \{0, 0, 0, 0, 1, 1, 1, 1\}$
Constraints	• Geometric bounds (52)	Bounds on the angle $\Phi_1 \in [-\pi, \pi]$
	Shape	Anisotropy
Number of design variables	12	48

Fig. 6 Example 1: optimal shape for isotropic material

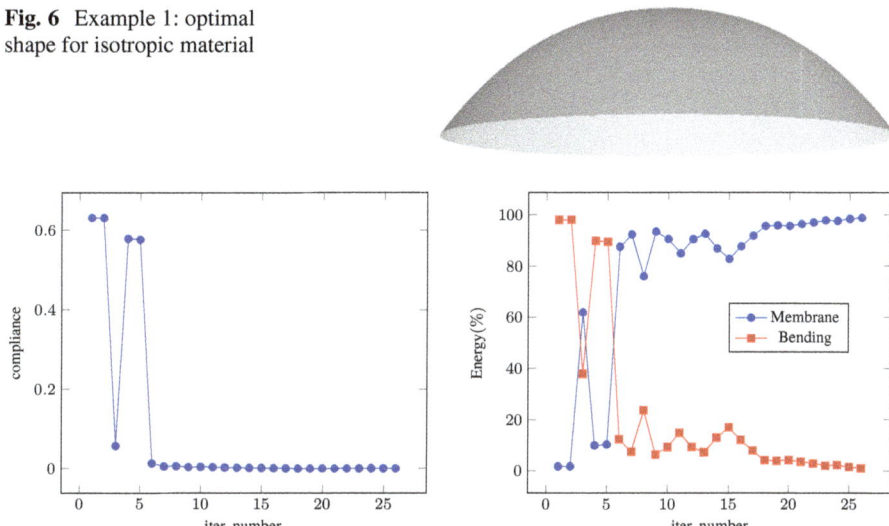

Fig. 7 Example one, shape optimization with isotropic material: evolution of the compliance through iterations (left) and ratio of the membrane and bending energy to the total strain energy (right)

7.1.1 First Case: Shape Design with Fixed Isotropic Material

The optimal shape found for the shell is represented in Fig. 6. We have checked that the optimal structure is a shell whose meridional profile is a catenary. It is well known that the catenary is the form of equilibrium of an arch of constant thickness under the action of its own weight in which all the internal actions reduce to a pure normal force (this result is due to R. Hooke, [63], see also [64] or [65]). Such an arch shape optimizes also its stiffness, because all the strain energy is stored in the structure under the form of extension energy, while the bending one is reduced to zero; this condition, as well known, corresponds to the maximum stiffness.

The optimal shell so obtained is, in the same way, submitted to only membrane internal actions, so that the bending energy vanishes. Actually, the membrane energy part for the optimal solution is $E_m = 98.35\%$. Figure 7 shows the evolution of the compliance, i.e. of the strain energy, and the contributions of the membrane and bending parts to the strain energy, along the optimization procedure. We can remark the migration of the strain energy from the dominant initial form of bending energy to the final prevailing form of membrane energy.

7.1.2 Second Case: Anisotropy and Shape Design

The second case that we consider for this example is the joint optimization of the polar parameters and of the shape. The polar parameters are subjected to geometric

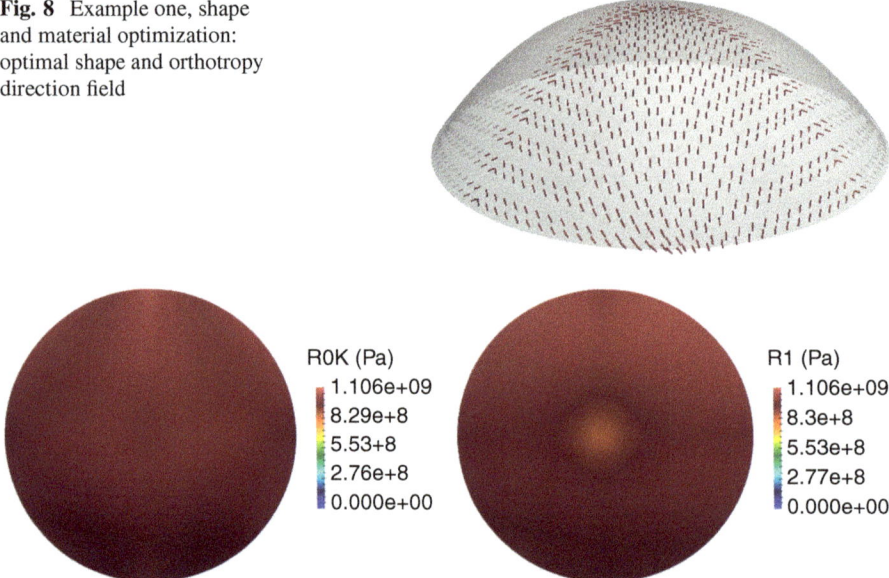

Fig. 8 Example one, shape and material optimization: optimal shape and orthotropy direction field

Fig. 9 Example one, shape and anisotropy optimization: optimal polar moduli fields

bounds on their control points and are defined as B-spline of degrees $d = 3$. The optimal shape and the orthotropy direction are plotted in Fig. 8. Also in this case the optimal shape is that of a shell with meridional sections in the form of a catenary. The polar parameters moduli are plotted in Fig. 9. One remarks that these parameters are not only uniform throughout the structure, but also that both of them take the highest possible value: the optimal shell is that with the highest possible degree of anisotropy of the two anisotropic phases. We also remark that the only orthotropy type is the one with $K = 0$.

Figure 10 shows the variation of the compliance and once again the migration of the strain energy from the bending to the membrane form throughout the optimization: for the optimal design, the membrane part is 93.34% of the total strain energy (Table 2).

7.2 Optimization of a Conical Shell

As second example, we consider the optimization of a conical shell: the starting structure is a holed circular plate which is simply supported at its external boundary and subjected to a uniformly distributed vertical load at its inner circular boundary. The radius of the circular plate is $R = 0.8$ m while that of the hole is $r = \frac{R}{3}$. The geometry and conditions are shown in Fig. 11 and Table 3.

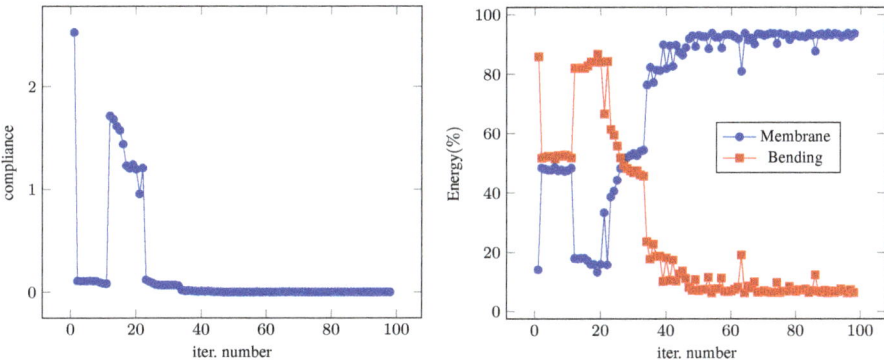

Fig. 10 Example one, shape and material optimization: evolution of the compliance through iterations (left) and ratio of the membrane and bending energy to the total strain energy (right)

Table 2 Summary of the global results concerning the first example

Design	Material			
	Anisotropic		Isotropic	
	Init	Final	Init	Final
Compliance (%)	100	0.174	100	0.167
$E_m(\%)$	14.11	93.33	1.82	98.95

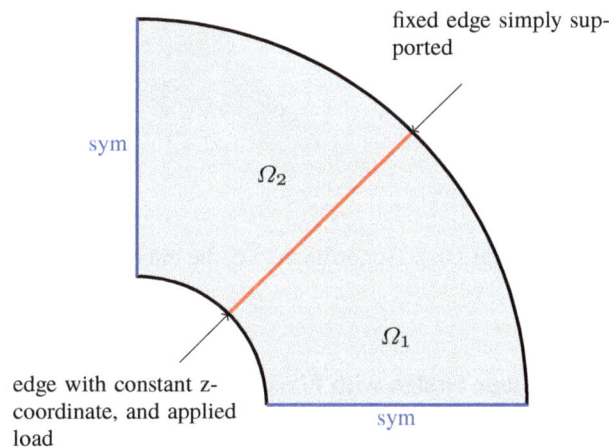

Fig. 11 Example two: geometry and boundary conditions

For symmetry reasons, the optimization is performed on a quarter of the structure. The geometry is defined by two sub-structures joined together as described in Fig. 11 and the optimization problem set up is reported in Table 3. The control points associated with the circular boundary with simply supported condition are kept fixed while those defining the internal crown, which carries the applied load,

Table 3 Description of the conical shell design problem

Geometry & loading:	Holed circular plate of outer radius $R = 0.8$ m and inner $r = \frac{R}{3}$	
	Loaded at its inner boundary	
Constraints:	• Symmetries with C^1 regularity throughout the planes of symmetry	
	• Fix place constraint on the simply supported boundary	
	• The inner loaded circle remains a circle	
	• Box constraints	
	• Bounded area constraints $\epsilon_{tol} = \frac{3}{2}$	
	Anisotropy	
	R_0^K, R_1	Φ_1
Knots	$\Sigma^\alpha = \{0, 0, 0, 1, 1, 1\}$	$\Sigma^\alpha_{\Phi_1} = \{0, 0, 0, 1, 1, 1\}$
Constraints	• Geometric bounds (52)	Bounds on the angle $\Phi_1 \in [-\pi, \pi]$
	Shape	Anisotropy
Number of design variables	25	45

Fig. 12 Example two: optimal shape for the isotropic case

are constrained to have the same z-coordinate, i.e. the internal circular boundary can move rigidly.

7.2.1 First Case: Shape Design with Fixed Isotropic Material

The optimal shape found is plotted in Fig. 12, while the variation of the compliance and of membrane and bending energy parts during the optimization procedure are plotted in Fig. 13.

Also in this case we can remark the migration of the strain energy from the bending to the membrane form: at the end of the calculation, this last is 98.88% of the whole strain energy.

For this example, the optimal shape is a conical surface, to which corresponds a strain energy completely stored under the form of membrane energy. The result shown in Fig. 12 is not exactly a conical surface, but it is close to it. This is due to the fact that the solution has not yet perfectly converged, which is attested by the

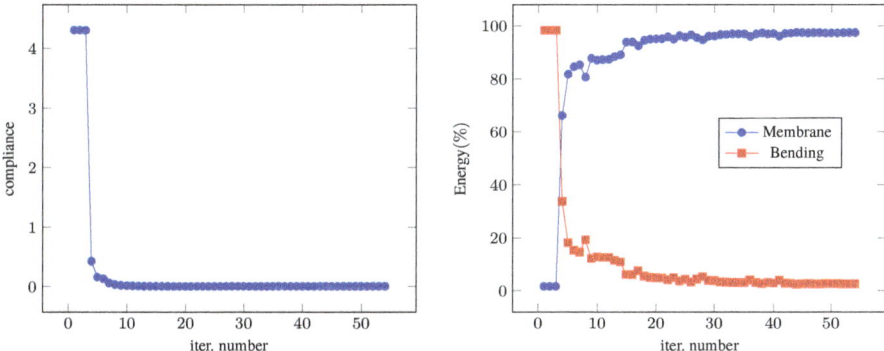

Fig. 13 Example two, shape optimization with isotropic material: evolution of the compliance through iterations (left) and ratio of the membrane and bending energy to the total strain energy (right)

fact that the membrane energy is not 100% of the whole elastic energy stored in the shell.

7.2.2 Second Case: Shape Design with Fixed Anisotropic Material

We consider now the optimization of the shape made with the anisotropic material specified in Sect. 7 under the same constraints on the geometry as in the previous case. In addition, we fix the angle of orthotropy: $\Phi_1 = \frac{\pi}{4}$. The optimal shape so found is shown in Fig. 14. In this case we do not obtain a shell with circular cross section, but a wrinkled surface, which is the apparent consequence of using a fixed anisotropic material with a fixed orientation throughout the shell. Unlike in the previous case, however, the wrinkles have an almost rectilinear profile, closer to a conical shape than before. We remark also that, as well known, wrinkled membranes are very stiff structures and it is interesting to notice that when anisotropy enters the design, the optimal shape becomes a wrinkled surface. Figure 15 shows the variation of the strain energy and of the membrane and bending parts throughout optimization. In this case, though the tendency is the same of the previous cases, the final membrane energy is only 87.3% of the whole strain energy stored in the shell; the remaining 12.7% is in the form of bending energy, which is due to the presence of the wrinkles.

7.2.3 Third Case: Shape and Anisotropy Optimal Design

We presently consider the joint optimization of the shape and material properties for this second example. The parameterization considered for the polar parameters corresponds to $d = 2$, thus, considering also the continuity condition between the

Fig. 14 Example two: optimal shape for fixed anisotropic material

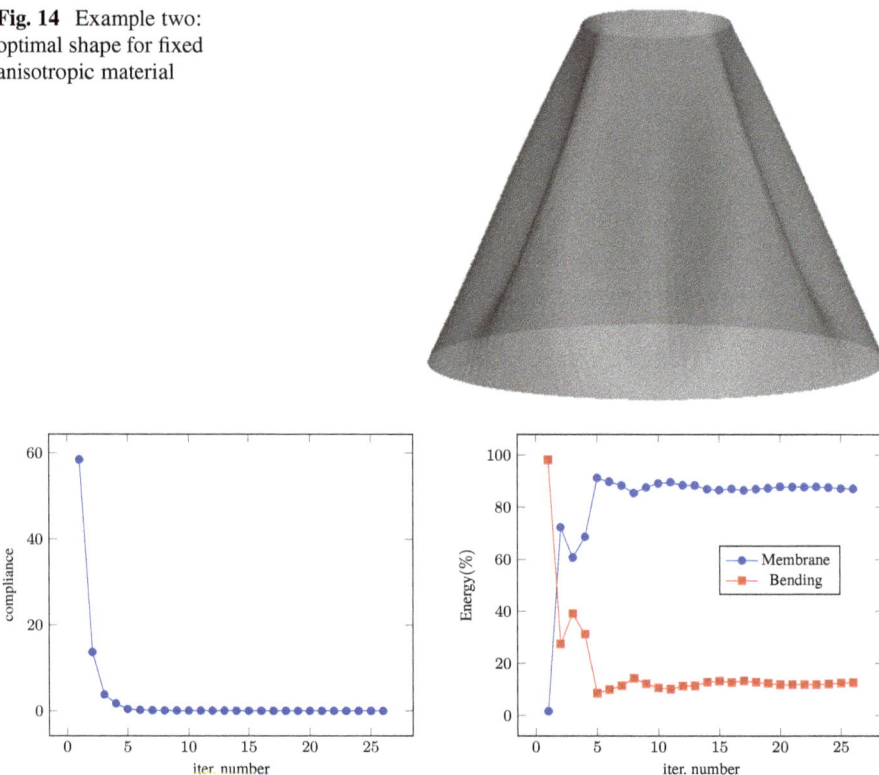

Fig. 15 Example two: optimal shape and material orientation for fixed anisotropic material. Evolution of the compliance through iterations (left) and ratio of the membrane and bending energy to the total strain energy (right)

Fig. 16 Example two, shape and anisotropy optimization: optimal shape and material orientation

two patches, there are 15 design variables for each polar parameter. Figure 16 shows the optimal shape and orthotropy direction. We remark that the optimal shape is,

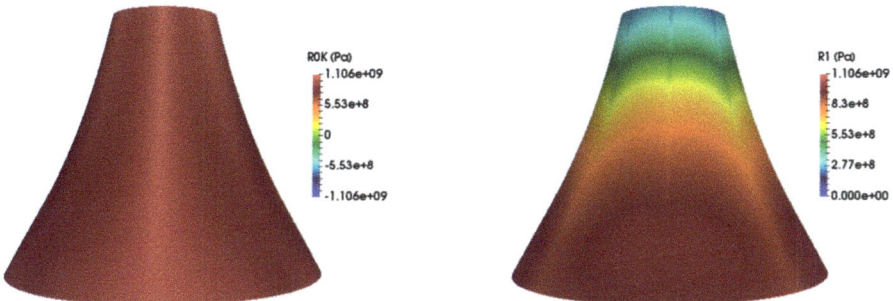

Fig. 17 Example two, shape and anisotropy optimization. Distribution of the optimal polar moduli R_0^K and R_1

Table 4 Summary of the different results concerning the second example (final 1: final value for the first case, final 2 for the second one and final 3 for the third one)

	Material				
	Anisotropic			Isotropic	
Design	Init	Final 2	Final 3	Init	Final 1
Compliance (%)	100	0.036	0.004	100	0.105
E_m (%)	1.74	87.3	97.2	1.65	98.9

like in the first case, a shell close to a conical surface; in particular, in this case the final surface is closer to a conical one than in the first case. Also, we notice that the optimal orientation is the same everywhere: the highest elastic modulus is in the meridional direction, which seems a logical result. Unlike the optimal material orientation, which is constant throughout the shell, the optimal distribution of the polar anisotropic moduli R_0^K and R_1 is not constant, see Fig. 17.

In particular, the field of the parameter R_0^K is uniform and almost constant over all the shell with the presence of only one kind of orthotropy $K = 0$, while R_1 changes, in particular it increases from top to bottom.

We remark also that the optimal material orientation is different from that, fixed a priori, of the previous case. The consequence of this is the different optimal shape of the shell that now is not wrinkled.

Figure 18 shows also for this case the variation along iterations of the strain energy and of its membrane and bending parts. The variation is analogous to that of the previous cases; in particular, one can notice that the convergence is practically reached after 60 iterations and at the end the membrane energy amounts to 97.2% of the whole elastic energy stored in the shell.

The overall results for the three cases considered for this second example are shown in Table 4. We can remark that the most effective case is the third one, with a final compliance which is just the 0.004% of the initial one. This shows that acting simultaneously on the geometry and on the material distribution is advantageous.

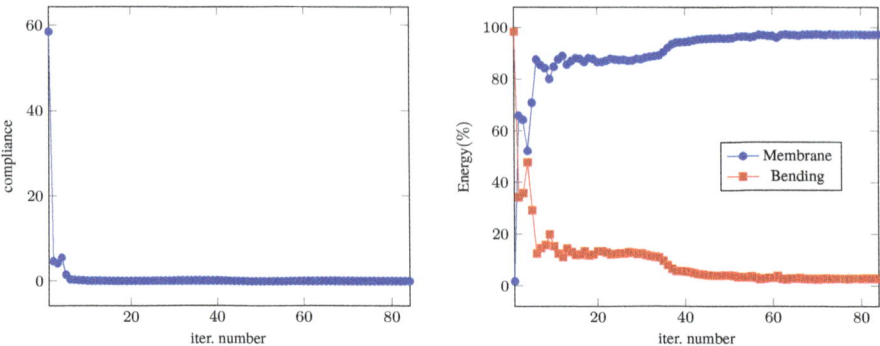

Fig. 18 Example two: shape and material optimization. Evolution of the compliance through iterations (left) and ratio of the membrane and bending energy to the total strain energy (right)

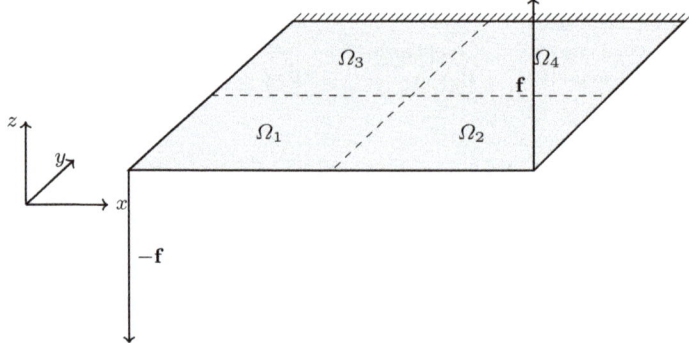

Fig. 19 Example three: geometry, actions, and boundary conditions

7.3 Plate Submitted to a Torsional Load

The last example that we consider is a square plate with a side of 1 m, clamped at one side and subjected to two equal but opposite loads applied at the free corners. On the whole, the plate is hence submitted to a torsional action. The geometry of the plate is sketched in Fig. 19, while in Table 5 we show the data of the problem. The two concentrated loads producing the torsion of the plate have a value $f = 1000$ N. The plate is defined by an assembling of 2×2 square plates of length 0.5 m, each being parameterized through a cubic B-splines with open knot vectors and four control points in each parametric direction. Finally a C^1-regularity is imposed at the junction of the patches. The design is constrained to preserve the boundary: the final shape of the optimal shell must have the same boundary of the original square plate. The orthotropy is initially oriented with an angle $\Phi_1 = 0$ with respect to the x-axis direction.

Table 5 Example three: data of the optimum problem

Geometry & loading:	Square plate of unit length	
	Torsion applied through two concentrated loads	
Constraints:	• C^1 regularity constraints between the patches	
	• Fix place constraint on the boundary	
	• Box constraints	
	• Bounded area constraints $\epsilon_{tol} = 0.2$	
	Anisotropy	
	R_0^K, R_1	Φ_1
Knot	$\Sigma^\alpha = \{0, 0, 0, 0, 1, 1, 1, 1\}$	$\Sigma^\alpha_{\Phi_1} = \{0, 0, 0, 0, 1, 1, 1, 1\}$
Constraints	• Geometric bounds (52)	Bounds on the angle $\Phi_1 \in [-\pi, \pi]$
	Shape	Anisotropy
Number of design variables	36	49

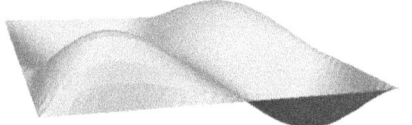
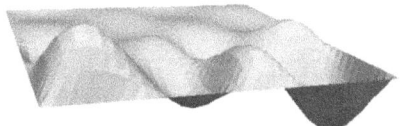

Fig. 20 Example three: optimal shapes corresponding to the first and second case (from the left)

7.3.1 First Case: Shape Design with Fixed Isotropic Material

As usual, we first consider that the plate is made with the isotropic material obtained putting $R_0^K = R_1 = 0$. Figure 20 shows, on the left, the optimal shape so found. It is interesting to notice that this shape corresponds with one of the fundamental vibration modes of the plate when it is simply supported along all its boundary.

7.3.2 Second Case: Shape Design with Fixed Anisotropic Material

Like for the previous example, we consider now the optimal shape design with the anisotropic material given in Sect. 7 and with the material orientation that is fixed everywhere to $\Phi_1 = 0$. The final optimal shape is shown in Fig. 20, on the right. Now, the optimal shape has remarkably changed with respect to the previous case: the waves are more numerous and they decrease going towards the clamped edge. It is interesting to remark that the greatest curvatures in the shell shape are located near the loaded corners.

Fig. 21 Example three, shape and material optimization: optimal shape and material orientation

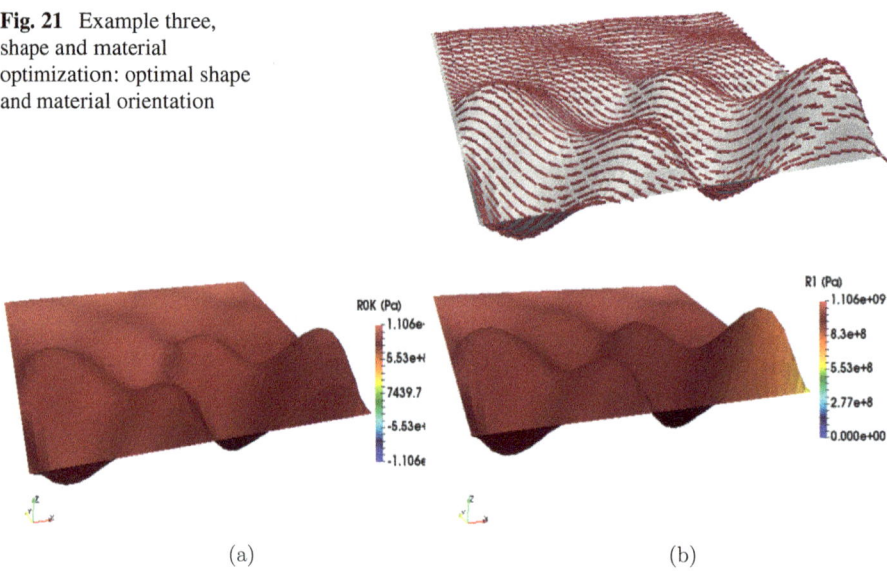

Fig. 22 Example three, shape and anisotropy optimization: optimal distribution of the anisotropy polar parameters. (**a**) Optimal distribution of R_0^K. (**b**) Optimal distribution of R_1

Table 6 Summary of the different results concerning the third example (final 1: final value for the first case, final 2 for the second one and final 3 for the third one)

	Material				
	Anisotropic			Isotropic	
Design	Init	Final 2	Final 3	Init	Final 1
Compliance (%)	100	1.000	0.256	100	5.150
E_m (%)	3.28	91.71	94.98	16.48	93.05

7.3.3 Third Case: Shape and Anisotropy Optimal Design

Finally, we consider the joint shape and anisotropy optimal design. Figure 21 shows the optimal shape and material orientation. The optimal shape changes again with respect to the two previous cases; the optimal distributions of the polar moduli R_0^K and R_1 are represented in Fig. 22. Like in the previous examples, they are practically constant throughout the structure and equal to their maximal allowed value. Also in this case, there is a unique final type of orthotropy, that with $K = 0$.

Table 6 summarizes the different optimization results for this third example. Like in the previous example, also in this case the best result is found for the third case, i.e. when shape and anisotropy are optimized. The optimal shapes so found for this example are rather unexpected and surprising is their similarity with the vibration modes of a simply supported plate. To this purpose, it is interesting to remark how the presence of anisotropy modifies the shape, changing the number of waves in the shell shape.

8 Conclusions

In this paper we have proposed a problem in a new field of structural optimization: the joint optimization of shape and material distribution for a shell-like structure. We have also proposed to tackle such a kind of problem by an approach that we have called polar-isogeometric, because it marries two distinct mathematical techniques: the polar formalism for the representation of plane anisotropy and an isogeometric-like approach for the parametrization of both the shell shape and the fields of the polar parameters.

Starting from a flat shape, i.e. from a plate, we have shown through different examples that the proposed method is able to drive the computation towards the optimal solution, the one where the elastic energy of the structure tends to be stored as membrane energy.

The use of the isogeometric approach, which is almost classical in shape design, has been used also for the parametrization of the variables describing the anisotropy of the shell. This has been done in order to reduce the number of design variables that become, in this approach, the control points of the parameterizations. Using the properties of the spline functions, we have also given a sufficient condition for ensuring the satisfaction of the bounds on the elastic parameters, to be fulfilled everywhere in the shell.

This is just one of the first works in this field and of course different improvements can be imagined, we discuss some of them. As a first point, we indicate the possibility of describing complex shapes using more than one patch. This problem has already been tackled and partially solved in [25]; the matter is delicate, because continuity conditions must be specified for the shape and for the elastic parameter fields. If for the first case, the shape, such conditions are delicate to be written, but rather well defined, for the second one, the elastic parameters, the definition of such continuity conditions among the patches is questionable, different possibilities can be imagined, driving towards different optimal problems.

A second point is the design of laminated shells. This point is rather well solved for plates and it just needs few amendments to the proposed approach: besides taking into account for the geometric instead of the elastic bounds, we need to use a specific procedure for finding a laminate able to have all the properties of the optimal shell, i.e. orthotropic, quasi-homogeneous and with the optimal distribution of e everywhere. Such a problem has been already solved satisfactorily for different problems concerning plates, see e.g. [12, 14, 15, 48], using metaheuristics, e.g. a genetic algorithm. However, a step further in this direction should be the introduction in the proposed optimization process of some technological constraints, e.g. on the trajectories and densities of the reinforcing fibers, using, for instance, the technique already proposed in [27]. For the case of laminated shells, the use of quasi-homogeneous laminates is almost essential, because the structure is, at least in some parts, submitted to extension and bending. To avoid this assumption means, on the one hand, to introduce separately the polar parameters for extension and bending, which doubles the number of design variables for the elastic part, and, on the other hand, to use some approximation of the feasible domain, because the exact bounds of the polar parameters for extension and bending, that are not independent, are still unknown, see [41].

The proposed approach can be used also for other objective functions rather than compliance. In particular, for buckling loads and vibration frequencies, i.e. for problems concerning eigenvalues, the proposed method can be applied almost directly; more difficult is the case of strength, because this is represented by a local functional; a possible way is that introduced in [12], because the use of the isogeometric approach used in this paper can be easily adapted to the polar parameters describing strength in the case of a tensorial strength criterion. A dual problem of the one considered in this study is the minimization of the mass of a shell that must have a minimal requirement in terms of stiffness, a problem of interest especially in aircraft construction.

Finally, the problem considered in this paper has been suggested by industrial applications: automotive, aeronautics, space, and sport engineering are more and more interested in the optimal design of composite structures, most of them being in the form of shells. Thanks to the new technologies of fiber placement, the possibility of tailoring anisotropy is today a true reality; so, the goal is to dispose of mathematical methods able to drive the design towards optimal solutions.

However, to our opinion, this approach has a wider interest and significance; in fact, as already said in the text, this approach allows to investigate the reciprocal influence of material properties and geometrical shape in the morphogenesis of 2D space structures; we think, namely to natural forms, like leaves or some skeletal organisms. This approach could be of interest in trying to investigate how Nature acts when it can dispose of geometry and material.

Acknowledgment The authors sincerely acknowledge RENAULT SAS for its support to this research through the granting of the PhD thesis of D. F. Kpadonou.

References

1. Banichuk, N.V.: Problems and Methods of Optimal Structural Design. Springer US, Berlin (1983). https://doi.org/10.1007/978-1-4613-3676-1
2. Allaire, G.: Shape Optimization by the Homogenization Method. Springer, New York (2002). https://doi.org/10.1007/978-1-4684-9286-6
3. Bendsøe, M.P., Sigmund, O.: Topology Optimization. Springer, Berlin (2004). https://doi.org/10.1007/978-3-662-05086-6
4. Gurdal, Z., Haftka, R.T., Hajela, P.: Design and Optimization of Laminated Composite Materials. Wiley, New York (1999)
5. Vannucci, P.: Anisotropic Elasticity. Springer, Singapore (2018). https://doi.org/10.1007/978-981-10-5439-6
6. Montemurro, M., Vincenti, A., Vannucci, P.: Design of the elastic properties of laminates with a minimum number of plies. Mech. Compos. Mater. **48**(4), 369–390 (2012). https://doi.org/10.1007/s11029-012-9284-4

7. Vannucci, P.: Designing the elastic properties of laminates as an optimisation problem: a unified approach based on polar tensor invariants. Struct. Multidiscip. Optim. **31**(5), 378–387 (2005). https://doi.org/10.1007/s00158-005-0566-5
8. Vannucci, P., Vincenti, A.: The design of laminates with given thermal/hygral expansion coefficients: a general approach based upon the polar-genetic method. Compos. Struct. **79**(3), 454–466 (2007). https://doi.org/10.1016/j.compstruct.2006.02.004
9. Vannucci, P., Barsotti, R., Bennati, S.: Exact optimal flexural design of laminates. Compos. Struct. **90**(3), 337–345 (2009). https://doi.org/10.1016/j.compstruct.2009.03.017
10. Vincenti, A., Desmorat, B.: Optimal orthotropy for minimum elastic energy by the polar method. J. Elast. **102**(1), 55–78 (2010). https://doi.org/10.1007/s10659-010-9262-9
11. Vincenti, A., Vannucci, P., Ahmadian, M.R.: Optimization of laminated composites by using genetic algorithm and the polar description of plane anisotropy. Mech. Adv. Mater. Struct. **20**(3), 242–255 (2013). https://doi.org/10.1080/15376494.2011.563415
12. Catapano, A., Desmorat, B., Vannucci, P.: Stiffness and strength optimization of the anisotropy distribution for laminated structures. J. Optim. Theory Appl. **167**(1), 118–146 (2015). https://doi.org/10.1007/s10957-014-0693-5.
13. Jibawy, A., Julien, C., Desmorat, B., Vincenti, A., Léné, F.: Hierarchical structural optimization of laminated plates using polar representation. Int. J. Solids Struct. **48**(18), 2576–2584 (2011). https://doi.org/10.1016/j.ijsolstr.2011.05.015
14. Montemurro, M., Vincenti, A., Vannucci, P.: A two-level procedure for the global optimum design of composite modular structures—application to the design of an aircraft wing. Part 1: theoretical formulation. J. Optim. Theory Appl. **155**(1), 1–23 (2012). https://doi.org/10.1007/s10957-012-0067-9
15. Montemurro, M., Vincenti, A., Vannucci, P.: A two-level procedure for the global optimum design of composite modular structures-application to the design of an aircraft wing. Part 2: numerical aspects and examples. J. Optim. Theory Appl. **155**, 24–53 (2012)
16. Vannucci, P.: Strange laminates. Math. Methods Appl. Sci. **35**(13), 1532–1546 (2012). https://doi.org/10.1002/mma.2539
17. Hughes, T., Cottrell, J., Bazilevs, Y.: Isogeometric analysis: CAD, finite elements, NURBS, exact geometry and mesh refinement. Comput. Methods Appl. Mech. Eng. **194**(39–41), 4135–4195 (2005). https://doi.org/10.1016/j.cma.2004.10.008
18. Bazilevs, Y., Calo, V.M., Hughes, T.J.R., Zhang, Y.: Isogeometric fluid-structure interaction: theory, algorithms, and computations. Comput. Mech. **43**(1), 3–37 (2008). https://doi.org/10.1007/s00466-008-0315-x
19. Bazilevs, Y., Calo, V.M., Zhang, Y., Hughes, T.J.R.: Isogeometric fluid–structure interaction analysis with applications to arterial blood flow. Comput. Mech. **38**(4), 310–322 (2006). https://doi.org/10.1007/s00466-006-0084-3.
20. Bazilevs, Y., Hsu, M.C., Scott, M.: Isogeometric fluid–structure interaction analysis with emphasis on non-matching discretizations, and with application to wind turbines. Comput. Methods Appl. Mech. Eng. **249–252**, 28–41 (2012). https://doi.org/10.1016/j.cma.2012.03.028
21. Cho, S., Ha, S.H.: Isogeometric shape design optimization: exact geometry and enhanced sensitivity. Struct. Multidiscip. Optim. **38**(1), 53–70 (2008). https://doi.org/10.1007/s00158-008-0266-z
22. Qian, X.: Full analytical sensitivities in NURBS based isogeometric shape optimization. Comput. Methods Appl. Mech. Eng. **199**(29–32), 2059–2071 (2010). https://doi.org/10.1016/j.cma.2010.03.005
23. De Nazelle, P.: Paramétrage de formes surfaciques pour l'optimisation. Ph.D. thesis, Ecole Centrale Lyon (2013)
24. Julisson, S.: Shape optimization of thin shell structures for complex geometries. Ph.D. thesis, Paris-Saclay University (2016). https://tel.archives-ouvertes.fr/tel-01503061
25. Kpadonou, D.F.: Shape and anisotropy optimization by an isogeometric-polar approach. Ph.D. thesis, Paris-Saclay University (2017)

26. Montemurro, M., Catapano, A.: A new paradigm for the optimum design of variable angle tow laminates. In: Variational analysis and aerospace engineering: mathematical challenges for the aerospace of the future. Springer Optimization and Its Applications, vol. 116, pp. 375–400. Springer, Berlin (2016). http://dx.doi.org/10.1007/978-3-319-45680-5
27. Montemurro, M., Catapano, A.: On the effective integration of manufacturability constraints within the multi-scale methodology for designing variable angle-tow laminates. Compos. Struct. **161**, 145–159 (2017)
28. Banichuk, N.V.: Introduction to Optimization of Structures. Springer, New York (1990). https://doi.org/10.1007/978-1-4612-3376-3
29. Love, A.E.H.: The small free vibrations and deformation of a thin elastic shell. Philos. Trans. R. Soc. A Math. Phys. Eng. Sci. **179**(0), 491–546 (1888). https://doi.org/10.1098/rsta.1888.0016
30. Koiter, W.: Foundations and Basic Equations of Shell Theory: A Survey of Recent Progress. Afdeling der Werktuigbouwkunde: WTHD. Labor. voor Techn. Mechanica (1968). https://books.google.fr/books?id=74REHQAACAAJ
31. Naghdi, P.: Foundations of Elastic Shell Theory. North-Holland, Amsterdam (1963)
32. Reissner, E.: On the theory of transverse bending of elastic plates. Int. J. Solids Struct. **12**(8), 545–554 (1976). https://doi.org/10.1016/0020-7683(76)90001-9
33. Ciarlet, P.: An Introduction to Differential Geometry with Application to Elasticity. Springer, Berlin (2005)
34. Bernadou, M.: Mèthodes d'éléments finis pour les problèmes de coques minces. Recherches en Mathématiques Appliquées, Masson (1994)
35. Bézier, P.: Essai de définition numérique des courbes et des surfaces expérimentales: contribution à l'étude des propriétés des courbes et des surfaces paramétriques polynomiales à coefficients vectoriels, vol. 1 (1977). https://books.google.fr/books?id=vK4POAAACAAJ
36. Rogers, D.F.: An Introduction to NURBS with Historical Perspective. Elsevier (2001). https://doi.org/10.1016/b978-1-55860-669-2.x5000-3
37. Risler, J.J.: Méthodes Mathématiques pour la CAO. Masson (1991)
38. Piegl, L., Tiller, W.: The NURBS Book, 2nd edn. Springer, Berlin (1997)
39. Verchery, G.: Les invariants des tenseurs d'ordre 4 du type de l'élasticité. In: Boehler, J.P. (ed.) Mechanical Behavior of Anisotropic Solids/Comportment Méchanique des Solides Anisotropes, pp. 93–104. Springer Netherlands (1982). https://doi.org/10.1007/978-94-009-6827-1_7
40. Vannucci, P.: Plane anisotropy by the polar method. Meccanica **40**, 437–454 (2005). https://doi.org/10.1007/s11012-005-2132-z
41. Vannucci, P.: A note on the elastic and geometric bounds for composite laminates. J. Elast. **112**, 199–215 (2013). https://doi.org/10.1007/s10659-012-9406-1
42. Vannucci, P., Verchery, G.: Anisotropy of plane complex elastic bodies. Int. J. Solids Struct. **47**, 1154–1166 (2010). https://doi.org/10.1016/j.ijsolstr.2010.01.002. http://www.sciencedirect.com/science/article/pii/S002076831000003X
43. Valot, E., Vannucci, P.: Some exact solutions for fully orthotropic laminates. Compos. Struct. **69**(2), 157–166 (2005). https://doi.org/10.1016/j.compstruct.2004.06.007
44. Vannucci, P., Pouget, J.: Laminates with given piezoelectric expansion coefficients. Mech. Adv. Mater. Struct. **13**(5), 419–427 (2006). https://doi.org/10.1080/15376490600777699
45. Vannucci, P.: Influence of invariant material parameters on the flexural optimal design of thin anisotropic laminates. Int. J. Mech. Sci. **51**, 192–203 (2009). https://doi.org/10.1016/j.ijmecsci.2009.01.005. http://www.sciencedirect.com/science/article/pii/S0020740309000198
46. Vannucci, P.: A new general approach for optimizing the performances of smart laminates. Mech. Adva. Mater. Struct. **18**(7), 548–558 (2011). https://doi.org/10.1080/15376494.2011.605015
47. Montemurro, M., Koutsawa, Y., Belouettar, S., Vincenti, A., Vannucci, P.: Design of damping properties of hybrid laminates through a global optimisation strategy. Compos. Struct. **94**(11), 3309–3320 (2012). https://doi.org/10.1016/j.compstruct.2012.05.003

48. Vannucci, P.: The design of laminates as a global optimization problem. J. Optim. Theory Appl. **157**(2), 299–323 (2012). https://doi.org/10.1007/s10957-012-0175-6
49. Montemurro, M., Vincenti, A., Koutsawa, Y., Vannucci, P.: A two-level procedure for the global optimization of the damping behavior of composite laminated plates with elastomer patches. J. Vib. Control. **21**(9), 1778–1800 (2013). https://doi.org/10.1177/1077546313503358
50. Vannucci, P.: The polar analysis of a third order piezoelectricity-like plane tensor. Int. J. Solids Struct. **44**(24), 7803–7815 (2007). https://doi.org/10.1016/j.ijsolstr.2007.05.012
51. Vannucci, P.: On special orthotropy of paper. J. Elast. **99**(1), 75–83 (2009). https://doi.org/10.1007/s10659-009-9232-2
52. Catapano, A., Desmorat, B., Vannucci, P.: Invariant formulation of phenomenological failure criteria for orthotropic sheets and optimisation of their strength. Math. Methods Appl. Sci. **35**(15), 1842–1858 (2012). https://doi.org/10.1002/mma.2530
53. Barsotti, R., Vannucci, P.: Wrinkling of orthotropic membranes: an analysis by the polar method. J. Elast. **113**(1), 5–26 (2012). https://doi.org/10.1007/s10659-012-9408-z
54. Vannucci, P.: General theory of coupled thermally stable anisotropic laminates. J. Elast. **113**(2), 147–166 (2012). https://doi.org/10.1007/s10659-012-9415-0
55. Desmorat, B., Vannucci, P.: An alternative to the kelvin decomposition for plane anisotropic elasticity. Math. Methods Appl. Sci. **38**(1), 164–175 (2013). https://doi.org/10.1002/mma.3059
56. Vannucci, P., Desmorat, B.: Analytical bounds for damage induced planar anisotropy. Int. J. Solids Struct. **60–61**, 96–106 (2015). https://doi.org/10.1016/j.ijsolstr.2015.02.017. http://www.sciencedirect.com/science/article/pii/S0020768315000578
57. Vannucci, P.: A note on the computation of the extrema of young's modulus for hexagonal materials: an approach by planar tensor invariants. Appl. Math. Comput. **270**, 124–129 (2015). https://doi.org/10.1016/j.amc.2015.08.025
58. Vannucci, P., Desmorat, B.: Plane anisotropic rari-constant materials. Math. Methods Appl. Sci. **39**(12), 3271–3281 (2015). https://doi.org/10.1002/mma.3770
59. Vannucci, P.: A special planar orthotropic material. J. Elast. **67**, 81–96 (2002). https://doi.org/10.1023/A:1023949729395
60. Vannucci, P., Verchery, G.: Stiffness design of laminates using the polar method. Int. J. Solids Struct. **38**(50–51), 9281–9294 (2001). https://doi.org/10.1016/S0020-7683(01)00177-9. http://www.sciencedirect.com/science/article/pii/S0020768301001779
61. Patrikalakis, N.M., Maekawa, T.: Shape interrogation for computer aided design and manufacturing. Springer, Berlin (2009)
62. de Boor, C.: A Practical Guide to Splines. Applied Mathematical Sciences. Springer, New York (2001). https://books.google.fr/books?id=m0QDJvBI_ecC
63. Hooke, R.: A Description of Helioscopes, and Some Other Instruments. John and Martyn Printer, London (1675)
64. Heyman, J.: The Stone Skeleton. Cambridge University Press, Cambridge (1995)
65. Cowan, H.J.: The Masterbuilders. Wiley, New York (1977)

Gradient Polyconvexity and Modeling of Shape Memory Alloys

Martin Horák, Martin Kružík, Petr Pelech, and Anja Schlömerkemper

1 Introduction

Gradient polyconvex functionals, introduced originally in [10], depend on the gradients of nonlinear minors of the deformation gradient, i.e., they involve not only the first but also the second spatial derivatives of the deformation field. Materials having such a broader energy dependence are generally called non-simple [54] and their idea can be traced back to 1901 when Korteweg [32] considered a gradient of the density in his model of fluid capillarity. Considering more than only the first deformation gradient in the description of elastic behavior of solids goes back to the 1960s and appeared in the work of Toupin [52, 53], and Green and Rivlin [29]. Such materials are usually called N-grade materials, where N refers to the highest deformation gradient appearing in the model. This approach has brought questions on thermodynamical consistency of such models, treated in [13, 22], for instance. Since then, it has been used and analyzed in many works; see, e.g.,

---

M. Horák
Faculty of Civil Engineering, Czech Technical University, Prague, Czechia
e-mail: martin.horak@fsv.cvut.cz

M. Kružík (✉)
Czech Academy of Sciences, Institute of Information Theory and Automation, Prague, Czechia

Faculty of Civil Engineering, Czech Technical University, Prague, Czechia
e-mail: kruzik@utia.cas.cz

P. Pelech
Weierstrass Institute, Berlin, Germany
e-mail: petr.pelech@wias-berlin.de

A. Schlömerkemper
Institute of Mathematics, University of Würzburg, Würzburg, Germany
e-mail: anja.schloemerkemper@mathematik.uni-wuerzburg.de

[7, 20, 21, 24, 26, 27, 34, 46, 50, 51]. From the material point of view, the more general energy functionals in higher grade continua lead to an additional force interaction in a form of an edge traction or the so-called couple-stress or double force acting on the boundary; see [38, 42, 48, 49].

Mathematically, the presence of higher-order gradients in the model brings additional compactness properties for the set of admissible functions and ensures the existence of minimizers. We refer to recent related results on the mathematical treatment of shape memory materials and solid-to-solid interfaces: [1, 4, 6, 18, 19]. We also refer to [9] for an overview of recent mathematical results in the calculus of variations. For computational results on NiMnGa see, e.g., [1].

The aim of this contribution (cf. [37]) is to apply a new class of non-simple material models introduced in [10] (called *gradient polyconvex materials*) to evolutionary problems of shape memory alloys and to consider a computational experiment. The novelty consists in considering only gradients on nonlinear minors in the stored energy density of the material. It is shown there, and also in Example 2 below, that corresponding deformations do not necessarily have integrable second weak derivatives. Nevertheless, it is possible to prove existence of an energetic solution.

The plan of the paper is as follows. We first introduce necessary notation and tools in Sect. 2. The notion of gradient polyconvexity is thoroughly discussed in Sect. 3 and the quasistatic evolution in Sect. 4. Finally, in Sect. 5 we consider a bar made of a specific shape memory material (NiMnGa) and provide first computational results on the evolution of a solid-to-solid phase transformation in a tension experiment.

2 Preliminaries

Hyperelasticity is a special area of Cauchy elasticity, where one assumes that the first Piola-Kirchhoff stress tensor P possesses a potential (called stored energy density) $W : (0, +\infty) \times \mathbb{R}^{3\times3} \to (-\infty, \infty]$. In other words,

$$P(\theta, F) := \frac{\partial W(\theta, F)}{\partial F} \quad (1)$$

on its domain, where $F \in \mathbb{R}^{3\times3}$ is such that $\det F > 0$ and θ stands for the absolute temperature. This concept emphasizes that all work done by external loads on the specimen is stored in it. The principle of frame-indifference requires that W satisfies, for all $F \in \mathbb{R}^{3\times3}$ and all proper rotations $R \in \mathrm{SO}(3)$,

$$W(\theta, F) = W(\theta, RF) = \tilde{W}(\theta, F^\top F) = \tilde{W}(\theta, C),$$

where $C := F^\top F$ is the right Cauchy-Green strain tensor and $\tilde{W} : (0, +\infty) \times \mathbb{R}^{3\times3} \to (-\infty, \infty]$.

Additionally, every elastic material is assumed to resist extreme compression, which is modeled by

$$W(\theta, F) \to +\infty, \text{ if } \det F \searrow 0. \tag{2}$$

Let the reference configuration be a bounded Lipschitz domain $\Omega \subset \mathbb{R}^3$. Deformation $y : \bar{\Omega} \to \mathbb{R}^3$ maps the points in the closure of the reference configuration $\bar{\Omega}$ to their positions in the deformation configuration. Solutions to the corresponding elasticity equations can then be formally found by minimizing the energy functional

$$I(\theta, y) := \int_{\Omega} W(\theta, \nabla y(x)) \, dx - \ell(y) \tag{3}$$

over the class of admissible deformations. Here, ℓ is a functional on the set of deformations, expressing (in a simplified way) the work of external loads on the specimen, and ∇y is the deformation gradient, which quantifies the strain. We only allow for deformations, which are orientation-preserving, i.e., if $a, b, c \in \mathbb{R}^3$ satisfy $(a \times b) \cdot c > 0$, then $(Fa \times Fb) \cdot Fc > 0$ for every $F := \nabla y(x)$ and $x \in \Omega$, which means that $\det F > 0$. This condition can be expressed by extending W by infinity on matrices with non-positive determinants, i.e.,

$$W(\theta, F) := +\infty, \text{ if } \det F \leq 0. \tag{4}$$

In view of (1), (2), and (4), we see that $W : (0, +\infty) \times \mathbb{R}^{3\times 3} \to (-\infty, +\infty]$, is continuous in the sense that if $F_k \to F$ in $\mathbb{R}^{3\times 3}$ for $k \to +\infty$, then $\lim_{k \to +\infty} W(\theta, F_k) = W(\theta, F)$. Furthermore, $W(\theta, \cdot)$ is differentiable on the set of matrices with positive determinants.

Relying on the direct method of the calculus of variations, the usual approach to prove the existence of minimizers is to study (weak) lower semicontinuity of the functional I on appropriate Banach spaces containing the admissible deformations. For definiteness, we assume that $y \mapsto -\ell(y)$ is weakly sequentially lower semicontinuous. Thus, the question reduces to a discussion of the assumptions on W. It is well-known that (2) prevents us from assuming convexity of W. See, e.g., [17] or the recent review for a detailed exposition of weak lower semicontinuity. Following earlier work by C.B. Morrey, Jr., [43], J.M. Ball [2] defined a polyconvex stored energy density W by assuming that there is a convex and lower semicontinuous function $\overline{W}(\theta, \cdot) : \mathbb{R}^{19} \to (-\infty, +\infty]$ such that

$$W(\theta, F) := \overline{W}(\theta, F, \text{Cof } F, \det F) \quad \forall F \in \mathbb{R}^{3\times 3}.$$

Here, Cof F denotes the cofactor matrix of F, which, for F being invertible, satisfies Cramer's rule:

$$\text{Cof } F = (\det F)(F^{-1})^\top.$$

Hence, $\det \operatorname{Cof} F = \det^2 F$ and because we assume that $\det F > 0$ we have that

$$F = \left(\frac{\operatorname{Cof} F}{\sqrt{\det \operatorname{Cof} F}}\right)^{-\top},$$

i.e., we can reconstruct F from $\operatorname{Cof} F$. It is well-known that polyconvexity is satisfied for a large class of constitutive functions and allows for the existence of minimizers of I under (2) and (4). On the other hand, there are still situations where polyconvexity cannot be adopted. A prominent example is shape memory alloys, where W has the so-called multi-well structure; see, e.g., [5, 11, 44]. Namely, there is a high-temperature phase, called austenite, which is usually of cubic symmetry, and a low-temperature phase, called martensite, which is less symmetric and exists in more variants, e.g., in three for the tetragonal structure (NiMnGa) or in twelve for the monoclinic one (NiTi). We can assume that

$$W(\theta, F) := \min_{0 \leq i \leq M} W_i(\theta, F), \tag{5}$$

where $W_i : (0, +\infty) \times \mathbb{R}^{3\times 3} \to (-\infty, +\infty]$ is the stored energy density of the i-th variant of martensite if $i > 0$, and W_0 is the stored energy density of the austenite. For every admissible i, we have $W_i(\theta, \cdot)$ is minimized if and only if $F = RF_i$ for a given matrix $F_i \in \mathbb{R}^{3\times 3}$ and an arbitrary proper rotation $R \in SO(3)$. This means that each variant of the martensite and the austenite is modeled as a hyperelastic material with its own stored energy density W_i. We also assume that each $W_i(\theta, \cdot)$ is differentiable on the set of matrices with positive determinants. Thus the variants can be described independently of each other, i.e., the elastic constants can be chosen differently. The drawback is obviously the non-smoothness of W, however, physically realistic elastic strain values do not occur in the set where W is not differentiable. We refer, e.g., to [39] for other models of the stored energy density of shape memory alloys.

Given a deformation gradient F, we need to decide if the corresponding deformation is in the well of the austenite, or in a martensitic variant. In order to do so, we define a volume fraction $\lambda(F)$ as follows: Let $\lambda : \mathbb{R}^{3\times 3} \to \mathbb{R}^{M+1}$. Set

$$\lambda^j(F) := \frac{1}{M}\left(1 - \frac{\operatorname{dist}(C, \mathcal{N}_j(C_j))}{\sum_{i=0}^M \operatorname{dist}(C, \mathcal{N}_i(C_i))}\right) \forall C = F^T F \in \mathbb{R}^{3\times 3},\ j = 0, \ldots, M, \tag{6}$$

where $\{\mathcal{N}_i(C_i)\}_i$ are pairwise disjoint neighborhoods of the right Cauchy-Green strain tensors $C_i = F_i^\top F_i$, for $i \in \{0, \ldots, M\}$. Notice that $\sum_{j=0}^M \lambda^j(F) = 1$ for every F, which, together with $\lambda^j \geq 0$, allows us to interpret λ as a volume fraction.

Moreover, note that λ is continuous and frame-indifferent in the sense that $\lambda(F) = \lambda(RF)$ for every proper rotation R. Volume fractions will play an important role in the definition of our evolutionary model in Sect. 4.

Remark 1 Note that this particular choice of λ allows for some elastic behavior close to the wells $SO(3)F_i$, $i = 0, \ldots, M$, since the volume fraction remains constant on the neighborhoods $\mathcal{N}_i(C_i)$, $i = 0, \ldots, M$.

Let us emphasize that (5) ruins even generalized notions of convexity as, e.g., rank-one convexity. (We recall that rank-one convex functions are convex on line segments with endpoints differing by a rank-one matrix and that rank-one convexity is a necessary condition for polyconvexity; cf. [17], for instance.) Namely, it is observed (see, e.g., [5, 11]) that there is a proper rotation R_{ij} such that $\text{rank}(R_{ij}F_i - F_j) = 1$. if $0 < i \neq j > 0$. Hence, generically, $W(\theta, R_{ij}F_i) = W(\theta, F_j) = -w_i(\theta)$, but $W(\theta, F) > -w_i(\theta)$ if F is on the line segment between $R_{ij}F_i$ and F_j. Nevertheless, not having a convexity property at hand that implied existence of minimizers is in accordance with experimental observations for these alloys.

Indeed, nonexistence of a minimizer corresponds to the formation of microstructure of strain-states. This is mathematically manifested via a faster and faster oscillation of deformation gradients in minimizing sequences, driving the functional I to its infimum. One can then formulate a minimization problem for a lower semicontinuous envelope of I, the so-called relaxation, see, e.g., [17]. Such a relaxation yields information of the effective behavior of the material and on the set of possible microstructures. Thus relaxation is not only an important tool for mathematical analysis, but also for applications. For numerical considerations it is a challenging problem, because the relaxation formula is generically not obtained in a closed form. Further difficulties come from the fact that a sound mathematical relaxation theory is developed only if W has p-growth; that is, for some $c(\theta), c > 1$, $p \in \,]1, +\infty[$ and all $F \in \mathbb{R}^{3\times3}$, the inequality

$$\frac{1}{c}(|F| - c(\theta)) \leq W(\theta, F) \leq c(1 + |F|^p + c(\theta))$$

is satisfied. This in particular implies that $W < +\infty$. We refer, however, to [8, 16, 33] for results allowing for infinite energies. Nevertheless, these works include other assumptions that severely restrict their usage. Let us point out that the right Cauchy-Green strain tensor $F^\top F$ maps $SO(3)F$ as well as $(O(3)\backslash SO(3))F$ to the same point. Here, $O(3)$ are the orthogonal matrices with determinant ± 1. Thus, for example, $F \mapsto |F^\top F - \mathbb{I}|$ is minimized on two energy wells, on $SO(3)$ and also on $O(3)\backslash SO(3)$. However, the latter set is not acceptable in elasticity, because the corresponding minimizing affine deformation is a mirror reflection. In order to distinguish between these two wells, it is necessary to incorporate $\det F$ in the model properly.

Besides relaxation, another approach guaranteeing existence of minimizers is to resort to non-simple materials, i.e., materials, whose stored energy density depends also on higher-order derivatives. Simple examples are functionals of the form

$$I(\theta, y) := \int_\Omega W(\theta, \nabla y(x)) + \varepsilon |\nabla^2 y(x)|^p \, dx - \ell(y),$$

where $\varepsilon > 0$. Obviously, the second-gradient term brings additional compactness to the problem, which allows to require only strong lower semicontinuity of the term

$$\nabla y \mapsto \int_\Omega W(\theta, \nabla y(x))\, dx$$

for existence of minimizers.

Here, we follow a different approach suggested in [10], which is a natural extension of polyconvexity exploiting weak continuity of minors in Sobolev spaces. Instead of the full second gradient, it is assumed that the stored energy density of the material depends on the deformation gradient ∇y and on gradients of nonlinear minors of ∇y, i.e., on $\nabla[\text{Cof}\,\nabla y]$ and on $\nabla[\det \nabla y]$. The corresponding functionals are then called gradient polyconvex. While we assume convexity of the stored energy density in the two latter variables, this is not assumed in the ∇y variable. The advantage is that minimizers are elements of Sobolev spaces $W^{1,p}(\Omega, \mathbb{R}^3)$, and no higher regularity is required.

The following example is inspired by a similar one in [10]. It shows that there are maps with smooth nonlinear minors whose deformation gradient is *not* a Sobolev map. Hence, gradient polyconvex energies are more general than second-gradient ones.

Example

Let $\Omega =]0, 1[^3$. For functions $f, g :]0, 1[\to]0, +\infty[$ to be specified later, let us consider the deformation

$$y(x_1, x_2, x_3) := (x_1, x_2 f(x_1), x_3 g(x_1)).$$

Then,

$$\nabla y(x_1, x_2, x_3) = \begin{pmatrix} 1 & 0 & 0 \\ x_2 f'(x_1) & f(x_1) & 0 \\ x_3 g'(x_1) & 0 & g(x_1) \end{pmatrix},$$

$$\text{Cof}\,\nabla y(x_1, x_2, x_3) = \begin{pmatrix} f(x_1)g(x_1) & -x_2 f'(x_1)g(x_1) & -x_3 f(x_1)g'(x_1) \\ 0 & g(x_1) & 0 \\ 0 & 0 & f(x_1) \end{pmatrix}$$

and

$$\det \nabla y(x_1, x_2, x_3) = f(x_1)g(x_1) > 0.$$

Finally, the nonzero entries of $\nabla^2 y(x_1, x_2, x_3)$ are

$$x_2 f''(x_1), \quad f'(x_1), \quad x_3 g''(x_1), \quad g'(x_1). \tag{7}$$

Note that we have in particular

$$|\nabla^2 y(x_1, x_2, x_3)| \geq |x_2||f''(x_1)|.$$

Any functions f, g such that $y \in W^{1,p}(\Omega; \mathbb{R}^3)$, $\operatorname{Cof} \nabla y \in W^{1,q}(\Omega; \mathbb{R}^{3\times 3})$, $0 < \det \nabla y \in W^{1,r}(\Omega)$, $(\det \nabla y)^{-s} \in L^1(\Omega)$ for some $p, q, r \geq 1$ and $s > 0$, but such that one of the quantities in (7) is not a function in $L^p(\Omega)$ yield a useful example since then $y \notin W^{2,p}(\Omega; \mathbb{R}^3)$. To be specific, we choose, for $1 > \varepsilon > 0$,

$$f(x_1) = x_1^{1-\varepsilon} \quad \text{and} \quad g(x_1) = x_1^{1+\varepsilon}.$$

Hence

$$f'(x_1) = (1-\varepsilon)x_1^{-\varepsilon}, \qquad g'(x_1) = (1+\varepsilon)x_1^{\varepsilon},$$
$$f''(x_1) = -\varepsilon(1-\varepsilon)x_1^{-1-\varepsilon}, \qquad g''(x_1) = \varepsilon(1+\varepsilon)x_1^{-1+\varepsilon}.$$

Since $x_2 f''(x_1)$ is not integrable, we have $\nabla^2 y \notin L^1(\Omega; \mathbb{R}^{3\times 3\times 3})$ and thus $y \notin W^{2,1}(\Omega; \mathbb{R}^3)$. We have only $y \in W^{1,p}(\Omega; \mathbb{R}^3) \cap L^\infty(\Omega; \mathbb{R}^3)$ for every $1 \leq p < 1/\varepsilon$. Moreover, direct computation shows that both $\operatorname{Cof} \nabla y$ and $\det \nabla y$ lie in $W^{1,\infty}$. Finally, $\det \nabla y = x_1^2 > 0$ and $(\det \nabla y)^{-s} \in L^1(\Omega)$ for all $0 < s < 1/2$.

Therefore, for any $r, q \geq 1, s > 0$, requiring a deformation $y : \Omega \to \mathbb{R}^3$ to satisfy $\det \nabla y \in W^{1,r}(\Omega)$, $(\det \nabla y)^{-s} \in L^1(\Omega)$ and $\operatorname{Cof} \nabla y \in W^{1,q}(\Omega; \mathbb{R}^{3\times 3})$ is a weaker assumption than $y \in W^{2,1}(\Omega; \mathbb{R}^3)$.

3 Gradient Polyconvexity

We start with a definition of gradient polyconvexity.

Definition 1 (See [10, 36]) Let $\hat{W} : (0, +\infty) \times \mathbb{R}^{3\times 3} \times \mathbb{R}^{3\times 3\times 3} \times \mathbb{R}^3 \to \mathbb{R} \cup \{+\infty\}$ be a lower semicontinuous function, and let $\Omega \subset \mathbb{R}^3$ be a bounded open domain. The functional

$$J(\theta, y) = \int_\Omega \hat{W}(\theta, \nabla y(x), \nabla[\operatorname{Cof} \nabla y(x)], \nabla[\det \nabla y(x)]) dx, \qquad (8)$$

defined for any measurable function $y : \Omega \to \mathbb{R}^3$ for which the weak derivatives ∇y, $\nabla[\operatorname{Cof} \nabla y]$, $\nabla[\det \nabla y]$ exist and which are integrable, is called *gradient polyconvex* if the function $\hat{W}(F, \cdot, \cdot)$ is convex for every $F \in \mathbb{R}^{3\times 3}$.

With J defined as in (8) and a functional $y \mapsto -\ell(y)$ expressing the work of external loads, we set

$$I(\theta, y) := J(\theta, y) - \ell(y). \tag{9}$$

Besides convexity properties, the results of weak lower semicontinuity of $I(\theta, \cdot)$ on $W^{1,p}(\Omega; \mathbb{R}^3)$, in the case $1 \leq p < +\infty$, rely on suitable coercivity properties. Here we assume that there are numbers $q, r > 1$ and $c, c(\theta), s > 0$ such that for every $F \in \mathbb{R}^{3\times 3}$, $\Delta_1 \in \mathbb{R}^{3\times 3\times 3}$, and every $\Delta_2 \in \mathbb{R}^3$

$$\hat{W}(\theta, F, \Delta_1, \Delta_2)$$
$$\geq \begin{cases} c\big(|F|^p + |\operatorname{Cof} F|^q + (\det F)^r + (\det F)^{-s} + |\Delta_1|^q + |\Delta_2|^r\big) - c(\theta), & \text{if } \det F > 0, \\ +\infty, & \text{otherwise.} \end{cases}$$
$$\tag{10}$$

The following existence result is taken from [10] where it is stated without the explicit dependance on θ. For the reader's convenience, we provide a proof below.

Proposition 1 *Let $\theta > 0$ be fixed. Let $\Omega \subset \mathbb{R}^3$ be a bounded Lipschitz domain, and let $\Gamma = \Gamma_0 \cup \Gamma_1$ be an $\mathcal{H}^2$-measurable partition of $\Gamma = \partial\Omega$ with the area of $\Gamma_0 > 0$. Let further $-\ell : W^{1,p}(\Omega; \mathbb{R}^3) \to \mathbb{R}$ be a weakly lower semicontinuous functional satisfying, for some $\tilde{C} > 0$ and $1 \leq \bar{p} < p$,*

$$\ell(y) \leq \tilde{C} \|y\|_{W^{1,p}(\Omega;\mathbb{R}^3)}^{\bar{p}}, \qquad \text{for all } y \in W^{1,p}(\Omega; \mathbb{R}^3). \tag{11}$$

Further, let J, as in (8), be gradient polyconvex on Ω and such that there is a $\hat{W}$ as in Definition 1 which in addition satisfies (10) for $p > 2$, $q \geq \frac{p}{p-1}$, $r > 1$, $s > 0$. Moreover, assume that, for some given measurable function $y_0 : \Gamma_0 \to \mathbb{R}^3$, the following set

$$\mathcal{A} := \big\{ y \in W^{1,p}(\Omega; \mathbb{R}^3) : \operatorname{Cof} \nabla y \in W^{1,q}(\Omega; \mathbb{R}^{3\times 3}), \det \nabla y \in W^{1,r}(\Omega),$$
$$(\det \nabla y)^{-s} \in L^1(\Omega), \det \nabla y > 0 \text{ a.e. in } \Omega, \ y = y_0 \text{ on } \Gamma_0 \big\}$$

is nonempty. If $\inf_{\mathcal{A}} I(\theta, \cdot) < \infty$ for I from (9), then the functional I has a minimizer on $\mathcal{A}$.

Proof Our proof closely follows the approach in [10]. Let $\{y_k\} \subset \mathcal{A}$ be a minimizing sequence of I. Due to coercivity assumption (10), the bound on the loading (11), the Poincaré inequality, and the Dirichlet boundary conditions on Γ_0, we obtain that

$$\sup_{k\in\mathbb{N}} \left(\|y_k\|_{W^{1,p}(\Omega;\mathbb{R}^3)} + \|\operatorname{Cof}\nabla y_k\|_{W^{1,q}(\Omega;\mathbb{R}^{3\times 3})} + \|\det\nabla y_k\|_{W^{1,r}(\Omega)} \right.$$
$$\left. + \|(\det\nabla y_k)^{-s}\|_{L^1(\Omega)} \right) < \infty. \qquad (12)$$

Hence, by standard results on weak convergence of minors, see, e.g., [14, Thm. 7.6-1], there are (not explicitly labeled) subsequences such that

$$y_k \rightharpoonup y \text{ in } W^{1,p}(\Omega;\mathbb{R}^3), \quad \operatorname{Cof}\nabla y_k \rightharpoonup \operatorname{Cof}\nabla y \text{ in } L^q(\Omega;\mathbb{R}^{3\times 3}),$$
$$\det\nabla y_k \rightharpoonup \det\nabla y \text{ in } L^r(\Omega)$$

for $k \to \infty$. Moreover, since bounded sets in uniformly convex Sobolev spaces are weakly sequentially compact,

$$\operatorname{Cof}\nabla y_k \rightharpoonup H \text{ in } W^{1,q}(\Omega;\mathbb{R}^{3\times 3}), \quad \det\nabla y_k \rightharpoonup D \text{ in } W^{1,r}(\Omega) \qquad (13)$$

for some $H \in W^{1,q}(\Omega;\mathbb{R}^{3\times 3})$ and $D \in W^{1,r}(\Omega)$. Since the weak limit is unique, we have $H = \operatorname{Cof}\nabla y$ and $D = \det\nabla y$. By compact embedding, also $\operatorname{Cof}\nabla y_k \to H$ in $L^q(\Omega;\mathbb{R}^{3\times 3})$ and hence we obtain a (not explicitly labeled) subsequence such that, for $k \to \infty$,

$$\operatorname{Cof}\nabla y_k \to \operatorname{Cof}\nabla y \quad \text{a.e. in } \Omega. \qquad (14)$$

Since, by Cramer's formula, $\det(\operatorname{Cof}\nabla y) = (\det\nabla y)^2$, we have, for $k \to \infty$, that

$$\det\nabla y_k \to \det\nabla y \quad \text{a.e. in } \Omega. \qquad (15)$$

Next we show that y belongs to the set of admissible functions $\mathcal{A}$. Notice that $\det\nabla y \geq 0$ since $\det\nabla y_k > 0$ for any $k \in \mathbb{N}$. Further, the conditions (10), (11), (12), and the Fatou lemma imply that

$$+\infty > \liminf_{k\to\infty} I(\theta, y_k) + \ell(y_k) \geq \liminf_{k\to\infty} \int_\Omega \frac{1}{(\det\nabla y_k(x))^s} \, dx$$
$$\geq \int_\Omega \frac{1}{(\det\nabla y(x))^s} \, dx.$$

Hence, inevitably, $\det\nabla y > 0$ almost everywhere in Ω and $(\det\nabla y)^{-s} \in L^1(\Omega)$. Since the trace operator is continuous, we obtain that $y \in \mathcal{A}$.

By Cramer's rule, the inverse of the deformation gradient satisfies, for almost all $x \in \Omega$ and $k \to \infty$, that

$$(\nabla y_k(x))^{-1} = \frac{(\text{Cof}\, \nabla y_k(x))^\top}{\det \nabla y_k(x)} \longrightarrow \frac{(\text{Cof}\, \nabla y(x))^\top}{\det \nabla y(x)} = (\nabla y(x))^{-1}. \tag{16}$$

Notice that, for almost all $x \in \Omega$,

$$\sup_{k \in \mathbb{N}} |\nabla y_k(x)| = \sup_{k \in \mathbb{N}} \det \nabla y_k(x) \, |((\text{Cof}(\nabla y_k(x)))^{-1}))^\top|$$

$$\leq \sup_{k \in \mathbb{N}} \frac{3}{2} \det \nabla y_k(x) \, |(\nabla y_k(x))^{-1}|^2 < \infty$$

because of the pointwise convergence of $\{\det \nabla y_k\}$ and (16).

Due to (16), we have, for almost all $x \in \Omega$ and $k \to \infty$, that

$$\nabla y_k(x) = ((\text{Cof}(\nabla y_k(x)))^{-1})^\top \det \nabla y_k(x) \longrightarrow ((\text{Cof}(\nabla y(x)))^{-1})^\top \det \nabla y(x)$$
$$= \nabla y(x),$$

where we have used that the cofactor of some matrix is invertible whenever the matrix itself is invertible too. As the Lebesgue measure on Ω is finite, we get by the Egoroff theorem, c.f. [23, Thm. 2.22],

$$\nabla y_k \to \nabla y \text{ in measure.} \tag{17}$$

Since $\hat{W}(\theta, \cdot)$ is bounded from below and continuous on matrices with positive determinants and $\hat{W}(\theta, F, \cdot, \cdot)$ is convex, we may use [23, Cor. 7.9] to conclude, from (17) and (13), that

$$\int_\Omega \hat{W}(\theta, \nabla y(x), \nabla \text{Cof}\, \nabla y(x), \nabla \det \nabla y(x))\, dx$$
$$\leq \liminf_{k \to \infty} \int_\Omega \hat{W}(\theta, \nabla y_k(x), \nabla \text{Cof}\, \nabla y_k(x), \nabla \det \nabla y_k(x))\, dx \, .$$

To pass to the limit in the functional $-\ell$, we exploit its weak lower semicontinuity. Therefore, the whole functional I is weakly lower semicontinuous along $\{y_k\} \subset \mathcal{A}$ and hence $y \in \mathcal{A}$ is a minimizer of $I(\theta, \cdot)$.

Remark 2 Note that the pointwise convergence (15) of the determinant, necessary for obtaining the crucial convergence in (17), was not achieved by compact embedding, as it was done for $\text{Cof}\, \nabla y$ in (14). Hence, the coercivity in $\nabla[\det \nabla y]$ is of minor importance and can be relaxed, provided the function $\hat{W}$ from (8) does not depend on its last argument, c.f. [10, Prop. 5.1]. On the other hand, although only $\nabla[\text{Cof}\, \nabla y]$ is necessary for regularizing the whole problem, making the functional in (8) dependent also on $\nabla[\det \nabla y]$ may be interesting from the applications point of view.

Let $\mathcal{L}^3$ denote the Lebesgue measure in $\mathbb{R}^3$. If $p > 3$ and $y \in W^{1,p}(\Omega; \mathbb{R}^3)$ is such that $\det \nabla y > 0$ almost everywhere in Ω, then the so-called Ciarlet-Nečas condition

$$\int_\Omega \det \nabla y(x)\, dx \leq \mathcal{L}^3(y(\Omega)), \tag{18}$$

derived in [15], ensures almost everywhere injectivity of deformations. We also refer to [28, Sec. 6, Thm.2] and to [3] for other conditions ensuring injectivity of deformations, requiring, however, a prescribed Dirichlet boundary datum on the whole $\partial \Omega$, which is difficult to ensure in a physical lab. If

$$\frac{|\nabla y|^3}{\det \nabla y} \in L^\delta(\Omega) \tag{19}$$

for some $\delta > 2$ and (18) holds, then we even get invertibility everywhere in Ω due to [30, Theorem 3.4]. Namely, this then implies that y is an open map. Hence, we get the following corollary of Proposition 1.

Corollary 1 *Let $\Omega \subset \mathbb{R}^3$ be a bounded Lipschitz domain, and let $\Gamma = \Gamma_0 \cup \Gamma_1$ be an $\mathcal{H}^2$-measurable partition of $\Gamma = \partial \Omega$ with the area of $\Gamma_0 > 0$. Let further $\ell : W^{1,p}(\Omega; \mathbb{R}^3) \to \mathbb{R}$ be a weakly upper semicontinuous functional and J as in (8) be gradient polyconvex on Ω such that $\hat{W}$ satisfies (10). Finally, let $p > 6$, $q \geq \frac{p}{p-1}$, $r > 1$, $s > 2p/(p-6)$, and assume that, for some given measurable function $y_0 : \Gamma_0 \to \mathbb{R}^3$, the following set*

$$\mathcal{A} := \{y \in W^{1,p}(\Omega; \mathbb{R}^3) : \operatorname{Cof} \nabla y \in W^{1,q}(\Omega; \mathbb{R}^{3 \times 3}),\ \det \nabla y \in W^{1,r}(\Omega),$$
$$(\det \nabla y)^{-s} \in L^1(\Omega),\ \det \nabla y > 0 \text{ a.e. in } \Omega,\ y = y_0 \text{ on } \Gamma_0,\ (18)\ \textit{holds}\}$$

is nonempty. If $\inf_\mathcal{A} I < \infty$ for I from (9), then the functional I has a minimizer on $\mathcal{A}$ which is injective everywhere in Ω.

A simple example of an energy density which satisfies the assumptions of Proposition 1 and Corollary 1 is

$\hat{W}(\theta, F, \Delta_1, \Delta_2)$

$$= \begin{cases} W(\theta, F) + \varepsilon\big(|F|^p + |\operatorname{Cof} F|^q + (\det F)^r + (\det F)^{-s} + |\Delta_1|^q + |\Delta_2|^r\big), & \text{if } \det F > 0, \\ +\infty, & \text{otherwise,} \end{cases}$$

for W defined in (5).

Remark 3 (Gradient Polyconvex Materials and Smoothness of Stress) Gradient polyconvex materials enable us to control regularity of the first Piola-Kirchhoff stress tensor by means of smoothness of the Cauchy stress. Assume that the Cauchy stress tensor $T^y : y(\Omega) \to \mathbb{R}^{3\times 3}$ is Lipschitz continuous, for instance. If $\operatorname{Cof} \nabla y : \Omega \to \mathbb{R}^{3\times 3}$ is Lipschitz continuous too, then the first Piola-Kirchhoff stress tensor P inherits the Lipschitz continuity from T^y because

$$P(x) := T^y(x^y) \operatorname{Cof} \nabla y(x),$$

where $x^y := y(x)$. In a similar fashion, one can transfer Hölder continuity of T^y to P via Hölder continuity of $x \mapsto \operatorname{Cof} \nabla y(x)$.

4 Evolution

If the loading changes in time or if the boundary condition becomes time-dependent, then the specimen evolves as well. We consider here the case, in which evolution is connected with energy dissipation. Experimental evidence shows that considering a rate-independent dissipation mechanism is a reasonable approximation in a wide range of rates of external loads. We hence need to define a suitable dissipation function.

Since we consider a rate-independent processes, this dissipation will be positively one-homogeneous. We associate the dissipation with the magnitude of the time derivative of the dissipative variable $z \in \mathbb{R}^{M+1}$, where $M \in \mathbb{N}$, i.e., with $|\dot z|_{M+1}$, where $|\cdot|_{M+1}$ denotes a norm on $\mathbb{R}^{M+1}$ (in our setting, the internal variable z can be seen as a vector of volume fractions of austenite and M variants of martensite). Therefore, the specific dissipated energy associated with a change from state z^1 to z^2 is postulated as

$$D(z^1, z^2) := |z^1 - z^2|_{M+1}. \tag{20}$$

Hence, for $z^i : \Omega \to \mathbb{R}^{M+1}, i = 1, 2$, the total dissipation reads

$$\mathcal{D}(z^1, z^2) := \int_\Omega D(z^1(x), z^2(x)) \, dx,$$

and the total $\mathcal{D}$-dissipation of a time-dependent curve $z : t \in [0, T] \mapsto z(t)$, where $z(t) : \Omega \to \mathbb{R}^{M+1}$ is defined as

$$\operatorname{Diss}_\mathcal{D}(z, [s, t]) := \sup \left\{ \sum_{j=1}^N \mathcal{D}(z(t_{i-1}), z(t_i)) : N \in \mathbb{N}, s = t_0 \leq \ldots \leq t_N = t \right\}.$$

Let $\mathcal{Z}$ denote the set of all admissible states of internal variables $z : \Omega \to \mathbb{R}^{M+1}$ and $\mathcal{A}$ be the set of admissible deformations as before. For a given triple $(t, y, z) \in [0, T] \times \mathcal{A} \times \mathcal{Z}$, we define the total energy of the system by

$$\mathcal{E}(t, \theta, y, z) = \begin{cases} J(\theta, y) - L(t, y), & \text{if } z = \lambda(\nabla y) \text{ a.e. in } \Omega, \\ +\infty, & \text{otherwise,} \end{cases}$$

where $L(t, \cdot)$ is a functional on deformations expressing time-dependent loading of the specimen, and λ is defined in (6).

4.1 Energetic Solution

Suppose, that we look for the time evolution of $t \mapsto y(t) \in \mathcal{A}$ and $t \mapsto z(t) \in \mathcal{Z} := L^\infty(\Omega, \mathbb{R}^{M+1})$ during a process on a time interval $[0, T]$, where $T > 0$ is the time horizon. We use the following notion of solution from [25], see also [40, 41].

Definition 2 (Energetic Solution) Let an energy $\mathcal{E} : [0, T] \times (0, +\infty) \times \mathcal{A} \times \mathcal{Z} \to \mathbb{R} \cup \{+\infty\}$ and a dissipation distance $\mathcal{D} : \mathcal{Z} \times \mathcal{Z} \to \mathbb{R} \cup \{+\infty\}$ be given. The set of admissible configurations is defined as

$$\mathcal{Q} := \{(y, z) \in \mathcal{A} \times \mathcal{Z} : \lambda(\nabla y) = z \text{ a.e. in } \Omega\}.$$

We say that $(y, z) : [0, T] \to \mathcal{Q}$ is an energetic solution to $(\mathcal{Q}, \mathcal{E}, \mathcal{D})$, if the mapping $t \mapsto \partial_t \mathcal{E}(t, \theta, y(t), z(t))$ is in $L^1(0, T)$ and if, for all $t \in [0, T]$, the stability condition

$$\mathcal{E}(t, \theta, y(t), z(t)) \leq \mathcal{E}(t, \theta, \tilde{y}, \tilde{z}) + \mathcal{D}(z(t), \tilde{z}) \qquad \forall (\tilde{y}, \tilde{z}) \in \mathcal{Q} \qquad (S)$$

and the energy balance

$$\mathcal{E}(t, \theta, y(t), z(t)) + \text{Diss}_\mathcal{D}(z; [s, t]) = \mathcal{E}(s, \theta, y(s), z(s)) \\ + \int_s^t \partial_t \mathcal{E}(a, \theta, y(\theta), z(\theta)) \, da \qquad (E)$$

are satisfied for any $0 \leq s < t \leq T$.

An important role is played by the set of so-called stable states, defined for each $t \in [0, T]$ as

$$\mathbb{S}(t) := \{(y, z) \in \mathcal{Q} : \mathcal{E}(t, \theta, y, z) < +\infty \text{ and } \mathcal{E}(t, \theta, y, z) \leq \mathcal{E}(t, \theta, \tilde{y}, \tilde{z}) \\ + \mathcal{D}(z, \tilde{z}) \ \forall (\tilde{y}, \tilde{z}) \in \mathcal{Q}\}.$$

4.2 Existence of an Energetic Solution

A standard way how to prove the existence of an energetic solution is to construct time-discrete minimization problems and then to pass to the limit. Before we give the existence proof we need some auxiliary results. For given $N \in \mathbb{N}$ and for $0 \leq k \leq N$, we define the time increments $t_k := kT/N$. Furthermore, we use the abbreviation $q := (y, z) \in \mathcal{Q}$. We assume that there exists an admissible deformation y^0 being compatible with the initial volume fraction z^0, i.e., $q^0 := (y^0, z^0) \in \mathbb{S}(0)$. For $k = 1, \ldots, N$, we define a sequence of minimization problems

$$\text{minimize } \mathcal{I}_k(\theta, y, z) := \mathcal{E}(t_k, \theta, y, z) + \mathcal{D}(z, z^{k-1}), \quad (y, z) \in \mathcal{Q}. \tag{21}$$

We denote a minimizer of (21), for a given k, as $q_k^N := (y^k, z^k) \in \mathcal{Q}$ for $1 \leq k \leq N$. The following lemma shows that a minimizer always exists if the elastic energy is not identically infinite on $\mathcal{Q}$:

Lemma 1 *Let $\Omega \subset \mathbb{R}^3$ be a bounded Lipschitz domain, and let $\Gamma = \Gamma_0 \cup \Gamma_1$ be an $\mathcal{H}^2$-measurable partition of $\Gamma = \partial\Omega$ with the area of $\Gamma_0 > 0$. Let J, of the from (8), be gradient polyconvex on Ω and such that the stored energy density $\hat{W}$ satisfies (10). Moreover, let $L \in C^1[0, T] \times W^{1,p}(\Omega; \mathbb{R}^3)$ be such that, for some $C > 0$ and $1 \leq \alpha < p$,*

$$L(t, y) \leq C \|y\|_{W^{1,p}}^\alpha, \quad \text{for all } t \in [0, T]$$

and $y \mapsto -L(t, y)$ is weakly lower semicontinuous on $W^{1,p}(\Omega; \mathbb{R}^3)$ for all $t \subset [0, T]$. Finally, let $p > 6$, $q \geq \frac{p}{p-1}$, $r > 1$, $s > 2p/(p-6)$.

If there is $(y, z) \in \mathcal{Q}$ such that $\mathcal{I}_k(y, z) < \infty$ for $\mathcal{I}_k$ from (21), then the functional $\mathcal{I}_k$ has a minimizer $q_k^N = (y^k, z^k) \in \mathcal{Q}$ such that y_k is injective everywhere in Ω. Moreover, $q_k^N \in \mathbb{S}(t_k)$ for all $1 \leq k \leq N$.

Proof Since the discretized problem (21) has a purely static character, we can follow the proof of Proposition 1. Let $\{(y_j^k, z_j^k)\}_{j \in \mathbb{N}} \subset \mathcal{Q}$ be a minimizing sequence. As

$$\nabla y_j^k \longrightarrow \nabla y^k \quad \text{strongly in } L^{\tilde{p}}(\Omega, \mathbb{R}^{3 \times 3}) \text{ as } j \to \infty$$

for every $1 \leq \tilde{p} < p$ and $\lambda \in C(\mathbb{R}^{3 \times 3}, \mathbb{R}^{M+1})$ is bounded, we obtain that

$$z_j^k = \lambda(\nabla y_j^k) \longrightarrow \lambda(\nabla y^k) \quad \text{strongly in } L^{\tilde{p}}(\Omega, \mathbb{R}^{M+1}) \text{ as } j \to \infty.$$

Since $\|z_j^k\|_{L^1(\Omega, \mathbb{R}^{M+1})}$ is uniformly bounded in j, there is a subsequence (not explicitly relabeled) such that $z_j^k \overset{*}{\rightharpoonup} \mu^k$ in Radon measures on Ω. This shows that $z^k := \mu^k = \lambda(\nabla y^k)$ and hence $q_k^N = (y^k, z^k) \in \mathcal{Q}$. Since $\mathcal{D}(\cdot, z^{k-1})$ is convex, we obtain that q_k^N is indeed a minimizer of $\mathcal{I}_k$. Moreover, y_k is injective everywhere by the reasoning used for proving Corollary 1. The stability $q_k^N \in \mathbb{S}(t_k)$ follows by standard arguments; see, e.g., [25].

Denoting by $B([0, T]; \mathcal{A})$ the set of bounded maps $t \in [0, T] \mapsto y(t) \in \mathcal{A}$, we have the following result showing the existence of an energetic solution to the problem $(\mathcal{Q}, \mathcal{E}, \mathcal{D})$:

Theorem 1 *Let $\theta > 0$ be fixed. Let $T > 0$ and let the assumptions in Lemma 1 be satisfied. Moreover, let the initial condition be stable, i.e., $q^0 := (y^0, z^0) \in \mathbb{S}(0)$. Then there is an energetic solution to $(\mathcal{Q}, \mathcal{E}, \mathcal{D})$ satisfying $q(0) = q^0$ and such that $y \in B([0, T]; \mathcal{A})$, $z \in \mathrm{BV}\left([0, T]; L^1(\Omega; \mathbb{R}^{M+1})\right) \cap L^\infty(0, T; \mathcal{Z})$, and such that for all $t \in [0, T]$ the identity $\lambda(\nabla y(t, \cdot)) = z(t, \cdot)$ holds a.e. in Ω. Moreover, for all $t \in [0, T]$, the deformation $y(t)$ is injective everywhere in Ω.*

Proof Let $q_k^N := (y^k, z^k)$ be the solution of (21), which exists by Lemma 1, and let $q^N : [0, T] \to \mathcal{Q}$ be given by

$$q^N(t) := \begin{cases} q_k^N, & \text{if } t \in [t_k, t_{k+1}[\text{ if } k = 0, \ldots, N-1, \\ q_N^N, & \text{if } t = T. \end{cases}$$

Following [25], we get, for some $C > 0$ and for all $N \in \mathbb{N}$, the estimates

$$\|z^N\|_{BV(0,T;L^1(\Omega;\mathbb{R}^{M+1}))} \leq C, \quad \|z^N\|_{L^\infty(0,T;BV(\Omega;\mathbb{R}^{M+1}))} \leq C, \tag{22a}$$

$$\|y^N\|_{L^\infty(0,T;W^{1,p}(\Omega;\mathbb{R}^3))} \leq C, \tag{22b}$$

as well as the following two-sided energy inequality

$$\int_{t_{k-1}}^{t_k} \partial_t \mathcal{E}(a, \theta, q_k^N)\, da \leq \mathcal{E}(t_k, \theta, q_k^N) + \mathcal{D}(z^k, z^{k-1}) - \mathcal{E}(t_{k-1}, \theta, q_{k-1}^N)$$

$$\leq \int_{t_{k-1}}^{t_k} \partial_t \mathcal{E}(a, \theta, q_{k-1}^N)\, da. \tag{23}$$

The second inequality in (23) follows since q_k^N is a minimizer of (21) and by comparison of its energy with $q := q_{k-1}^N$. The lower estimate is implied by the stability of $q_{k-1}^N \in \mathbb{S}(t_{k-1})$, see Lemma 1, when compared with $\tilde{q} := q_k^N$. By this inequality, the a-priori estimates and a generalized Helly's selection principle [41, Cor. 2.8], we get that there is indeed an energetic solution obtained as a limit for $N \to \infty$.

Let us comment more on the two main properties of the minimizer, namely that it is orientation-preserving and injective everywhere in Ω. The condition $\det \nabla y > 0$ a.e. in Ω follows from the fact that if $t_j \to t$, $(y_{(j)}, z_{(j)}) \in \mathbb{S}(t_j)$ and $(y_{(j)}, z_{(j)}) \rightharpoonup$

(y, z) in $W^{1,p}(\Omega; \mathbb{R}^3) \times BV(\Omega; \mathbb{R}^{M+1})$, then $(y, z) \in \mathbb{S}(t)$. Indeed, we have $z_{(j)} \to z$ in $L^1(\Omega; \mathbb{R}^{M+1})$ in our setting and hence for all $(\tilde{y}, \tilde{z}) \in \mathcal{Q}$, we get

$$\mathcal{E}(t, \theta, y, z) \leq \liminf_{j \to \infty} \mathcal{E}(t_j, \theta, y_{(j)}, z_{(j)}) \leq \liminf_{j \to \infty} (\mathcal{E}(t_j, \theta, \tilde{y}, \tilde{z}) + \mathcal{D}(z_{(j)}, \tilde{z}))$$
$$= \mathcal{E}(t, \theta, \tilde{y}, \tilde{z}) + \mathcal{D}(z, \tilde{z}).$$

In particular, as $\mathcal{E}(t_j, \theta, \tilde{y}, \tilde{z})$ is finite for some $(\tilde{y}, \tilde{z}) \in \mathcal{Q}$, we get $\mathcal{E}(t, \theta, y, z) < +\infty$ and thus $\det \nabla y > 0$ a.e. in Ω in view of (10).

To prove injectivity, we profit again from the fact that quasistatic evolution of energetic solutions is very close to a purely static problem. In view of (22b), we obtain, for each $t \in [0, T]$, all necessary convergences that were used in the proof of Corollary 1 to pass to the limit in the conditions (18) and (19).

5 Computational Experiments

In this section, we demonstrate computational performance of the above model on a numerical experiment. We will use a *St. Venant-Kirchhoff*-like form of the stored energy of each particular phase variant, which allows for an explicit reference to measured data and can easily be applied to various materials. We consider that the material can occur in $M + 1$ stress-free configurations that are determined by *distortion matrices* F_i, $i = 0, \ldots, M$, which are independent of θ, i.e., thermal expansion is neglected. The austenite well is defined by $F_0 = \mathbb{I}$.

The frame-indifferent free energy of particular phase (variant) is considered as a function of *Green strain* tensor ε^ℓ related to the distortion of this phase(variant). In the simplest case (cf. [47, Sect. 6.6], or [35], e.g.), one can consider a function quadratic in terms of ε^ℓ of the form (if $\det F > 0$)

$$W_\ell(F, \theta) = \sum_{i,j,k,l=1}^{d} \varepsilon_{ij}^\ell \mathcal{C}_{ijkl}^\ell \varepsilon_{kl}^\ell + d_\ell(\theta) + \alpha((\det F)^{-2} + |\nabla[\operatorname{Cof} F]|^2),$$

$$\varepsilon^\ell = \frac{(F_\ell^\top)^{-1} F^\top F F_\ell^{-1} - \mathbb{I}}{2}, \qquad (24)$$

where $\mathcal{C}^\ell = \{\mathcal{C}_{ijkl}^\ell\}$ is the fourth-order tensor of elastic moduli satisfying the usual symmetry relations depending also on symmetry of the specific phase(variant) ℓ and d_ℓ is some offset. The overall stored energy is assembled as in (5).

The data required for the potential are available for many alloys, except perhaps the measurements of the elastic tensor $\mathcal{C}^\ell$, which are standardly done (with few exceptions) only for the austenite so that elastic response of the martensitic variants has to be extrapolated. The heat capacities c_ℓ are usually obtained experimentally, while the offsets d_ℓ are then to be fitted to get the agreement with energetical equilibrium between martensite and austenite at a specific temperature. Typically, heat capacity of austenite is larger than that of martensite, which is just what causes the shape memory effect.

We performed our computation on a prismatic single crystal of Ni_2MnGa in a specific orientation, mostly (1,0,0). This alloy (or, more precisely, intermetallic) undergoes a cubic/tetragonal transformation, which is relatively easy to model because the martensite forms only 3 variants, i.e., $M = 3$.

Following [12] we describe the variants of martensite by $F_1 = \text{diag}(\eta_2, \eta_1, \eta_1)$, $F_2 = \text{diag}(\eta_1, \eta_2, \eta_1)$ and $F_3 = \text{diag}(\eta_1, \eta_1, \eta_2)$ where $\eta_1 = 0.9512$ and $\eta_2 = 1.130$. The stretch tensor of the austenite is the identity, i.e., $F_0 = \text{diag}(1, 1, 1)$. The Euclidean distance between any two variants of the martensite is about 0.253 while the distance between the austenite and any variant of the martensite is 0.147. The distances here are calculated as the Frobenius norms of the corresponding right Cauchy-Green strains. Hence, we can define $\mathcal{N}_i(C_i) = \{C \in \mathbb{R}^{3 \times 3} : |C - C_i| < \epsilon_i\}$ for some $\epsilon_i > 0$. Then

$$\text{dist}(C, \mathcal{N}_i(C_i)) = \begin{cases} 0 & \text{if } |C - C_i| < \epsilon_i, \\ |C - C_i| - \epsilon_i & \text{otherwise.} \end{cases}$$

We can take $\epsilon_i = 0.07$ for every $0 \leq i \leq 3$. This formula is then used in (6). As the elastic moduli are much bigger than the transformation strains, the volume fraction λ will have one dominant component because ∇y must be pointwise in a small vicinity of some energy well. Using [5] we can see that the martensitic variants are rank-one connected with each other while none of them is rank-one connected with the austenite. Rank-one connection allows for the formation of a planar interface between two martensitic variants.

We prescribe the dissipation energy density as 0.35 MPa for transformations between the austenite and any martensitic variant [1] and almost no dissipation is assumed for transformations among martensitic variants. This can be done by setting $|z|_4 := \sum_{i=0}^{3} \gamma_i |z_i|$ in (20) and taking $\gamma_0 = 35 \times 10^4$ Pa and $\gamma_i = 1$ Pa if $i \neq 0$. The equilibrium temperature θ_0 of the austenite and the martensite is about 288 K. The Clausius-Clapeyron constant describing the rate of the increase of the bottoms of the martensitic wells with respect to the austenite is about 0.2 MPa/K. Therefore, we can take $d_\ell(\theta) = 0.2$ MPa $(\theta - 288$ K$)$ for $\ell > 0$ and $d_0(\theta) = 0$.

Elastic moduli of the austenite are taken zero but $\mathcal{C}_{1111}^0 = 136$ GPa, $\mathcal{C}_{1122}^0 = \mathcal{C}_{2211}^0 = 92$ GPa, $\mathcal{C}_{2323}^0 = \mathcal{C}_{2332}^0 = \mathcal{C}_{3223}^0 = \mathcal{C}_{3232}^0 = 102$ GPa.

We consider a simple problem of uniaxial tension of a three-dimensional bar, i.e., the horizontal displacements are fixed at the left end and all the nodes at the right end are loaded by increasing horizontal displacements, while the vertical displacements at the both ends are prescribed such as the rigid body modes are removed but the bar is free to deform laterally. In the case the bar is considered as perfectly uniform, the onset of phase transition from austenite to martensite is reached for all the points at the same time. This situation can be studied analytically, assuming zero dissipation for simplicity. First, we know that the only nonzero component of the second Piola-

Kirchhoff stress tensor S^ℓ is S^ℓ_{33} calculated as

$$S^\ell_{33} = C_{33}\varepsilon^\ell_{33} + C_{23}\varepsilon^\ell_{22} + C_{13}\varepsilon^\ell_{11}. \tag{25}$$

The condition of zero stress components S^ℓ_{11} and S^ℓ_{22} can be written as

$$S^\ell_{11} = C_{11}\varepsilon^\ell_{11} + C_{12}\varepsilon^\ell_{22} + C_{13}\varepsilon^\ell_{33} = 0 \tag{26}$$

$$S^\ell_{22} = C_{12}\varepsilon^\ell_{11} + C_{22}\varepsilon^\ell_{22} + C_{23}\varepsilon^\ell_{33} = 0, \tag{27}$$

where C_{ij} are components of the stiffness tensor in Voigt notation, i.e., $C_{12} = C_{21} = C_{2211} = C_{1122}$, $C_{22} = C_{2222}$, $C_{23} = C_{23} = C_{2233} = C_{3322}$, etc. Solution of the above system of two equations is given as

$$\varepsilon^\ell_{11} = \varepsilon^\ell_{22} \tag{28}$$

$$\varepsilon^\ell_{22} = -\frac{C_{23}}{C_{22} + C_{12}}\varepsilon^\ell_{33}. \tag{29}$$

Substituting back to (25) we arrive at

$$S^\ell_{33} = \underbrace{\left(C_{33} - 2\frac{C^2_{23}}{C_{22} + C_{23}}\right)}_{K}\varepsilon^\ell_{33}. \tag{30}$$

The transformation from austenite to the first variant of martensite happens when the energy of both phases is the same

$$W_0(F) = W_3(F) \tag{31}$$

which can be written in terms of strain as

$$K(\varepsilon^0_{33})^2 = K(\varepsilon^3_{33})^2, \tag{32}$$

where the strains are calculated as

$$\varepsilon^0_{33} = \frac{1}{2}\left(F^2_{33} - 1\right) \tag{33}$$

$$\varepsilon^3_{33} = \frac{1}{2}\left(\frac{F^2_{33}}{\eta^2_2} - 1\right). \tag{34}$$

Therefore, the critical stretch F_c of the bar at the onset of transformation from austenite to martensite can be determined as

$$F_c = \sqrt{\frac{2\eta_2^2}{\eta_2^2 + 1}} \qquad (35)$$

for the given value of $\eta_2 = 1.13$, the stretch is $F_c = 1.059$, and the strains are

$$\varepsilon_{33}^0 = \frac{1}{2}\left(F_c^2 - 1\right) = 0.0608 \qquad (36)$$

$$\varepsilon_{33}^3 = \frac{1}{2}\left(\frac{F_c^2}{\eta_2^2} - 1\right) = -0.0608. \qquad (37)$$

The solution is represented graphically in Fig. 1.

Moreover, also remaining nonzero components of the strain tensor before and after transformation can be calculated as

$$\varepsilon_{22}^0 = -\frac{C_{23}}{C_{33} + C_{23}}\varepsilon_{33}^0 = -0.608\frac{92}{136 + 92} = -0.0245 \qquad (38)$$

$$\varepsilon_{22}^3 = -\frac{C_{23}}{C_{33} + C_{23}}\varepsilon_{11}^1 = 0.608\frac{92}{136 + 92} = 0.0245 \qquad (39)$$

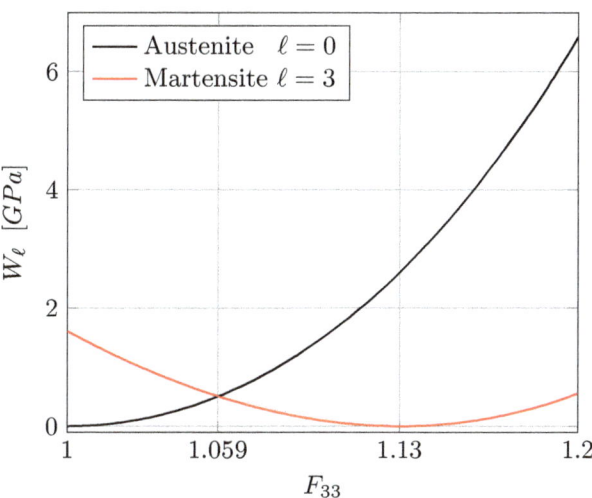

Fig. 1 Uniaxial tension: free energy of particular phase(variant)s, namely W_0 and W_3 in terms of F_{33}

and the stretches in the lateral direction before and after deformation are therefore given as

$$F_{22}^0 = \sqrt{2\varepsilon_{22}^0 + 1} = 0.9752 \tag{40}$$

$$F_{22}^3 = \sqrt{2\varepsilon_{22}^3 + 1} = 1.0242. \tag{41}$$

Let us now calculate also the stress at the point of transition from austenite to martensite. The first Piola-Kirchhoff stresses right before and after the transformation, i.e., P_{33}^0 and P_{33}^1 are calculated as

$$P_{33}^0 = F_c S_{33}^0 = F_c \left(C_{33} - 2\frac{C_{23}^2}{C_{22} + C_{23}} \right) \varepsilon_{33}^0 \tag{42}$$

$$= 1.059 \left(136 - 2\frac{92}{136 + 92} \right) 0.0608 = 8.705 \text{ GPa}. \tag{43}$$

$$P_{33}^3 = F_c S_{33}^3 = F_c \left(C_{33} - 2\frac{C_{23}^2}{C_{22} + C_{23}} \right) \varepsilon_{33}^1 \tag{44}$$

$$= 1.059 \left(136 - 2\frac{92}{136 + 92} \right) (-0.0608) = -8.705 \text{ GPa}. \tag{45}$$

Interestingly, jump from tension to compression occurs during the transformation, see Fig. 2 for the dependence of the first Piola-Kirchhoff stress on the stretch.

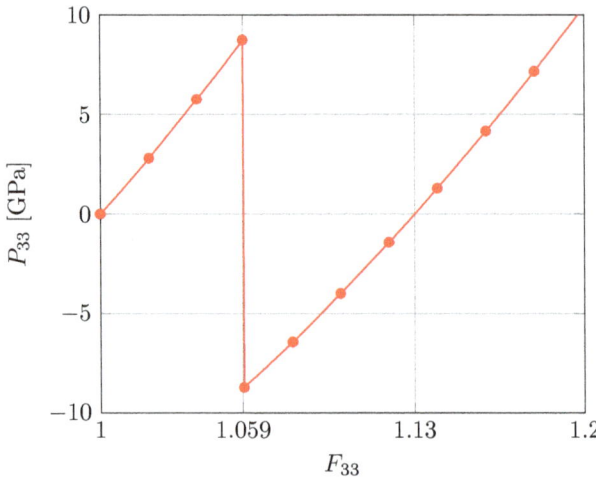

Fig. 2 Uniaxial tension: first Piola-Kirchoff stress—stretch graph

However, in reality, the material is never homogeneous and uniform but shows certain variation in material properties. Such a variation can trigger the transformation from austenite to martensite only in a small part of the bar. Nonetheless, such a uniaxial state would violate the equilibrium condition as well as the compatibility condition since the distortion matrices F_0 and F_3 are not rank-1 connected. Therefore, the bar must deform in a more complex way that is in general not possible to study analytically. Therefore, we simulate this case by the finite element method.

The proposed material model enhanced by gradient polyconvexity has been implemented into a finite element code OOFEM [45]. The implementation of gradient polyconvexity was based on the so-called micromorphic approach, see [31] for more details. Thus, in the present example we perform a uniaxial tension test of a bar with η_2 considered as a random variable with a Gaussian distribution, specified by mean $\mu = 1.13$ and standard deviation parameter $\sigma = 0.01$. As expected, the martensite transformation starts in several separated parts of the bar leading to violation of uniaxial stress state resulting into bending of the bar. Moreover, since the variants $\ell = 0$ and $\ell = 3$ are not rank-1 connected, an interface consisting of the other two variants of martensite is created. The transformation process is depicted in Fig. 3 where gradual change from the initial austenite state to the final state of martensite variant $\ell = 3$ is shown.

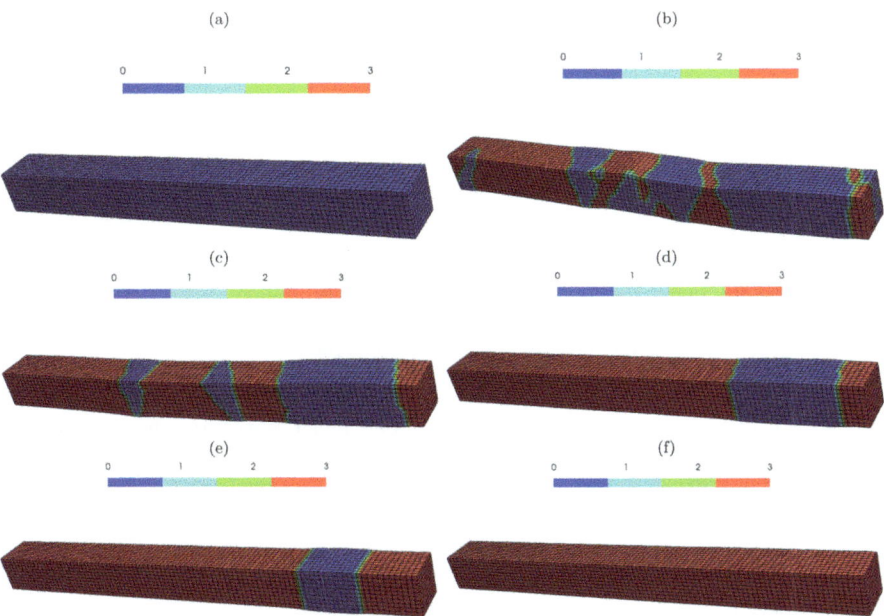

Fig. 3 Uniaxial tension test: evolution of a austenite-martensite transformation form (**a**) to (**f**). Blue color represents the austenite variant, while the remaining colors represent different variants of martensite according to the color bar

Note that the solution was obtained by the Newton-Raphson procedure which generally leads to a critical point rather than the global minima. Since the present problem involves several local minima, a more robust technique will be further implemented into OOFEM to allow development of austenite-martensite laminates without perturbing material parameters.

Acknowledgment MH and MK acknowledge the support of the Czech Science Foundation (project No. 19-26143X). P. P. was supported by Deutsche Forschungsgemeinschaft via SPP 2256 (project ID 441470105), project Analysis for thermo-mechanical models with internal variables.

References

1. Arndt, M., Griebel, M., Novák, V., Roubíček, T., Šittner, P.: Martensitic transformation in NiMnGa single crystals: numerical simulation and experiments. Int. J. Plasticity **22**, 1943–1961 (2006)
2. Ball, J.M.: Convexity conditions and existence theorems in nonlinear elasticity. Arch. Ration. Mech. Anal. **63**, 337–403 (1977)
3. Ball, J.M.: Global invertibility of Sobolev functions and the interpenetration of matter. Proc. Roy. Soc. Edinb. **88A**, 315–328 (1981)
4. Ball, J.M., Crooks, E.C.M.: Local minimizers and planar interfaces in a phase-transition model with interfacial energy. Calc. Var. **40**, 501–538 (2011)
5. Ball, J.M., James, R.D.: Fine phase mixtures as minimizers of energy. Arch. Ration. Mech. Anal. **100**, 13–52 (1988)
6. Ball, J.M., Mora-Corral, C.: A variational model allowing both smooth and sharp phase boundaries in solids. Commun. Pure Appl. Anal. **8**, 55–81 (2009)
7. Ball, J.M., Currie, J.C., Olver, P.L.: Null Lagrangians, weak continuity, and variational problems of arbitrary order. J. Funct. Anal. **41**, 135–174 (1981)
8. Benešová, B., Kružík, M.: Characterization of gradient Young measures generated by homeomorphisms in the plane. ESAIM Control Optim. Calc. Var. **22**, 267–288 (2016)
9. Benešová, B., Kružík, M.: Weak lower semicontinuity of integral functionals and applications. SIAM Rev. **59**, 703–766 (2017)
10. Benešová, B., Kružík, M., Schlömerkemper, A.: A note on locking materials and gradient polyconvexity. Math. Mod. Meth. Appl. Sci. **28**, 2367–2401 (2018)
11. Bhattacharya, K.: Microstructure of Martensite. Why It Forms and How It Gives Rise to the Shape-Memory Effect. Oxford Univ. Press, New York (2003)
12. Bhattacharya, K., James, R.D.: A theory of thin films of martensitic materials with applications to microactuators. J. Mech. Phys. Solids **47**, 531–576 (1999)
13. Capriz, G.: Continua with latent microstructure. Arch. Ration. Mech. Anal. **90**, 43–56 (1985).
14. Ciarlet, P.G.: Mathematical Elasticity Vol. I: Three-dimensional Elasticity. North-Holland, Amsterdam (1988)
15. Ciarlet, P.G., Nečas, J.: Injectivity and self-contact in nonlinear elasticity. Arch. Ration. Mech. Anal. **97**, 171–188 (1987)
16. Conti, S., Dolzmann, G.: On the theory of relaxation in nonlinear elasticity with constraints on the determinant. Arch. Ration. Mech. Anal. **217**, 413–437 (2015)
17. Dacorogna, B.: Direct Methods in the Calculus of Variations, 2nd edn. Springer, Berlin (2008)
18. Davoli, E., Friedrich, M.: Two-well rigidity and multidimensional sharp-interface limits for solid–solid phase transitions. Calc. Var. PDE **59**, 44 (2020)
19. Davoli, E., Friedrich, M.: Two-well linearization for solid-solid phase transitions. Preprint, arXiv:2005.03892 (2020)

20. Dell'Isola, F., Steigmann, D.: A two-dimensional gradient-elasticity theory for woven fabrics. J. Elast. **118**, 113–125 (2014)
21. Dell'Isola, F., Sciarra, G., Vidoli, S.: Generalized Hooke's law for isotropic second gradient materials. Proc. R. Soc. Lond. A **465**, 2177–2196 (2009)
22. Dunn, J.E., Serrin, J.: On the thermomechanics of interstitial working. Arch Ration. Mech. Anal. **88**, 95–133 (1985)
23. Fonseca, I., Leoni, G.: Modern Methods in the Calculus of Variations: L^p Spaces. Springer, New York (2007)
24. Forest, S.: Micromorphic approach for gradient elasticity, viscoplasticity, and damage. J. Eng. Mech. **135**, 117 (2009)
25. Francfort, G., Mielke, A.: Existence results for a class of rate-independent material models with nonconvex elastic energies. J. Reine Angew. Math. **595**, 55–91 (2006)
26. Friedrich, M., Kružík, M.: On the passage from nonlinear to linearized viscoelasticity. SIAM J. Math. Anal. **50**, 4426–4456 (2018)
27. Friedrich, M., Kružík, M.: Derivation of von Kármán plate theory in the framework of three-dimensional viscoelasticity. Arch. Ration. Mech. Anal. **238**, 489–540 (2020)
28. Giaquinta, M., Modica, M., Souček, J.: Cartesian currents, weak diffeomorphisms and existence theorems in nonlinear elasticity. Arch. Ration. Mech. Anal. **105**, 97–159 (1990). Erratum and Addendum **109**, 385–392 (1990)
29. Green, A.E., Rivlin, R.S.: Multipolar continuum mechanics. Arch. Ration. Mech. Anal. **17**, 113–147 (1964)
30. Hencl, S., Koskela, P.: Lectures on Mappings of Finite Distortion. LNM 2096, Springer, Cham (2014)
31. Horák, M., Kružík, M.: Gradient polyconvex material models and their numerical treatment. Int. J. Solid Struct. **195**, 57–65 (2020)
32. Korteweg, D.J.: Sur la forme que prennent les équations du mouvement des fuides si lón tient compte des forces capillaires causées par des variations de densité considérables mais continues et sur la théorie de la capillarité dans l'hypothèse d'une variation continue de la densité. Arch. Néerl. Sci. Exactes Nat. **6**, 1–24 (1901)
33. Koumatos, K., Rindler, F., Wiedemann, E.: Orientation-preserving Young measures. Q. J. Math. **67**, 439–466 (2016)
34. Kouranbaeva, S., Shkoller, S.: A variational approach to second-order multisymplectic field theory. J. Geom. Phys. **35**, 333–366 (2000)
35. Kružík, M., Roubíček, T.: Mesoscopic model of microstructure evolution in shape memory alloys with applications to NiMnGa. Preprint IMA No. 2003, Univ. of Minnesota, Minneapolis (2004)
36. Kružík, M., Roubíček, T.: Mathematical Methods in Continuum Mechanics of Solids. Springer Nature, Cham (2019)
37. Kružík, M., Pelech, P., Schlömerkemper, A.: Gradient polyconvexity in evolutionary models of shape-memory alloys. J. Optim. Theory Appl. **184**, 5–20 (2020)
38. Mariano, P.M.: Geometry and balance of hyperstresses. Rendiconti Lincei Mat. Appl. **18**, 311–331 (2007)
39. Mielke, A., Roubíček, T.: Rate-Independent Systems: Theory and Applications. Springer, New York (2015)
40. Mielke, A., Theil, F.: On rate-independent hysteresis models. NoDEA Nonlinear Differ. Equ. Appl. **11**, 151–189 (2004)
41. Mielke, A., Theil, F., Levitas, V.I.: A variational formulation of rate-independent phase transformations using an extremum principle. Arch. Ration. Mech. Anal. **162**, 137–177 (2002)
42. Mindlin, R.D.: Micro-structure in linear elasticity. Arch. Ration. Mech. Anal. **16**, 51–78 (1964)
43. Morrey, C.B.: Quasi-convexity and the lower semicontinuity of multiple integrals. Pac. J. Math. **2**, 25–53 (1952)
44. Müller, S.: Variational Models for Microstructure and Phase Transitions. Lecture Notes in Mathematics 1713, pp. 85–210. Springer, Berlin (1999)

45. Patzák, B.: OOFEM—an object-oriented simulation tool for advanced modeling of materials and structures. Acta Polytech. **52**, 6 (2012)
46. Pideri, C., Seppecher, P.: A second gradient material resulting from the homogenization of an heterogeneous linear elastic medium. Contin. Mech. Thermodyn. **9**, 241–257 (1997)
47. Pitteri, M., Zanzotto, G.: Continuum Models for Phase Transitions and Twinning in Crystals. Chapman & Hall, Boca Raton (2003)
48. Podio-Guidugli, P., Caffarelli, G.V.: Surface interaction potentials in elasticity. Arch. Ration. Mech. Anal. **109**, 343–383 (1990)
49. Segev, R.: Geometric analysis of hyper-stresses. Int. J. Eng. Sci. **120**, 100–118 (2017)
50. Seppecher, P., Alibert, J.-J., dell'Isola, F.: Linear elastic trusses leading to continua with exotic mechanical interactions. J. Phys. Conf. Ser. **319**, 13 pp. (2011)
51. Šilhavý, M.: Phase transitions in non-simple bodies. Arch. Ration. Mech. Anal. **88**, 135–161 (1985)
52. Toupin, R.A.: Elastic materials with couple stresses. Arch. Ration Mech. Anal. **11**, 385–414 (1962)
53. Toupin, R.A.: Theory of elasticity with couple stress. Arch. Ration. Mech. Anal. **17**, 85–112 (1964)
54. Truesdell, C., Noll, W.: The Non-Linear Field Theories of Mechanics. Springer, Berlin (2004)

Placement of an Obstacle for Optimizing the Fundamental Eigenvalue of Divergence Form Elliptic Operators

Anisa M. H. Chorwadwala and Souvik Roy

AMS Subject Classifications 35J05, 35J10, 35P15, 49R05, 58J50

1 Introduction

Shape optimization problems deal with finding optimal shapes of certain objects or structures that optimize some objective cost functional with associated constraints. Such problems find applications in various fields like aerodynamics, medical imaging, engineering, and structural design [6, 22, 26, 35]. For example, in the designing of musical instruments like guitars, one needs to consider the optimal shape and placement of the hole in relation to the entire instrument to obtain the optimal frequency, thus enabling better sound output [22]. Another shape optimization problem arises in the field of medical imaging, where one needs to determine the shape of inhomogeneities inside the object of interest so as to minimize the Mumford–Shah functional containing the Radon data [26]. The placement of foreign inclusions inside a disk filled with liquid crystals (LCs) is also a shape optimization problem, where inclusions of different shapes and symmetries inside a LC-filled disk could be modeled as obstacles (see [7, 28], and [8]).

A commonly considered shape optimization problem is the optimization of the Dirichlet eigenvalues of the Laplace operator with a volume constraint. The origin of such problems dates back to 1800s when Rayleigh conjectured the famous isoperimetric inequality [31], which was proved by Faber [17] in 1923 and by Krahn [27] in 1925, independently. Since then, there have been numerous notable researches on the eigenvalue optimization problems involving various constraints. For a review of such results, we refer the readers to [4, 5, 23, 29]. In addition,

---

A. M. H. Chorwadwala
Indian Institute of Science Education and Research Pune, Pune, India
e-mail: anisa@iiserpune.ac.in

S. Roy (✉)
University of Texas at Arlington, Arlington, TX, USA
e-mail: souvik.roy@uta.edu

an introduction to the problems called "shape optimization problems" and the motivation to study general shape optimization problems can be found in [11]. For another mini review of various kinds of mathematical shape optimization problems, one may also refer to [10].

The shape optimization problem involving the placement of an obstacle inside a given planar domain was first studied by Hersch [24]. In this chapter, the optimal configuration for the fundamental Dirichlet eigenvalue for the Laplacian was characterized for the case, where a circular obstacle is placed inside a disk (please see also [30]). The result of [24] and [30] was subsequently extended to higher dimensional Euclidean spaces by the authors in [21, 25]. In [21], the case of multiple circular obstacles of possibly different sizes was also considered. In all these results, except in the more general family of domains in [24], the obstacles were balls in a Euclidean space, and thus only translation of the obstacle/s affected the eigenvalues. Therefore, these obstacle placement problems reduce to positioning of the center/s of the obstacle/s inside the outer disk. These results were further extended from the Euclidean case to all the three space forms in [12] and later to all rank one symmetric spaces of non-compact type in [15]. These results were extended to the case of the p-Laplacian in [13]. The mini review article [10] gives a brief explanation of the difficulties faced in proving these generalizations and about how the respective authors overcame these difficulties.

The case of obstacles that are not circular was first considered by the authors in [16]. In this work, it was assumed that the obstacle and the planar domain possessed a dihedral symmetry, with concentric centers. For such pairs of obstacles and enclosing domains, they considered a family of punctured domains, where the obstacle is removed from the enclosing domain. Among this family of punctured domains, the extremal configurations for the fundamental Dirichlet eigenvalue of the Laplacian were obtained by rotating the obstacle about its fixed center. The authors characterize both the minimizing and the maximizing configurations for the eigenvalue under consideration. The results in [16] motivated the work by the authors of this chapter in [14], where the planar domain is a disk punctured by the obstacle with dihedral symmetry, but the centers of the obstacles and the disk are non-concentric. The extremal configurations of the punctured disk with respect to rotation of the obstacle about its center were characterized for the fundamental Dirichlet eigenvalue of the Laplacian. Furthermore, the global extremal configurations of the punctured disks with respect to the combined rotations of the obstacle about its center and translations within the disk were also obtained. The proofs of the results rely mainly on the Hadamard perturbation formula and the reflection technique as in [34].

In this chapter, we present a generalization of the results obtained in [14]. We start with a generic second-order elliptic operator in the divergence form and consider the associated Dirichlet eigenvalue problem. Such divergence form elliptic operators frequently arise in hybrid imaging, fluid mechanics, and engineering [1, 18, 19, 32, 33]. The coefficients of the elliptic operator are a function of the spatial domain that is chosen such that the operator is invariant under rotations and translations in the plane. Our shape optimization problem is to place an obstacle with dihedral symmetry to optimize the fundamental eigenvalue. We use the techniques similar to the ones in [14] to obtain the extremal configurations with respect to rotations of the obstacle with even order dihedral symmetry and the global extremal configurations with respect to the rotations and translations of the obstacle in the disk. The crucial point in the theoretical derivation of the results is the generic Hadamard perturbation formula, which is different from the one used in [14]. We also provide results of several numerical simulations using the finite element method and with different obstacle shapes and coefficient functions. A generalization to domains with non-smooth boundaries can be made based on the work in [3] and [2].

The chapter is organized as follows: in the next section, we describe our generic eigenvalue shape optimization problem. Section 3 deals with the proof of a monotonicity property on the boundary of an arbitrary disk using its representation in polar coordinates with respect to a point different from its center. In addition, some elementary results regarding the geometries of the obstacle and the disk are proved. In Sect. 4, we state our main result that describes the extremal configurations for the fundamental eigenvalue. In Sect. 5, we give a proof of the extremal configurations for obstacles with even order dihedral symmetry. We further provide a partial result for obstacles with odd order symmetry. In Sect. 6, we provide a proof for the global extremal configurations. Section 7 presents some numerical results that validate the obtained extremal configurations. We end with a section of conclusions.

2 The Eigenvalue Optimization Problem

We now describe the elliptic eigenvalue optimization problem. For this purpose, we consider a family of domains $\Omega_t \subset \mathbb{R}^2$ with $t \in \mathbb{R}$. Let the fundamental elliptic eigenvalue corresponding to the domain Ω_t be denoted as $\lambda_1(t)$. Then, the corresponding optimization problem can be formulated as follows:

$$\min / \max_{t \in \mathbb{R}} \lambda_1(t)$$

$$\text{subject to } -Lu \equiv -\nabla \cdot (a(x)\nabla u) = \lambda_1(t) u, \ u > 0, \text{ in } \Omega_t$$

$$u(t) = 0, \text{ on } \partial \Omega_t \tag{1}$$

$$\int_{\Omega_t} u^2(x)\, dx = 1.$$

The coefficient $a(x)$ is chosen such that the following conditions are satisfied:

1. The operator L is invariant under rotations and translations in the plane. For example, one can choose $a(x)$ as constants, which gives L as a scalar multiple of the Laplacian. Another choice of $a(x)$ is $f(x_1^2 + x_2^2)$, which is a radial function.

2. The operator L is uniformly elliptic, i.e., $\langle a\xi, \xi \rangle \geq C\|\xi\|^2$ for all $\xi \in \mathbb{R}^2$ with $C > 0$, where $\langle \cdot, \cdot \rangle$ represents the standard Euclidean inner product and $\|\cdot\|$ represents the standard norm in $\mathbb{R}^2$.

We remark that for the case $a(x) = 1$, the operator L reduces to the standard Laplacian, and the corresponding analysis of (1) done below is already provided in [14].

In the next sections, we describe a family of admissible domains for the elliptic eigenvalue optimization problem. We then introduce some definitions to identify the different important configurations in this family. For the rest of the chapter, $n \in \mathbb{N}$, $n \geq 3$.

2.1 The Family of Admissible Domains

Let n be a positive integer, $n \geq 3$. Consider the dihedral group $\mathbb{D}_n$ generated by a rotation r of order n and a reflection s of order 2 such that $srs = r^{-1}$. Here, r is a rotation by an angle $2\pi/n$. Fix $A > 0$. Let the obstacle P denote a compact simply connected subset of the Euclidean plane $\mathbb{E}^2$ satisfying the following assumptions:

Assumption 2.1

(a) *The boundary ∂P of P is a simple closed C^2 curve in $\mathbb{R}^2$.*
(b) *The obstacle P possesses a $\mathbb{D}_n$ symmetry for some $n \geq 3$, n even, i.e., P is invariant under the action of a dihedral group $\mathbb{D}_n$ for some $n \geq 3$,*
(c) *The area of P is A.*

As a result of the above conditions, the axes of symmetry of P intersect in a unique point in the interior of P. We call this point the center $\underline{o}$ of P. Without loss of generality, we assume that $\underline{o}$ is the origin $(0, 0)$ of $\mathbb{R}^2$. The axes of symmetry of P divide $\mathbb{R}^2$ into $2n$ components. We call each of these $2n$ components as sectors and denote them by S_i, $1 \leq i \leq 2n$. We further make the following assumption on the monotonicity of the boundary ∂P (Fig. 1):

Assumption 2.2

(d) *The monotonicity of the boundary ∂P, that is, the distance $d(\underline{o}, x)$ between the center $\underline{o}$ of P and the point x on the boundary ∂P of P, is monotonic as a function of the argument ϕ in a sector delimited by two consecutive axes of symmetry of P.*

Assumptions 2.1 and 2.2 imply that P is a star-shaped domain with respect to its center $\underline{o}$.

Definition 2.1 (Incircle and Circumcircle) Let P be a compact simply connected subset of $\mathbb{R}^2$ satisfying Assumptions 2.1 and 2.2 and centered at $\underline{o}$. By an incircle of P we mean the largest circle in $\mathbb{R}^2$ centered at $\underline{o}$ that fits completely in P and

Placement of an Obstacle for Optimizing the Fundamental Eigenvalue of... 161

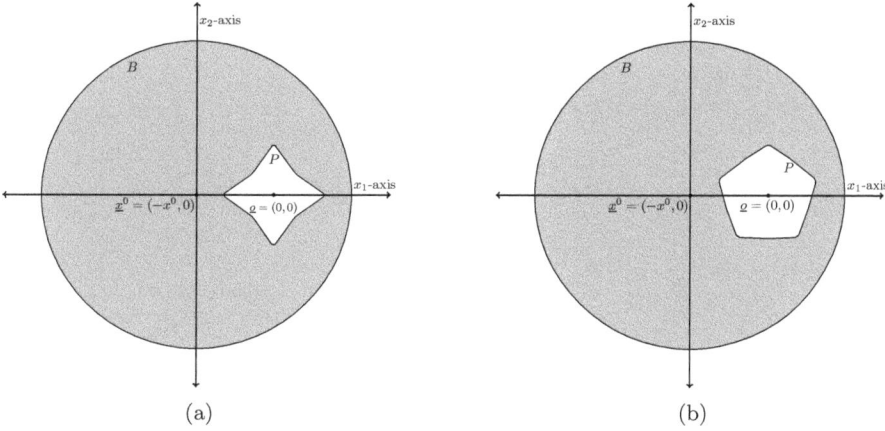

Fig. 1 Obstacles P with $\mathbb{D}_n$ symmetry. (**a**) $\mathbb{D}_4$ symmetry. (**b**) $\mathbb{D}_5$ symmetry

that is tangent to ∂P in each of its $2n$ sectors. By a circumcircle of P we mean the smallest circle in $\mathbb{R}^2$ centered at $\underline{o}$ that contains P and that is tangent to ∂P in each of its $2n$ sectors. Let $C_1(P)$ (respectively, $C_2(P)$) denote the incircle (respectively, the circumcircle) of P. When the set P is fixed, we will simply refer to the incircle as C_1 and the circumcircle as C_2. Please note here that $C_1(\rho(P)) = C_1(P)$ and $C_2(\rho(P)) = C_2(P)$ for each $\rho \in \mathbb{D}_n$.

Let $conv(A)$ denote the convex hull of a subset A in $\mathbb{R}^2$, and let $cl(conv(A))$ denote its closure. Clearly, for a compact simply connected subset P of the Euclidean plane $\mathbb{E}^2$ satisfying Assumptions 2.1 and 2.2, we have $P \subset cl(conv(C_2(P)))$, and hence $\rho(P) \subset cl(conv(C_2(P)))$ for each $\rho \in \mathbb{D}_n$. We take an open disk B in $\mathbb{R}^2$ with radius $r_1 > 0$ such that $B \supset cl(conv(C_2(P)))$ (Fig. 2).

2.2 The OFF and the ON Positions

Let C_1 and C_2 denote the incircle and the circumcircle, respectively, of an obstacle P satisfying Assumptions 2.1 and 2.2. We now define the *inner vertex set* V_{in} and the *outer vertex set* V_{out} of P as follows: $V_{in} := \partial P \cap C_1$ and $V_{out} := \partial P \cap C_2$. By a vertex set V we simply mean $V_{in} \cup V_{out}$. Elements of V_{in} (respectively, V_{out}) will be called *inner vertices* (respectively, *outer vertices*) *of* P. Elements of V will simply be referred to as *vertices of* P. A radial segment of the incircle C_1 of P containing an inner vertex will be referred to as an inradius of P, and likewise, a radial segment of the circumcircle C_2 of P containing an outer vertex of P will be referred to as a circumradius of P.

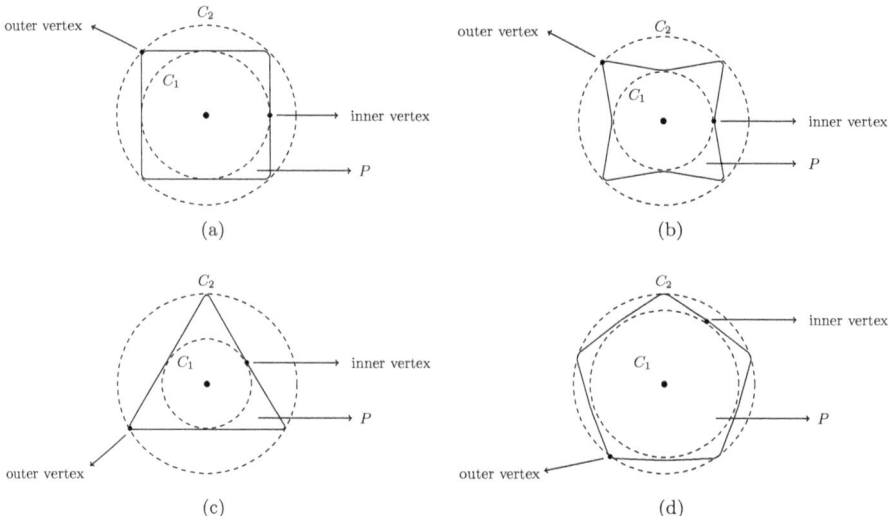

Fig. 2 Vertices of P. (**a**) P: square. (**b**) P: star. (**c**) P: triangle. (**d**) P: pentagon

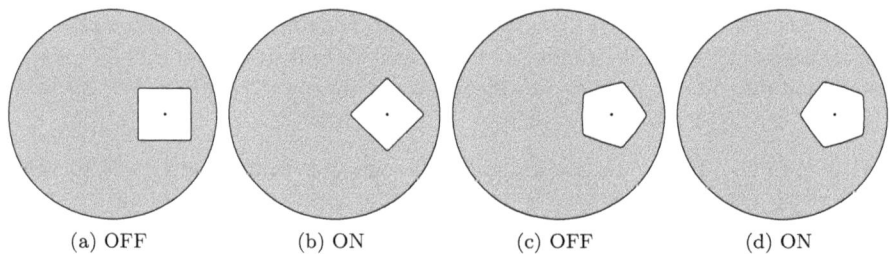

Fig. 3 OFF and ON configurations for obstacles having $\mathbb{D}_n$ symmetry. (**a**) OFF. (**b**) ON. (**c**) OFF. (**d**) ON

As a result of the fundamental eigenvalue λ_1 being invariant under isometries of $\mathbb{R}^n$, without loss of generality, we now make the following set of assumptions:

(a) The centers of B and P are on the x_1-axis.
(b) The center of P is at the origin.
(c) The center of B is on the negative x_1-axis.

We say that P is in an OFF position with respect to B if an inner vertex of P is on the negative x_1-axis and that P is in an ON position if an outer vertex of P is on the negative x_1-axis (Fig. 3).

We note that, for a given Ω_t, there are either two or no vertices of P lying on the axis of symmetry, the x_1-axis. When n is even, both the vertices are of the same type: either inner vertices or outer vertices. When n is odd, one of the two vertices is an inner vertex and the other is an outer vertex. So, alternate characterizations of the ON and the OFF positions are as follows:

(a) For n even, P is in an OFF position if two inner vertices lie on the x_1-axis and is in an ON position if two outer vertices lie on the x_1-axis,

(b) For n odd, P is in an OFF position if an outer vertex of P is on the positive x_1-axis and is in an ON position if an inner vertex of P is on the positive x_1-axis

3 Auxiliary Results

3.1 Boundary Monotonicity Property

In Lemma 3.1, we prove a monotonicity property on the boundary of an arbitrary disk B using the representation of B in polar coordinates with respect to a point other than its center.

Lemma 3.1 *Let $B((-x^0, 0), r_1)$ be a disk in $\mathbb{R}^2$ with center at $(-x^0, 0)$ and radius $r_1 > 0$ such that $0 < x^0 < r_1$. Let $\{re^{i\phi} : \phi \in [0, 2\pi[, 0 \leq r < g(\phi)\}$ be a representation B in polar coordinates, where $g : [0, 2\pi] \to [0, \infty[$ is a C^2 map with $g(0) = g(2\pi)$. Here, the polar coordinates (r, ϕ) are measured with respect to the origin $(0, 0)$ and the positive x_1-axis of $\mathbb{R}^2$. Then, the distance $\delta(\phi)$ of a point $g(\phi) e^{i\phi}$ on ∂B from $(0, 0)$ is a strictly increasing function of ϕ in $[0, \pi]$ and is a strictly decreasing function of ϕ in $[\pi, 2\pi]$.*

Proof Let ∂B^+ be defined as $\{g(\phi) e^{i\phi} \in \partial B \mid \phi \in [0, \pi]\} \subset \partial B$. Similarly, we define ∂B^- as the set $\{g(\phi) e^{i\phi} \in \partial B : \phi \in [\pi, 2\pi[\}$. We will prove that $\delta(\phi)$ is a strictly increasing function of ϕ in $[0, \pi]$. The proof for $\phi \in [\pi, 2\pi]$ is similar.

Let (x_1, x_2) denote the Cartesian coordinates of a point $g(\phi) e^{i\phi} \in \partial B^+$ as shown in Fig. 4. Then, $x_2 \geq 0$ and $(x_1 + x^0)^2 + x_2^2 = r_1^2$. We will first show that the Euclidean norm of the point $(x_1, x_2) \in \partial B^+$ is a monotonic function of x_1 for all $(x_1, x_2) \in \partial B^+$. Here, $x_1 \in [-x^0 - r_1, -x^0 + r_1]$. We thus consider $\|(x_1, x_2)\| = d((x_1, x_2), (0, 0))$ subject to $(x_1 + x^0)^2 + x_2^2 = r_1^2$. Now, $\|(x_1, x_2)\| = (x_1^2 + x_2^2)^{\frac{1}{2}} = (x_1^2 + r_1^2 - (x_1 + x^0)^2)^{\frac{1}{2}} = (r_1^2 - 2x_1 x^0 - (x^0)^2)^{\frac{1}{2}} =: h(x_1) > 0$. Therefore,

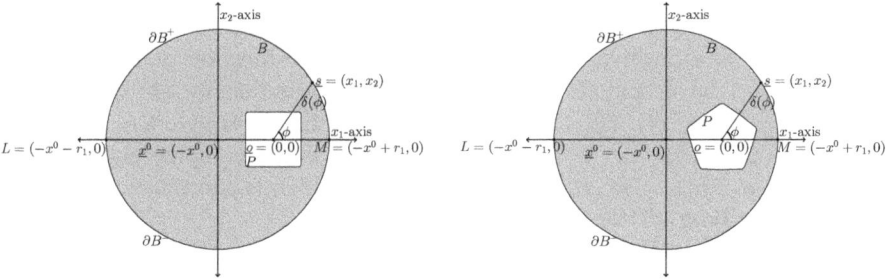

Fig. 4 The distance function δ for the boundary points

$h'(x_1) = \dfrac{-x^0}{h(x_1)} < 0$ for $(x_1, x_2) \in \partial B^+$. Hence, h is a strictly decreasing function of x_1 for $(x_1, x_2) \in \partial B^+$. We also note that $h(x_1) = \|(x_1, x_2)\| = |g(\phi)| = \delta(\phi)$ for $(x_1, x_2) = g(\phi)e^{i\phi} \in \partial B^+$, $\phi \in [0, \pi]$.

Next, we show that $x_1 = x_1(\phi)$ is a monotonic decreasing function of ϕ. We have $x_1 = \|(x_1, x_2)\| \cos \phi = h(x_1) \cos \phi$. Hence, $\cos(\phi) = \dfrac{x_1}{h(x_1)}$. Consider $\phi :]-x^0 - r_1, -x^0 + r_1[\to]0, \pi[$. Then,

$$\frac{d\phi}{dx_1} = -\frac{h(x_1)^2 + x^0 x_1}{h(x_1)^3} \frac{1}{\sin \phi} = -\frac{h(x_1)^2 + x^0 x_1}{x_2 \, h(x_1)^2} = -\frac{r_1^2 - x^0 x_1 - (x^0)^2}{x_2 \, h(x_1)^2}$$

$$= -\frac{r_1^2 - x^0 (x_1 + x^0)}{x_2 \, h(x_1)^2}.$$

Since $|x_1 + x^0| < r_1$ and $0 < x^0 < r_1$, we get $-r_1^2 < x^0(x^0 + x_1) < r_1^2$. This implies that $\dfrac{d\phi}{dx_1} < 0$ on $]-x^0 - r_1, -x^0 + r_1[$. Thus, ϕ as a function of x_1 is strictly decreasing and hence injective on $]-x^0 - r_1, -x^0 + r_1[$.

Finally, we show that $\phi :]-x^0 - r_1, -x^0 + r_1[\to]0, \pi[$ is surjective. Let $\theta \in]0, \pi[$. Define

$$x_1 = g(\theta) \cos \theta \in]-x^0 - r_1, -x^0 + r_1[.$$

Hence, $\phi :]-x^0 - r_1, -x^0 + r_1[\to]0, \pi[$ is a bijective and strictly decreasing function of x_1. Since the distance function $\delta(\phi)$ is decreasing with respect to x_1, it is increasing with respect to ϕ. This proves the lemma. □

3.2 Properties of a Planar Simply Connected Bounded Domain K

In this section, we derive some properties associated with a planar simply connected bounded domain K. In polar coordinates, the planar simply connected bounded domain K can be given by $K = \{re^{i\phi} : \phi \in [0, 2\pi[, 0 \leq r < h(\phi)\} \subset \mathbb{R}^2$, where h is a positive, bounded, and 2π-periodic function of class $\mathcal{C}^2$. Let $v \in C_0^\infty(\mathbb{R}^2)$ be a smooth vector field whose restriction to ∂K is given by $v(x_1, x_2) = (-x_2, x_1) \,\forall (x_1, x_2) \in \partial K$. This implies $v(h(\phi)(\cos \phi, \sin \phi)) = h(\phi)(-\sin \phi, \cos \phi) \,\forall \phi \in [0, 2\pi[$. Treating $\mathbb{R}^2$ as the complex plane $\mathbb{C}$, one can write v as $v(\zeta) = i\zeta \,\forall \zeta = h(\phi)e^{i\phi} \in \partial K$, which is equivalent to $v(\phi) := v\left(h(\phi)e^{i\phi}\right) = ih(\phi) \, e^{i\phi} \,\forall \phi \in \mathbb{R}$.

Denote by η the unit outward normal vector field to K on ∂K. For $\alpha \in [0, 2\pi]$, let $z_\alpha := \{re^{i\alpha} : r \in \mathbb{R}\}$ denote the line in $\mathbb{R}^2$ corresponding to angle $\phi = \alpha$

represented in polar coordinates. Clearly, $z_\alpha = z_{\alpha+\pi}$ for each $\alpha \in [0, 2\pi]$, where the addition is taken modulo 2π.

We now prove the following auxiliary lemma.

Lemma 3.2 *Let K, h, v, η, and z_α be as defined above. Then, at any point $h(\phi)e^{i\phi}$ of ∂K, we have the following:*

(i) $\eta(\phi) := \eta(h(\phi)e^{i\phi}) = \dfrac{h(\phi)e^{i\phi} - ih'(\phi)e^{i\phi}}{\sqrt{h^2(\phi) + (h'(\phi))^2}} \quad \forall \phi \in \mathbb{R}.$

(ii) $\langle \eta, v \rangle (\phi) := \langle \eta, v \rangle (h(\phi)e^{i\phi}) = \dfrac{-h(\phi)h'(\phi)}{\sqrt{h^2(\phi) + (h'(\phi))^2}} \quad \forall \phi \in \mathbb{R}.$ *Hence,* $\langle \eta, v \rangle$ *has a constant sign on an interval $I \subset \mathbb{R}$ iff h is monotonic in I.*

(iii) *If for some $\alpha \in [0, 2\pi[$, the domain K is symmetric with respect to the axis z_α, then, for each $\theta \in [0, \pi]$, $\langle \eta, v \rangle (\alpha + \theta) = - \langle \eta, v \rangle (\alpha - \theta)$.*

Proof

(i) Let $\gamma : [0, 2\pi[\to \mathbb{R}^2$ be defined as $\gamma(\phi) = h(\phi)e^{i\phi}$. That is, γ is a parametrization of the boundary curve ∂K. Then, the tangent vector field to the boundary ∂K is given by $\gamma'(\phi) = \big(h'(\phi) + ih(\phi)\big) e^{i\phi}$. Thus, the outward unit normal $\eta(\phi)$ to K at a point $\gamma(\phi) \in \partial K$ is given by $\dfrac{\big(h(\phi) - ih'(\phi)\big) e^{i\phi}}{\sqrt{h^2(\phi) + (h'(\phi))^2}}$.

(ii) Therefore, $\langle \eta, v \rangle (\phi) = \dfrac{h^2(\phi) \langle e^{i\phi}, ie^{i\phi} \rangle - h(\phi)h'(\phi)|ie^{i\phi}|^2}{\sqrt{h^2(\phi) + (h'(\phi))^2}} = -\dfrac{h(\phi)h'(\phi)}{\sqrt{h^2(\phi) + (h'(\phi))^2}}.$

(iii) Since K is symmetric with respect to the axis z_α, the function h satisfies $h(\alpha + \theta) = h(\alpha - \theta)$ for each $\theta \in [0, \pi]$. Moreover, $h'(\alpha - \theta) = -h'(\alpha + \theta)$ for each $\theta \in [0, \pi]$. Using (ii), we then have $\langle \eta, v \rangle (\alpha + \theta) = - \langle \eta, v \rangle (\alpha - \theta)$.

□

Remark 3.1 We note here that since h is a 2π-periodic function on $\mathbb{R}$, so are the functions v, η, and $\langle v, \eta \rangle$.

4 The Main Results: Extremal Configurations

We recall here that the obstacle P is a compact and simply connected subset of $\mathbb{R}^2$ satisfying Assumptions 2.1 and 2.2 and that B is an open disk in $\mathbb{R}^2$ of radius r_1 such that $B \supset cl(conv(C_2(P)))$. For $t \in \mathbb{R}$, let $\rho_t \in SO(2)$ denote the rotation in $\mathbb{R}^2$ about the origin $\underline{o}$ in the anticlockwise direction by an angle t, i.e., for $\zeta \in \mathbb{C} \cong \mathbb{R}^2$, we have $\rho_t \zeta := e^{it} \zeta$. Now, fix $t \in [0, 2\pi[$. Let $\Omega_t := B \setminus \rho_t(P)$ and $\mathcal{F} := \{\Omega_t : t \in [0, 2\pi)\}$. We now state the following theorem for n even, $n \geq 3$, about the extremal configurations with respect to rotations of the obstacle P.

Theorem 4.1 (Extremal Configurations) *The fundamental Dirichlet eigenvalue $\lambda_1(\Omega_t)$ for $\Omega_t \in \mathcal{F}$ is optimal precisely for those $t \in [0, 2\pi[$ for which an axis of symmetry of P_t coincides with a diameter of B. Among these optimal configurations,* the maximizing configurations are the ones corresponding to those $t \in [0, 2\pi[$ for which P_t is in an ON position with respect to B, and the minimizing configurations are the ones corresponding to those $t \in [0, 2\pi[$ for which P_t is in an OFF position with respect to B.

We next state the result pertaining to the global extremal configurations with respect to rotations and translations of the obstacle P. For this purpose, let r_0^1 and r_0^2 denote the radii of the incircle C_1 and the circumcircle C_2 of the obstacle P, respectively. Let $P_{(d,t)}$ be the obstacle P_t as in Theorem 4.1 with its center $\underline{o}$ at a distance $d < r_1 - r_0^2$ from the center of B. Please note that in Theorem 4.1, d is fixed and is always > 0. This is because, for the case $d = 0$, $t \longmapsto \lambda_1(\Omega_t)$ is a constant map. Since we want to study the behavior of λ_1 w.r.t. the translations of the obstacle too, we now allow d to be 0. Let $\Omega_{(d,t)} := B \setminus P_{(d,t)}$ for $d \in [0, r_1 - r_0^2)$, $t \in [0, 2\pi[$. Let $\lambda_1((d,t)) := \lambda_1(\Omega_{(d,t)})$. Let $\mathcal{G}$ be defined as $\{\Omega_{(d,t)} : (d,t) \in [0, r_1 - r_0^2[\times[0, 2\pi[\}$. We then have the following theorem.

Theorem 4.2 (Global Extremal Configurations, I.e., Extremal Configurations w.r.t. the Translations and Rotations of the Obstacle Within B**)** *Fix* $n \geq 3$, *even or odd. The concentric configuration, i.e.,* $\Omega_{(0,t)}$, *for any* $t \in [0, 2\pi[$, *is the maximizing configuration for* $\lambda_1((d,t))$ *over* $\mathcal{G}$. *At a minimizing configuration for* $\lambda_1(d,t)$ *over* $\mathcal{G}$, *the circumcircle of the obstacle must touch* ∂B.

For n even, $n \geq 3$, *we further have that, at the minimizing configuration over* $\mathcal{G}$, *the obstacle must be in an OFF position w.r.t.* B.

The concentric configurations are the global maximizing configurations w.r.t. all the translations and all rotations of the obstacle within B, *for* $n \geq 3$, n *even or odd. The OFF configurations with an outer vertex touching* ∂B, *i.e.,* $\Omega_{(r_1-r_0^1, 2k\frac{\pi}{n})}$, $k \in \mathbb{Z}$, *are the global minimizing configurations w.r.t. all the translations and rotations of the obstacle within* B, *for* $n \geq 3$, n *even.*

5 Proofs of Theorem 4.1

In this section, we prove our first main result, viz., Theorem 4.1 for $n \geq 3$, n even. We first show that, for any $n \geq 3$, even or odd, the fundamental eigenvalue λ_1 of the operator L for the family of domains under consideration is a function of just one real variable, which is an even, differentiable, and periodic function of period $2\pi/n$. The result is shown through equation (6) in Sect. 5.1.4. This helps in identifying the critical points of λ_1. Therefore, in order to determine the extremal configuration/s for λ_1, we study its behavior on the interval $[0, \frac{\pi}{n}]$. The Hadamard perturbation formula (4) becomes useful in this analysis.

We then prove the result about a sufficient condition for the existence of λ_1 in Proposition 5.1 for $n \geq 3$, even or odd. Next, in Proposition 5.2, for n even, $n \geq 3$, we state and prove the necessary conditions for the existence of λ_1. In view of

Eq. (6), Propositions 5.1 and 5.2 imply that, for n even, $n \geq 3$, (a) these are the only critical points for λ_1 and that (b) between every pair of consecutive critical points, λ_1 is a strictly monotonic function of the argument. Finally, the proof of Theorem 4.1 is a direct consequence of Propositions 5.1 and 5.2 and Eq. (6).

5.1 Sufficient Condition for the Critical Points of $\lambda_1(B \setminus P_t)$, $t \in [0, 2\pi[$

Fix $n \geq 3$, even or odd. Let $\lambda_1(t)$ denote the fundamental Dirichlet eigenvalue of the Laplacian on Ω_t i.e., $\lambda_1(t) := \lambda_1(\Omega_t)$. In this section, we establish a sufficient condition for the critical points of the $\mathcal{C}^1$ function $\lambda_1 : \mathbb{R} \to]0, \infty[$.

In polar coordinates, the open disk B can be represented as the set $\{re^{i\phi} : \phi \in [0, 2\pi[, 0 \leq r < g(\phi)\}$, where $g : [0, 2\pi] \to [0, \infty[$ is a $\mathcal{C}^2$ map with $g(0) = g(2\pi)$. Here, (r, ϕ) is measured with respect to the origin $\underline{o} = (0, 0)$ of $\mathbb{R}^2$. The boundary ∂B of B, then, is given by $g(\phi) e^{i\phi}$, $0 \leq \phi < 2\pi$. Let $\delta(\phi)$ denote the Euclidean norm of $g(\phi) e^{i\phi}$, that is, $\delta(\phi)$ is the distance of a point $g(\phi) e^{i\phi}$ on ∂B from the center $\underline{o}$ of the obstacle P. Then, by Lemma 3.1, δ is a strictly increasing function of ϕ on $[0, \pi]$.

5.1.1 The Initial Configuration

We start with the following initial configuration Ω_{init} of a domain $\Omega \in \mathcal{F}$. Let P and B be as described in Sect. 4. Let Ω_{init} denote the domain $B \setminus P \in \mathcal{F}$, where P is in an OFF position with respect to B. Recall that we assumed, without loss of generality, that (a) the centers of B and P are on the x_1-axis, (b) the center of P is at the origin, and (c) the center of B is on the negative x_1-axis. Let $\underline{x}^0 := (-x^0, 0)$ be the center of the disk B, where $0 < x^0 < r_1$. The initial configurations for obstacles with $\mathbb{D}_n$ symmetry are shown in Fig. 5.

We parametrize P in polar coordinates as follows:

$$P = \{re^{i\phi} : \phi \in [0, 2\pi[, 0 \leq r < f(\phi)\}, \tag{2}$$

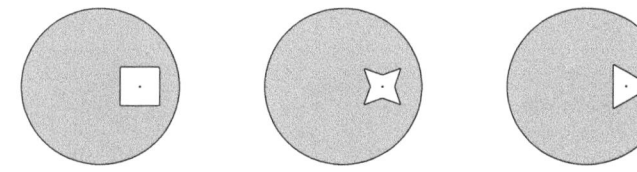
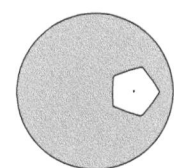

Fig. 5 The initial configurations

where $f : [0, 2\pi] \to [0, \infty)$ is a C^2 map with $f(0) = f(2\pi)$. Because of the initial configuration assumptions on $B \setminus P$, f is an increasing function of ϕ on $]0, \frac{\pi}{n}[$ for n even and is a decreasing function of ϕ on $]0, \frac{\pi}{n}[$ for n odd. The condition that the obstacle P can rotate freely about its center $\underline{o}$ inside B, that is, $\rho(P)$ is contained in $B \; \forall \rho \in SO(2)$, is guaranteed by assuming that the closure of the convex hull of the circumcircle $C_2(P)$ is contained in B. This gives us the following relation:

$$f\left(\frac{\pi}{n}\right) = \max_{0 \leq \phi \leq 2\pi} f(\phi) < \min_{0 \leq \phi \leq 2\pi} g(\phi) = g(0).$$

5.1.2 Configuration at Time t

Now, fix $t \in [0, 2\pi[$. We set

$$P_t := \rho_t(P), \qquad \Omega_t := B \setminus P_t. \tag{3}$$

Then, in polar coordinates, we have $\partial P_t := \{f(\phi - t)e^{i\phi} \mid \phi \in [0, 2\pi[\}$ (Fig. 6).

5.1.3 Hadamard Perturbation Formula

Let $\lambda_1(t)$ denote the fundamental Dirichlet eigenvalue of the Laplacian on Ω_t, i.e., $\lambda_1(t) := \lambda_1(\Omega_t)$. Let $y_1(t)$ denote the unique positive unit norm principal Dirichlet eigenfunction for the Laplacian on Ω_t, i.e., $y_1(t)$ is the eigenfunction corresponding to $\lambda_1(t)$ on Ω_t satisfying (1). By Proposition 3.1 in [12], the map $t \longmapsto \lambda_1(t)$ is a C^1 map in $\mathbb{R}$ from a neighborhood of 0 in $\mathbb{R}$. Then, the derivative $\lambda_1'(t)$ of λ_1 at a point $t \in \mathbb{R}$ is given by the Hadamard perturbation formula [20]

$$\lambda_1'(t) = -\int_{x \in \partial P_t} \langle a(x)\eta_t, \eta_t \rangle \left|\frac{\partial y_1(t)(x)}{\partial \eta_t}\right|^2 \langle \eta_t, v \rangle (x) \, d\Sigma(x), \tag{4}$$

where $d\Sigma$ is the line element on ∂P_t, $\eta_t(x)$ is the outward unit normal vector to Ω_t at $x \in \partial \Omega_t$, and $v \in C_0^\infty(\Omega_t)$ is the deformation vector field defined as

$$v(\zeta) = \rho(\zeta) \, i\zeta, \qquad \forall \zeta \in \mathbb{C} \cong \mathbb{R}^2. \tag{5}$$

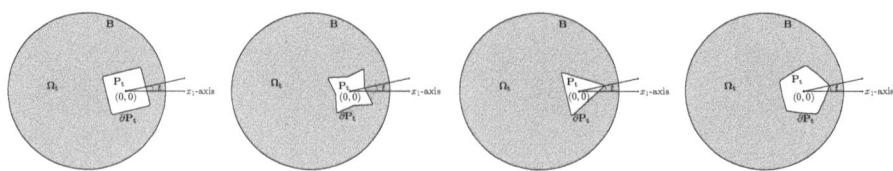

Fig. 6 Configuration at time t

Here, $\rho : \mathbb{R}^2 \to [0, 1]$ is a smooth function with compact support in B such that $\rho \equiv 1$ in a neighborhood of $cl(conv(C_2(P)))$.

Remark 5.1 We are interested in the outward unit normal to the domain Ω_t at points on the boundary $\partial P_t := \{f(\phi)e^{i\phi} : \phi \in [0, 2\pi[\}$ of the obstacle P_t. Therefore, the outward unit normal with respect to the domain Ω_t at a point $f(\phi)e^{i\phi}$ on ∂P_t will be the negative of the vector field $\eta(f(\phi)e^{i\phi})$, for $h = f$ in Lemma 3.2.

5.1.4 λ_1 Is an Even and Periodic Function with Period $\frac{2\pi}{n}$

Recall that $n \geq 3$ is a fixed integer, even or odd. Since P_t is invariant under the action of the dihedral group $\mathbb{D}_n$, it follows that $\Omega(t + \frac{2\pi}{n}) = \Omega_t$ for each $t \in \mathbb{R}$. Let $R_0 : \mathbb{R}^2 \to \mathbb{R}^2$ denote the reflection in $\mathbb{R}^2$ about the x_1-axis. That is, $R_0((x_1, x_2)) := (x_1, -x_2) \, \forall (x_1, x_2) \in \mathbb{R}^2$. Then, we have $\rho_{2\pi-t} = R_0 \circ \rho_t \circ R_0$ for each $t \in \mathbb{R}^2$. This gives $P_{2\pi-t} = R_0(P_t)$ and $\Omega_{2\pi-t} = R_0(\Omega_t)$. In $SO(2, \mathbb{R})$, $\rho_{s+t} = \rho_s \circ \rho_t = \rho_t \circ \rho_s \, \forall s, t \in \mathbb{R}$ and $\rho_{2\pi} = \mathrm{Id}$, the identity map. Therefore, we get $P_{-t} = R_0(P_t)$ and $\Omega_{-t} = R_0(\Omega_t)$ for all $t \in \mathbb{R}$. Moreover, since $\rho_{\frac{2\pi}{n}}(P_t) = P_t$ for all $t \in \mathbb{R}$, $\Omega_{\frac{2\pi}{n}+t} = \Omega_t$ for all $t \in \mathbb{R}$. This implies that $\lambda_1 : \mathbb{R} \to (0, \infty)$ is an even and periodic function with period $\frac{2\pi}{n}$. Thus, we have

$$\lambda_1\left(t + \frac{2\pi}{n}\right) = \lambda_1(t), \text{ and } \lambda_1(-t) = \lambda_1(t) \quad \forall \, t \in \mathbb{R}. \tag{6}$$

Therefore, it suffices to study the behavior of $\lambda_1(t)$ only on the interval $\left[0, \frac{\pi}{n}\right]$.

5.1.5 Sufficient Condition for the Critical Points of λ_1

The following theorem states a sufficient condition for the critical points of the function $\lambda_1 : \mathbb{R} \to (0, \infty)$.

Proposition 5.1 (Sufficient Condition for Critical Points of λ_1) *Let $n \geq 3$ be a fixed integer, even or odd. For each $k = 0, 1, 2, \ldots, 2n-1$, $\lambda'_1\left(k\frac{\pi}{n}\right) = 0$.*

Proof Fix $k \in \{0, 1, 2, \ldots, 2n-1\}$. Let $t_k := k\frac{\pi}{n}$. Then, the domain Ω_{t_k} is symmetric with respect to the x_1-axis. The first Dirichlet eigenfunction $y_1(t_k)$ satisfies $u \circ R_0 = u$, where $R_0 \in O(2, \mathbb{R})$ is the reflection about the x_1-axis. Clearly, for each $x \in \partial P_{t_k}$, where η is defined, $\eta(R_0(x)) = DR_0(\eta(x)) = R_0(\eta(x))$. Note also that

$$\frac{\partial(y_1(t_k) \circ R_0)}{\partial \eta}(x) = \frac{\partial(y_1(t_k))}{\partial \eta}(R_0(x)) \tag{7}$$

for each x on ∂P_{t_k} for which the normal derivative makes sense. By the Hadamard perturbation formula (4), we have

$$\lambda_1'(t_k) = -\int_{\partial P_{t_k}^+} \langle a(x)\eta_t, \eta_t \rangle \left| \frac{\partial (y_1(t_k))}{\partial \eta_{t_k}} \right|^2 (x) \langle \eta_{t_k}, v \rangle (x) \, d\Sigma(x)$$
$$-\int_{\partial P_{t_k}^-} \langle a(x)\eta_t, \eta_t \rangle \left| \frac{\partial (y_1(t_k))}{\partial \eta_{t_k}} \right|^2 (x) \langle \eta_{t_k}, v \rangle (x) \, d\Sigma(x), \quad (8)$$

where $\partial P_{t_k}^+$ and $\partial P_{t_k}^-$ represent the parts of ∂P_{t_k} above the x_1-axis and below the x_1-axis, respectively. Therefore, we have

$$\lambda_1'(t_k) = -\int_{\partial P_{t_k}^+} \langle a(x)\eta_t, \eta_t \rangle \left| \frac{\partial y_1(t_k)(x)}{\partial \eta_{t_k}} \right|^2 \langle \eta_{t_k}, v \rangle (x) \, d\Sigma(x)$$
$$-\int_{R_0(\partial P_{t_k}^+)} \langle a(x)\eta_t, \eta_t \rangle \left| \frac{\partial y_1(t_k)(x)}{\partial \eta_{t_k}} \right|^2 \langle \eta_{t_k}, v \rangle (x) \, d\Sigma(x).$$

Using Eq. (7) and property (iii) of Lemma 3.2, we get $\lambda_1'(t_k) = 0$. Thus, $k\frac{\pi}{n}$, $k \in \{0, 1, 2, \ldots, 2n-1\}$, are the critical points of λ_1. □

5.2 The Sectors of Ω_t

Fix $n \geq 3$, even or odd. For a fixed $t \in \mathbb{R}$ and $a, b \in \mathbb{Z}$, $a < b$, let $\sigma\left(t + \frac{a\pi}{n}, t + \frac{b\pi}{n}\right)$ denote the set $\{r e^{i\phi} \in \mathbb{R}^2 : \phi \in \left(t + \frac{a\pi}{n}, t + \frac{b\pi}{n}\right), r \in \mathbb{R}\}$. For convenience, we will simply write $\sigma_{(a,b)}$ to denote $\sigma\left(t + \frac{a\pi}{n}, t + \frac{b\pi}{n}\right)$. When we write $\sigma_{(k,k+1)}$, $k \in \mathbb{Z}$, we take addition modulo $2n$, that is, $k, k+1 \in (\mathbb{Z}_{2n}, +)$. From Eq. (4), we have

$$\lambda_1'(t) = -\sum_{k=0}^{2n-1} \int_{\partial P_t \cap \sigma]t+\frac{k\pi}{n}, t+\frac{(k+1)\pi}{n}[} \langle a(x)\eta_t, \eta_t \rangle \left| \frac{\partial y_1(t)(x)}{\partial \eta_t} \right|^2 \langle \eta_t, v \rangle (x) \, d\Sigma(x). \quad (9)$$

Equation (9) can be written as

$$\lambda_1'(t) = -\sum_{k=0}^{n-1} \int_{\partial P_t \cap \sigma_{(k,k+1)}} \langle a(x)\eta_t, \eta_t \rangle \left| \frac{\partial y_1(t)(x)}{\partial \eta_t} \right|^2 \langle \eta_t, v \rangle (x) \, d\Sigma(x)$$
$$-\sum_{k=n}^{2n-1} \int_{\partial P_t \cap \sigma_{(k,k+1)}} \langle a(x)\eta_t, \eta_t \rangle \left| \frac{\partial y_1(t)(x)}{\partial \eta_t} \right|^2 \langle \eta_t, v \rangle (x) \, d\Sigma(x). \quad (10)$$

We now fix a $t \in]0, \frac{2\pi}{n}[$ and note the following properties for the sectors $\sigma_{(k,k+1)}$:

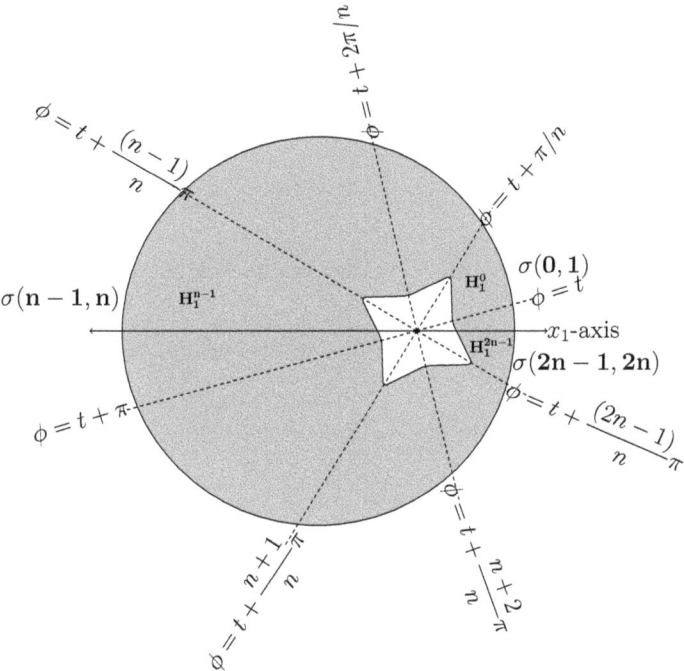

Fig. 7 Sectors of Ω_t for $n = 4$

1. For $k = 0, 1, 2, \ldots, n - 2$, each of the sectors $\sigma_{(k,k+1)}$ is completely above the x_1-axis.
2. For $k = n, \ldots, 2n - 2$, the sectors $\sigma_{(k,k+1)}$ are completely below the x_1-axis.
3. The sectors $\sigma_{(n-1,n)}$ and $\sigma_{(2n-1,2n)}$ are partially above the x_1-axis and partially below it.

These facts are illustrated in Fig. 7.

5.3 A Sector Reflection Technique

Here onward, we fix $n \geq 3$, n even. We recall here from Sect. 3.2 that, for $\alpha \in [0, 2\pi]$, the set $z_\alpha := \{re^{i\alpha} \mid r \in \mathbb{R}\}$ denotes the line in $\mathbb{R}^2$ corresponding to angle $\phi = \alpha$, represented in polar coordinates. Let $R_\alpha : \mathbb{R}^2 \to \mathbb{R}^2$, $\alpha \in \mathbb{R}$, denote the reflection map about the z_α-axis. For each $t \in \mathbb{R}$, the obstacle P_t is symmetric with respect to the line $z_{t + \frac{(k+1)\pi}{n}}$. We have, for $k = 0, 1, 2, \ldots, 2n - 1$,

$$R_{t + \frac{(k+1)\pi}{n}} (\partial P_t \cap \sigma_{(k,k+1)}) = \partial P_t \cap \sigma_{(k+1,k+2)}. \tag{11}$$

For $k = 0, 1, 2, \ldots, 2n - 1$, let $H_1^k(t) := \Omega_t \cap \sigma_{(k,k+1)}$. Now, let $\tilde{H}_1^k := cl(\Omega_t) \cap \sigma_{(k,k+1)}$. This implies $\tilde{H}_1^k(t) = H_1^k(t) \cup \left(cl(H_1^k(t)) \cap \partial \Omega_t\right)$.

We consider pairs of consecutive sectors of Ω_t, namely $\sigma_{(k,k+1)}$ and $\sigma_{(k+1,k+2)}$ for each even k such that $k = 0, 2 \ldots 2n - 2$. We now state and prove the following lemma.

Lemma 5.1 *Fix $n \geq 3$, n even. For all $t \in]0, \frac{\pi}{n}[$, we have the following:*

$$R_{t + \frac{(k+1)\pi}{n}}(H_1^k(t)) \subsetneq H_1^{k+1}(t) \quad \text{for } k = 0, 2, 4, \ldots, n - 2. \tag{12}$$

$$R_{t + \frac{(k+1)\pi}{n}}(\tilde{H}_1^k(t)) \subsetneq \tilde{H}_1^{k+1}(t) \setminus \partial B \quad \text{for } k = 0, 2, 4, \ldots, n - 2. \tag{13}$$

$$R_{t + \frac{(k+1)\pi}{n}}(H_1^{k+1}(t)) \subsetneq H_1^k(t) \quad \text{for } k = n, n + 2, \ldots, 2n - 2. \tag{14}$$

$$R_{t + \frac{(k+1)\pi}{n}}(\tilde{H}_1^{k+1}(t)) \subsetneq \tilde{H}_1^k(t) \setminus \partial B \quad \text{for } k = n, n + 2, \ldots, 2n - 2. \tag{15}$$

Proof We first prove (12)–(13) for $k = 0, 2, 4, \ldots, n - 4$, where the pair of sectors $\sigma_{(k,k+1)}$ and $\sigma_{(k+1,k+2)}$ are completely above the x_1-axis. A similar technique can be used to prove (14)–(15) for $k = n, n + 2, \ldots, 2n - 4$, where the sectors $\sigma_{(k,k+1)}$ and $\sigma_{(k+1,k+2)}$ are completely below the x_1-axis. We then prove (12)–(13) for $k = n - 2$ separately and similarly prove (14)–(15) for $k = 2n - 2$ separately.

Let $\beta \in [0, \frac{\pi}{n}]$ be arbitrary. The line L_1 containing the center $\underline{o}$ and the point

$$p_1 = g\left(t + (k+1)\frac{\pi}{n} - \beta\right)\left(\cos(t + (k+1)\frac{\pi}{n} - \beta), \sin(t + (k+1)\frac{\pi}{n} - \beta)\right) \in \partial B$$

is reflected about $z_{t+(k+1)\frac{\pi}{n}}$-axis to the line L_2 containing $\underline{o}$ and the point

$$p_2 = g\left(t + (k+1)\frac{\pi}{n} + \beta\right)\left(\cos(t + (k+1)\frac{\pi}{n} + \beta), \sin(t + (k+1)\frac{\pi}{n} + \beta)\right) \in \partial B$$

(see Fig. 8).

Since P_t is invariant under this reflection and B is star-shaped with respect to $\underline{o}$, to prove (12)–(13), it suffices to show that $g\left(t + \frac{(k+1)\pi}{n} - \beta\right) < g\left(t + \frac{(k+1)\pi}{n} + \beta\right)$ for $k = 0, 2, 4, \ldots, n - 2$.

Now, for $k = 0, 2, 4, \ldots, n - 4$, $\left(t + \frac{k\pi}{n}, t + \frac{(k+2)\pi}{n}\right) \subset]0, \pi[$. So, by Lemma 3.1, g is a strictly increasing function of the argument in $\left(t + \frac{k\pi}{n}, t + \frac{(k+2)\pi}{n}\right)$ for $k = 0, 2, 4, \ldots, n - 4$. Therefore, (12)–(13) for $k = 0, 2, 4, \ldots, n - 4$ follow from the fact that $t + \frac{(k+1)\pi}{n} - \beta < t + \frac{(k+1)\pi}{n} + \beta$.

Next, we consider the case $k = n - 2$. The sector $\sigma_{(n-2,n-1)}$ is completely above the x_1-axis, whereas the sector $\sigma_{(n-1,n)}$ is partially above and partially below the x_1-axis. If the point p_2 is above the x_1-axis, we have

Fig. 8 Reflection of sector H_1^k about the axis $z_{t+\frac{(k+1)\pi}{n}}$

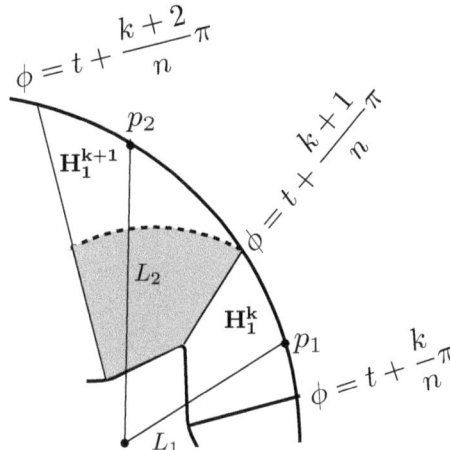

$$0 < t + \frac{(n-1)\pi}{n} - \beta < t + \frac{(n-1)\pi}{n} + \beta < \pi.$$

Since g is strictly increasing in $[0, \pi]$, we have the desired results (12)–(13) in this case.

Suppose the point p_2 is below the x_1-axis. Let $\theta > 0$ be the angle between L_2 and the positive x_1-axis. Then, since ∂B is symmetric with respect to the x_1-axis, we get $g\left(t + \frac{(n-1)\pi}{n} + \beta\right) = g\left(t + \frac{(n-1)\pi}{n} + (\beta - 2\theta)\right)$. Now, since $\beta > \theta$, we have $\left(t + \frac{(n-1)\pi}{n} + (\beta - 2\theta)\right) > \left(t + \frac{(n-1)\pi}{n} - \beta\right)$. Clearly, $\left(t + \frac{(n-1)\pi}{n} - \beta\right)$ $\in]0, \pi[$. Moreover, by the choice of θ, $\left(t + \frac{(n-1)\pi}{n} + (\beta - 2\theta)\right) \in]0, \pi[$. Since g is a strictly increasing function of the argument on $[0, \pi]$, we have the desired results (12)–(13) in this case.

For $k = 2n - 2$, we first note that we can write $\sigma_{(2n-2, 2n-1)}$ as $\sigma_{(-2,-1)}$ and $\sigma_{(2n-1, 2n)}$ as $\sigma_{(-1, 0)}$. We also note that the sector $\sigma_{(-2,-1)}$ is completely below the x_1-axis, whereas the sector $\sigma_{(-1, 0)}$ is partially above and partially below the x_1-axis. The line L_3 joining the center o of P_t to the point $p_3 = g\left(t - \frac{\pi}{n} + \beta\right)\left(\cos\left(t - \frac{\pi}{n} + \beta\right), \sin\left(t - \frac{\pi}{n} + \beta\right)\right) \in \partial B$ is reflected about $z_{t-\frac{\pi}{n}}$ to the line L_4 joining o to the point $p_4 = g\left(t - \frac{\pi}{n} - \beta\right)\left(\cos\left(t - \frac{\pi}{n} - \beta\right), \sin\left(t - \frac{\pi}{n} - \beta\right)\right) \in \partial B$ (see Fig. 9).

Thus, to prove (14, 15), it suffices to show that $g\left(t - \frac{\pi}{n} + \beta\right) <$ $g\left(t - \frac{\pi}{n} - \beta\right)$. Suppose the point p_3 is above the x_1-axis. Let $r > 0$ be the angle between L_3 and the positive x_1-axis. Then, $g\left(t - \frac{\pi}{n} + \beta\right) = g\left(t - \frac{\pi}{n} + (\beta - 2r)\right)$. Now, $r < \beta$ implies that $\left(t - \frac{\pi}{n} + (\beta - 2r)\right) > \left(t - \frac{\pi}{n} - \beta\right)$. Since g is a strictly decreasing function of the argument in $[\pi, 2\pi]$, we get the desired results (14), (15) in this case.

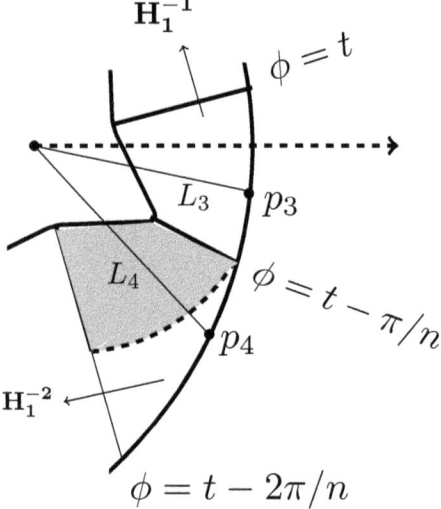

Fig. 9 Reflection of sector H_1^{-1} about the axis $z_{t-\frac{\pi}{n}}$

If the point p_3 is below the x_1-axis, then $2\pi > \left(t - \frac{\pi}{n} + \beta\right) > \left(t - \frac{\pi}{n} - \beta\right) > \pi$, and the fact that g is a strictly decreasing function of the argument in $[\pi, 2\pi]$ gives the desired results (14, 15) in this case. □

5.4 The Rotating Plane Method

Recall here that $n \geq 3$ is a fixed even integer. In order to study the behavior of λ_1 as a function of t, we now analyze the two terms appearing on the right-hand side of (10), which is an expression for $\lambda_1'(t)$. For each $\phi \in [0, \pi]$, by Lemma 3.2, we have

$$\langle \eta_t, v \rangle \left(t + \frac{(k+1)\pi}{n} + \phi\right)$$
$$= - \langle \eta_t, v \rangle \left(t + \frac{(k+1)\pi}{n} - \phi\right) \quad \text{for } k = 0, 2, 4, \ldots, n-2. \quad (16)$$

In particular, (16) holds for each $\phi \in [0, \frac{\pi}{n}]$. In other words, if $x' := R_{t+\frac{(k+1)\pi}{n}}(x)$, then by Eq. (11), for each $k = 0, 2, 4, \ldots, n-2$, $x' \in \partial P_t \cap \sigma_{(k+1,k+2)}$ for each $x \in \partial P_t \cap \sigma_{(k,k+1)}$, and we have $\langle \eta_t, v \rangle (x') = - \langle \eta_t, v \rangle (x) \ \forall \ x \in \partial P_t \cap \sigma_{(k,k+1)}$. Thus, for each $k = 0, 2, 4, \ldots, n-2$, we have the following:

$$\int_{\partial P_t \cap \sigma_{(k,k+1)}} \langle a(x)\eta_t, \eta_t \rangle \left|\frac{\partial y_1(t)}{\partial \eta_t}(x)\right|^2 \langle \eta_t, v \rangle (x) \, d\Sigma$$

$$+ \int_{\partial P_t \cap \sigma_{(k+1,k+2)}} \langle a(x)\eta_t, \eta_t \rangle \left| \frac{\partial y_1(t)}{\partial \eta_t}(x) \right|^2 \langle \eta_t, v \rangle (x) \, d\Sigma$$

$$= \int_{\partial P_t \cap \sigma_{(k,k+1)}} \langle a(x)\eta_t, \eta_t \rangle \left(\left| \frac{\partial y_1(t)}{\partial \eta_t}(x) \right|^2 - \left| \frac{\partial y_1(t)}{\partial \eta_t}(x') \right|^2 \right) \langle \eta_t, v \rangle (x) \, d\Sigma. \tag{17}$$

Now, we know that f is a positive and strictly increasing function of ϕ in $]t + \frac{k\pi}{n}, t + \frac{(k+1)\pi}{n}[$ for each $k = 0, 2, 4, \ldots, n-2$. Thus, applying Lemma 3.2 for $\eta_t = -n$, we get

$$\langle \eta_t, v \rangle > 0 \text{ on } \partial P_t \cap \sigma_{(k,k+1)} \text{ for each } k = 0, 2, 4, \ldots, n-2. \tag{18}$$

Using a similar argument, we have the following: for each $k = n, n+2, \ldots, 2n-2$,

$$\int_{\partial P_t \cap \sigma_{(k,k+1)}} \langle a(x)\eta_t, \eta_t \rangle \left| \frac{\partial y_1(t)}{\partial \eta_t}(x) \right|^2 \langle \eta_t, v \rangle (x) \, d\Sigma$$

$$+ \int_{\partial P_t \cap \sigma_{(k+1,k+2)}} \langle a(x)\eta_t, \eta_t \rangle \left| \frac{\partial y_1(t)}{\partial \eta_t}(x) \right|^2 \langle \eta_t, v \rangle (x) \, d\Sigma$$

$$= \int_{\partial P_t \cap \sigma_{(k+1,k+2)}} \langle a(x)\eta_t, \eta_t \rangle \left(\left| \frac{\partial y_1(t)}{\partial \eta_t}(x) \right|^2 - \left| \frac{\partial y_1(t)}{\partial \eta_t}(x') \right|^2 \right) \langle \eta_t, v \rangle (x) \, d\Sigma, \tag{19}$$

where $x' := R_{t + \frac{(k+1)\pi}{n}}(x)$. Then, for each $k = n, n+2, \ldots, 2n-2$, $x' \in \partial P_t \cap \sigma_{(k,k+1)}$ for each x in $\partial P_t \cap \sigma_{(k+1,k+2)}$. We note that the function f is a positive and strictly increasing function of ϕ in $]t + \frac{(k+2)\pi}{n}, t + \frac{(k+1)\pi}{n}[$ for each $k = n, n+2, \ldots, 2n-2$. Thus, applying Lemma 3.2 for $\eta_t = -n$, we get

$$\langle \eta_t, v \rangle > 0 \text{ on } \partial P_t \cap \sigma_{(k+1,k+2)} \text{ for each } k = n, n+2, \ldots, 2n-2. \tag{20}$$

5.5 Necessary Condition for the Critical Points of λ_1

Recall here that $n \geq 3$ is a fixed even integer. We finally show that $\left\{ \frac{k\pi}{n} \mid k = 0, 1, \ldots n-1 \right\}$ are the only critical points of λ_1 and that between every pair of consecutive critical points of λ_1, it is a strictly monotonic function of the argument. In view of Proposition 5.1 and Eq. (6), it now suffices to study the behavior of λ_1 only on the interval $\left(0, \frac{\pi}{n} \right)$.

Proposition 5.2 (Necessary Condition for Critical Points) *Fix $n \geq 3$, n even. For each $t \in]0, \frac{\pi}{n}[$, $\lambda'_1(t) > 0$.*

Proof Fix $t \in]0, \frac{\pi}{n}[$. Using (17) and (19), integral (10) can be written as

$$\lambda'_1(t) = - \sum_{\substack{0 \leq k \leq n-2 \\ k \text{ even}}} \int_{\partial P_t \cap \sigma_{(k,k+1)}} \langle a(x)\eta_t, \eta_t \rangle \left(\left| \frac{\partial y_1(t)(x)}{\partial \eta_t} \right|^2 - \left| \frac{\partial y_1(t)(x')}{\partial \eta_t} \right|^2 \right)$$

$$\langle \eta_t, v \rangle (x) \, d\Sigma(x)$$

$$- \sum_{\substack{n \leq k \leq 2n-2 \\ k \text{ even}}} \int_{\partial P_t \cap \sigma_{(k+1,k+2)}} \langle a(x)\eta_t, \eta_t \rangle \left(\left| \frac{\partial y_1(t)(x)}{\partial \eta_t} \right|^2 - \left| \frac{\partial y_1(t)(x')}{\partial \eta_t} \right|^2 \right)$$

$$\langle \eta_t, v \rangle (x) \, d\Sigma(x) \tag{21}$$

Let $H(t) := \bigcup_{\substack{0 \leq k \leq n-2 \\ k \text{ even}}} H_1^k(t)$. Let $w(x) := y_1(t)(x) - y_1(t)(x')$. By Lemma 5.1, the real-valued function w is well defined on $H(t)$. Moreover, $w \equiv 0$ on $\partial P_t \cap \partial H(t)$ and also on $\partial H(t) \cap z_{t+k\frac{\pi}{n}}$ for each $k = 1, 3, \ldots n - 1$. That is, $w(x) = 0 \; \forall \; x \in \partial H(t) \cap \left(\partial P_t \cup_{\substack{1 \leq k \leq n-1 \\ k \text{ odd}}} z_{t+\frac{k\pi}{n}} \right)$. Moreover, since $y_1(t)$ vanishes on ∂B and is positive inside $\Omega(t)$, and since for each $k = 0, 2, \ldots n - 2$, the reflection of $\partial H_1^k(t) \cap \partial B$ about the axis $z_{t+(k+1)\frac{\pi}{n}}$ lies completely inside $H_1^{k+1}(t) \subset \Omega(t)$, we obtain $w(x) < 0$ for each x in $(\partial H(t) \cap \partial B) \setminus \left(\bigcup_{\substack{1 \leq k \leq n-1 \\ k \text{ odd}}} z_{t+\frac{k\pi}{n}} \right)$. Now, we claim that

$$w(x) < 0 \; \forall \; x \in \partial H(t) \cap \bigcup_{\substack{0 \leq k \leq n-2 \\ k \text{ even}}} z_{t+\frac{k\pi}{n}}. \tag{22}$$

This is equivalent to saying that for each k, $0 \leq k \leq n - 2$, k even, $w(x) < 0$ for all $x \in \partial H_1^k(t) \cap z_{t+\frac{k\pi}{n}}$. Fix a k_0 such that $0 \leq k_0 \leq n - 2$, k_0 even. Now, the axis of symmetry $z_{t+\frac{(k_0+1)\pi}{n}}$ divides Ω_t into two unequal components. Let us denote the smaller component of the two by $\mathcal{O}_{k_0}(t)$. Then, we have $\mathcal{O}_{k_0}(t) := \Omega_t \cap \sigma_{(-(k_0+1+n),k_0+1)}$. Now, it can be shown that $R_{t+\frac{(k_0+1)\pi}{n}}\left(\mathcal{O}_{k_0}(t)\right) \subset \Omega_t \cap (cl(\mathcal{O}_{k_0}(t)))^c$. Therefore, if we define $w_{k_0}(x) := y_1(t)(x) - y_1(t)(x')$, then the real-valued function w_{k_0} is well defined on $\mathcal{O}_{k_0}(t)$. Here, $x' := R_{t+\frac{(k_0+1)\pi}{n}}(x)$ for $x \in \mathcal{O}_{k_0}(t)$. Moreover, $w \equiv 0$ on $\partial P_t \cap \partial \mathcal{O}_{k_0}(t)$ and also on $\partial \mathcal{O}_{k_0}(t) \cap z_{t+(k_0+1)\frac{\pi}{n}}$. That is, $w_{k_0}(x) = 0 \; \forall \; x \in \partial \mathcal{O}_{k_0}(t) \cap \left(\partial P_t \cup z_{t+\frac{(k_0+1)\pi}{n}} \right)$. Moreover, since $y_1(t)$ vanishes on ∂B and is positive inside Ω_t, and since the reflection of $\partial \mathcal{O}_{k_0}(t) \cap \partial B$ about the

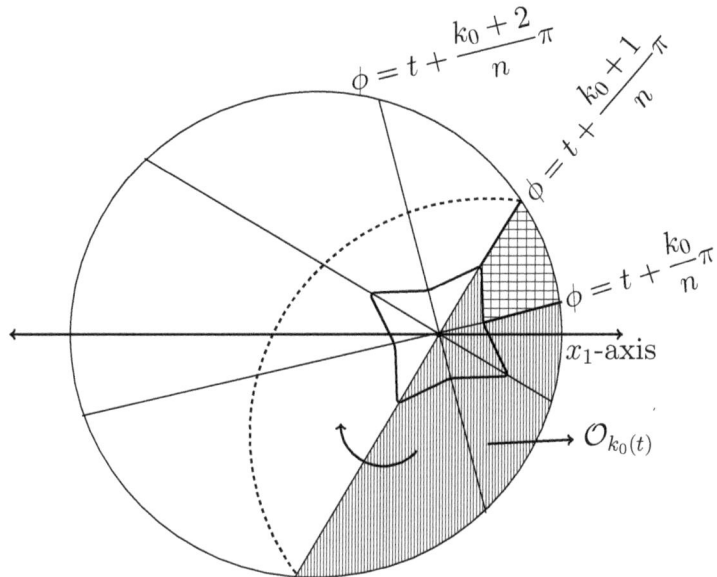

Fig. 10 $\mathcal{O}_{k_0}(t)$ for $n = 4$

axis $z_{t+(k_0+1)\frac{\pi}{n}}$ lies completely inside Ω_t, we have the following: $w_{k_0}(x) < 0 \; \forall \; x \in \left(\partial \mathcal{O}_{k_0}(t) \cap \partial B\right) \setminus z_{t+\frac{(k_0+1)\pi}{n}}$. Therefore, the non-constant function w_{k_0} satisfies

$$-\nabla \cdot (a(x)\nabla w_{k_0}) = \lambda_1(t) \, w_{k_0} \text{ in } \mathcal{O}_{k_0}(t), \; w_{k_0} \leq 0 \text{ on } \partial \mathcal{O}_{k_0}(t). \tag{23}$$

Hence, by the maximum principle, $w_{k_0} < 0$ in $\mathcal{O}_{k_0}(t)$. In particular, $w_{k_0} < 0$ in $\partial H_1^{k_0}(t) \cap z_{t+\frac{k_0\pi}{n}}$. Now, by definition, w and w_{k_0} coincide in $H_1^{k_0}$. Therefore, by continuity of both w and w_{k_0}, we get $w < 0$ in $\partial H_1^{k_0}(t) \cap z_{t+\frac{k_0\pi}{n}}$. But k_0 such that $0 \leq k_0 \leq n-2$, k_0 even, was chosen arbitrarily. This proves our claim (22) (Fig. 10).

Therefore, the non-constant function w satisfies

$$-\nabla \cdot (a(x)\nabla w) = \lambda_1(t) \, w \text{ in } H(t), \; w \leq 0, \text{ on } \partial H(t). \tag{24}$$

Hence, by the maximum principle, w is non-positive on the whole of $H(t)$. Therefore, from (24), we have $\nabla \cdot (a(x)\nabla w) \geq 0$ in $H(t)$. Since w achieves its maximal value zero on $\bigcup_{\substack{0 \leq k \leq n-2 \\ k \equiv 0 \bmod 2}} \left(\partial P_t \cap \sigma_{(k,k+1)}\right) \subset \partial H(t)$, by the Hopf maximum principle, one has $\frac{\partial w}{\partial \eta_t}(x) > 0 \; \forall \; x \in \bigcup_{\substack{0 \leq k \leq n-2 \\ k \equiv 0 \bmod 2}} \left(\partial P_t \cap \sigma_{(k,k+1)}\right)$; that is,

$$\frac{\partial y_1(t)}{\partial \eta_t}(x) - \frac{\partial y_1(t)}{\partial \eta_t}(x') > 0 \ \forall \ x \in \bigcup_{\substack{0 \leq k \leq n-2 \\ k \equiv 0 \bmod 2}} \left(\partial P_t \cap \sigma_{(k,k+1)} \right).$$

Also, by the application of the Hopf maximum principle to problem (1), it follows that

$$\frac{\partial y_1(t)}{\partial \eta_t}(x) < 0 \ \forall \ x \in \partial \Omega_t.$$

Thus,

$$\left| \frac{\partial y_1(t)}{\partial \eta_t}(x) \right|^2 - \left| \frac{\partial y_1(t)}{\partial \eta_t}(x') \right|^2 < 0 \ \forall \ x \in \bigcup_{\substack{0 \leq k \leq n-2 \\ k \equiv 0 \bmod 2}} \left(\partial P_t \cap \sigma_{(k,k+1)} \right). \tag{25}$$

Now, from (25) and (18), it follows that the first term in (21) is strictly positive. Similarly, one can prove using (20) that the second term in (21) is also strictly positive. This proves the proposition for n even. □

5.6 Proof of Theorem 4.1

Theorem 4.1, for n even, now follows from Propositions 5.1 and 5.2 and Eq. (6).

Remark 5.2 In the proof of Lemma 5.1, we considered two consecutive sectors in each of the two hemispheres of the disk B determined by the z_t-axis. We then took the reflection of the smaller sector of this pair into the bigger one about the axis of symmetry separating these two sectors. This was possible because the obstacle P we consider had a $\mathbb{D}_n$ symmetry, where $n \geq 3$ was chosen to be even. As a result, the axes of symmetry of P divide B into even number of sectors in each of these hemispheres. When n is odd, the axes of symmetry of P divide B into odd number of sectors in each of the hemispheres. Therefore, unlike the n even case, it is not possible to find a complete pairing of consecutive sectors within each of the hemispheres. Thus, the proof for n odd still remains an open question. However, we provide some numerical evidence that enables us to make a conjecture that Theorem 4.1 holds true for n odd too.

6 Proof of Theorem 4.2

We now prove our second main result viz., Theorem 4.2. Let $d \geq 0$ denote the distance between the center of the disk and the center of the obstacle P. Let $t \in [0, 2\pi[$ denote the angle by which the obstacle is rotated about its center in the anticlockwise direction starting from the initial configuration as described in Sect. 5.1.1. Clearly, λ_1 is a function of both d and t. Please note that when $d = 0$, the map $t \mapsto \lambda_1(t)$ is constant as the domain remains unaltered. In Theorem 4.1, we have studied the behavior of map $t \mapsto \lambda_1(t)$, for a fixed $d > 0$.

We next analyze the behavior of the map $d \mapsto \lambda_1(d)$ for a fixed $t \in [0, 2\pi[$. Using similar arguments as in [21, Theorem 2.1], it follows that, for a fixed $t \in [0, 2\pi[$, the eigenvalue maximizer is the concentric configuration, i.e., when $d = 0$. As a corollary of Theorem 2.1 in [21], we have the following proposition:

Proposition 6.1 *Fix $n \geq 3$, even or odd. Fix $t \in [0, 2\pi[$. Let x denote the center of the obstacle P_t. Then, at any maximizing x,*

(a) $\Omega = B \setminus P_t$ *has no hyperplane of interior reflection containing x. Moreover, at any maximizing x, either statement (a) above is true, or else*
(b) *the circumcircle C_2 of P_t intersects the small side of ∂B.*

Now, since the disk B enjoys the interior reflection property w.r.t. all secant lines that are not the diameters of B, as a consequence, we have the following result, similar to [21].

Corollary 6.1 *Fix $n \geq 3$, even or odd and $t \in [0, 2\pi[$. Then, (a) the concentric configuration, i.e., $d = 0$, is the only candidate for the maximizer of the map $d \longmapsto \lambda_1(d)$, and (b) at any minimizing configuration of the map $d \longmapsto \lambda_1(d)$, the circumcircle C_2 of the obstacle P_t must touch ∂B.*

Since L is invariant under the isometries of the domain, it follows that for a fixed $t \in [0, 2\pi[$, in order to study the behavior of $d \mapsto \lambda_1(d)$, it is enough to translate the center of the obstacle P_t along the positive x_1-axis. Using an analysis similar to that in [25], it follows that, $d \mapsto \lambda_1(d)$ is maximum for $d = 0$ and is a strictly decreasing function of d in $]0, r_1 - r_0^2[$. Now, Corollary 6.1 along with Theorem 4.1 implies Theorem 4.2 that characterizes the maximizing and the minimizing configurations over the family of domains $\mathcal{G}$. Applying the idea from [21] to the candidates for the minimizing configurations over $\mathcal{G}$ for n even, $n \geq 3$, we get that, at the global minimizing configurations, w.r.t. both the translations of the obstacle within B as well as the rotations of the obstacle about its center, the obstacle must be in an OFF position w.r.t. B with its outer vertex touching ∂B. The global maximizer for $n \geq 3$, even or odd, remains to be the concentric configuration.

7 Numerical Results

In this section, we provide the results of numerical experiments that validate the main results: Theorems 4.1 and 4.2. For this purpose, we consider two different sets of obstacles, a square and a pentagon, that have dihedral symmetry of order $n = 4, 5$. We also choose two different functions for $a(x)$ so that the operator L is invariant under translations and rotations: the first one is the constant function

$a(x) = 1$ and the second one is $a(x) = 0.5 + x^2 + y^2$. To solve the eigenvalue value problem (1) in the domain $\Omega = B \setminus P$, we use the finite element method with P^1 elements (see e.g., [9, 33]) on a mesh with element size $h = 0.018$.

In Test Case 1, we consider the square obstacle P with side length 0.5. To validate Theorem 4.1, we perform rotations of P without any translations, with a fixed value of $d = 0.5$. Furthermore, for validation of Theorem 4.2, we shift P along the x-axis from the point (0,0) to (0.7,0), where it actually touches the boundary of B. The results are shown in Fig. 11.

The first row in Fig. 11 shows the results for $a(x) = 1$, and the second row corresponds to the results for $a(x) = 0.5 + x^2 + y^2$. Figure 11a–d and e–d shows the OFF, intermediate, and ON configurations. The OFF and the ON configurations are the minimizer and the maximizer for λ_1, which is also reflected in Table 1. Furthermore, we see the global extremal configurations in Fig. 11d–e and i–j.

In Test Case 2, we demonstrate the results for a pentagon-shaped phantom P. The side length of the pentagon is 0.4. We follow a similar experimental setup as in the case for the square phantom. The extremal configurations and the values of λ_1 are shown in Fig. 12 and Table 2

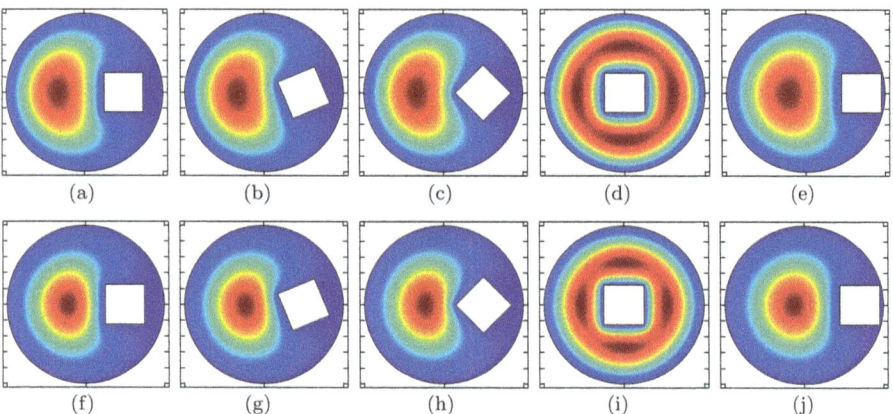

Fig. 11 Test Case 1: simulations of extremal configurations for a square obstacle. (**a**) OFF position. (**b**) Intermediate Position. (**c**) ON position. (**d**) Global maximizer. (**e**) Global minimizer. (**f**) OFF position. (**g**) Intermediate position. (**h**) ON position. (**i**) Global maximizer. (**j**) Global minimizer

Table 1 Values of λ_1 at different configurations for a square obstacle with different values of $a(x)$

Configuration	$\lambda_1(a = 1)$	$\lambda_1(a = 0.5 + x^2 + y^2)$
OFF	9.4137	8.3865
INTERMEDIATE	9.4385	8.4246
ON	9.464	8.4644
GLOBAL MAX	19.118	18.672
GLOBAL MIN	7.7142	6.9886

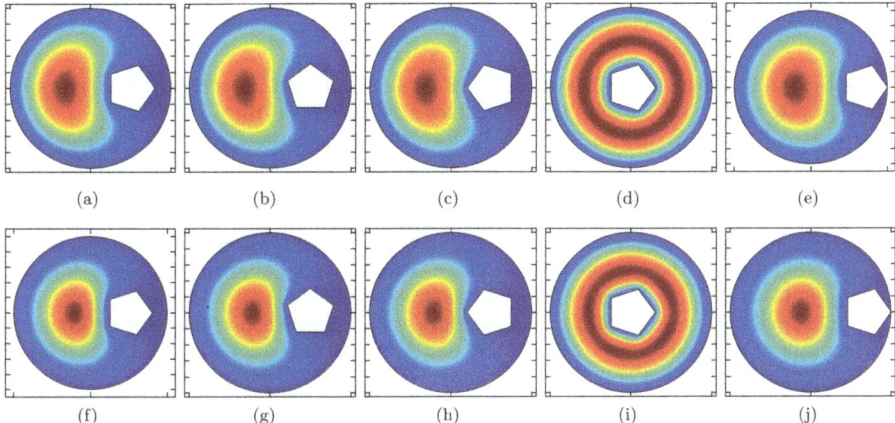

Fig. 12 Test Case 2: simulations of extremal configurations for a pentagon obstacle. (**a**) OFF position. (**b**) Intermediate Position. (**c**) ON position. (**d**) Global maximizer. (**e**) Global minimizer. (**f**) OFF position. (**g**) Intermediate position. (**h**) ON position. (**i**) Global maximizer. (**j**) Global minimizer

Table 2 Values of λ_1 at different configurations for a pentagon-shaped obstacle with different values of $a(x)$

Configuration	$\lambda_1(a=1)$	$\lambda_1(a=0.5+x^2+y^2)$
OFF	9.0895	8.127
INTERMEDIATE	9.0905	8.1303
ON	9.0915	8.1336
GLOBAL MAX	17.619	17.046
GLOBAL MIN	7.523	6.8521

8 Conclusions

In this chapter, we analyze an obstacle placement problem inside a disk, where the obstacle is invariant under the action of a dihedral group. We then characterize the local extremal configurations of the obstacle with respect to the disk for the fundamental eigenvalue of a general divergence form elliptic operator by rotating the obstacle, inside the disk about the fixed center of the obstacle. We prove this result for the case when the obstacle has an even order dihedral symmetry and formulate conjectures about such configurations for obstacles with odd order dihedral symmetry. We further characterize the global maximizing and the global minimizing configurations with respect to the rotations of the obstacle about its center as well as the translations of the obstacle within the disk. Several numerical experiments validate our theoretical findings.

References

1. Adesokan, B., Knudsen, K., Krishnan, V.P., Roy, S.: A fully non-linear optimization approach to acousto-electric tomography. Inverse Problems **34**, 104004 (2018)
2. Aithal, A.R., Raut, R.: On the extrema of Dirichlet's first eigenvalue of a family of punctured regular polygons in two dimensional space forms. Proc. Math. Sci. **122**(2), 257–281 (2012)
3. Aithal, A.R., Sarswat, A.: On a functional connected to the Laplacian in a family of punctured regular polygons in $\mathbb{R}^2$. Indian J. Pure Appl. Math. **45**, 861–874 (2014)
4. Ashbaugh, M.S.: Isoperimetric and universal inequalities for eigenvalues. In: Spectral Theory and Geometry (Edinburgh, 1998). London Mathematical Society Lecture Note Series, vol. 273, pp. 95–139. Cambridge University Press, Cambridge (1999)
5. Ashbaugh, M.S.: Open problems on eigenvalues of the Laplacian. In: Analytic and Geometric Inequalities and Applications. Mathematics and Its Applications, vol. 478, pp. 13–28. Kluwer Academic, Dordrecht (1999)
6. Azegami H.: Solution of shape optimization problem and its application to product design. In: Itou, H., Kimura, M., Chalupecký, V., Ohtsuka, K., Tagami, D., Takada, A. (eds.) Mathematical Analysis of Continuum Mechanics and Industrial Applications. Mathematics for Industry, vol 26. Springer, Singapore (2017)
7. Canevari, G., Majumdar, A., Spicer, A.: Order reconstruction for nematics on squares and hexagons: a Landau-de Gennes study. SIAM J. Appl. Math. **77**(1), 267–293 (2017)
8. Canevari, G., Majumdar, A., Wang, Y.: Order reconstruction for nematics on squares with isotropic inclusions: A Landau-de gennes study. SIAM J. Appl. Math. **79**(4), 1314–1340
9. Chandrashekar, P., Roy, S., Vasudeva Murthy, A.S.: A variational approach to estimate incompressible fluid flows. Proc. Math. Sci. **127**(1), 175–201 (2017)
10. Chorwadwala, A.M.H.: A glimpse of shape optimization problems. Curr. Sci. **112**(7), 1474–1477 (2017)
11. Chorwadwala, A.M.H., Aithal A.R.: Convex polygons and the Isoperimetric problem in simply connected space forms M_κ^2. The Mathematical Intelligencer, accepted
12. Chorwadwala, A.M.H., Aithal, A.R.: On two functionals connected to the Laplacian in a class of doubly connected domains in space-forms. Proc. Indian Acad. Sci. (Math. Sci.) **115**(1), 93–102 (2005)
13. Chorwadwala, A.M.H., Mahadevan, R.: An eigenvalue optimisation problem for the p-Laplacian. Proc. Roy. Soc. Edinburgh Sect. A Math. **145**(6), 1145–1151 (2015)
14. Chorwadwala, A.M.H., Roy, S.: How to place an obstacle having a dihedral symmetry centered at a given point inside a disk so as to optimize the fundamental Dirichlet eigenvalue. J. Optim. Theory Appl. **184**(1), 162–187 (2020)
15. Chorwadwala, A.M.H., Vemuri, M.K.: Two functionals connected to the Laplacian in a class of doubly connected domains of rank one symmetric spaces of non-compact type. Geometriae Dedicata **167**(1), 11–21 (2013)
16. El Soufi, A., Kiwan, R.: Extremal first Dirichlet eigenvalue of doubly connected plane domains and dihedral symmetry. SIAM J. Math. Anal. **39**(4), 1112–1119 (2007)
17. Faber, G., Beweis: Dass unter allen homogenen membranen von gleicher fläche und gleicherspannung die kreisförmige den tiefsten grundton gibt. Sitz. Ber. Bayer. Akad. Wiss. 169–172 (1923)
18. Gupta, M., Mishra, R.K., Roy, S.: Sparse reconstruction of log-conductivity in current density impedance tomography. J. Math. Imag. Vision **62**, 189–205 (2020)
19. Gupta, M., Mishra, R.K., Roy, S.: Sparsity-based nonlinear reconstruction of optical parameters in two-photon photoacoustic computed tomography. Inverse Probl. **37**, 044001 (2021)
20. Haddad, J., Montenegro, M.: On differentiability of eigenvalues of second order elliptic operators on non-smooth domains. J. Differ. Eq. **259**(1), 408–421 (2015)
21. Harrell II, E.M., Kröger, P., Kurata, K.: On the placement of an obstacle or a well so as to optimize the fundamental eigenvalue. SIAM J. Math. Anal. **33**(1), 240–259 (2001)

22. Henrique, L., Antunes, J., Carvalho, J.S.: Shape optimization techniques for musical instrument design. J. Acoust. Soc. Amer. **112**(5), 2210–2210 (2002)
23. Henrot, A.: Minimization problems for eigenvalues of the Laplacian. J. Evol. Equ. **3**, 443–461 (2003)
24. Hersch, J.: The method of interior parallels applied to polygonal or multiply connected membranes, Pacific J. Math. **13**, 1229–1238 (1963)
25. Kesavan, S.: On two functionals connected to the Laplacian in a class of doubly connected domains, Proc. Roy. Soc. Edinburgh **133A**, 617–624 (2003)
26. Klann, E., Ramlau, R., Ring, W.: A Mumford-Shah level-set approach for the inversion and segmentation of SPECT/CT data. Amer. Instit. Math. Sci. **5**(1), 137–166 (2011)
27. Krahn, E.: Über eine von Rayleigh formulierte Minimaleigenschaft des Kreises. Math. Ann. **94**, 97–100 (1925)
28. Majumdar, A., Lewis, A.: A theoreticia's approach to nematic liquid crystals and their applications. In: Variational Methods in Molecular Modeling. Springer, Berlin (2017)
29. Osserman, R.: The isoperimetric inequality. Bull. Amer. Math. Soc. **84**, 1182–1238 (1978)
30. Ramm, A.G., Shivakumar, P.N.: Inequalities for the minimal eigenvalue of the Laplacian in an annulus. Math. Inequalities Appl. **1**(4), 559–563 (1998)
31. Rayleigh, L.: The Theory of Sound, 1st edn. Macmillan, London (1877)
32. Roy, S., Borzì, A.: A new optimisation approach to sparse reconstruction of log-conductivity in acousto-electric tomography. SIAM J. Imag. Sci. **11**(2), 1759–1784 (2018)
33. Roy, S., Chandrashekar, P., Vasudeva Murthy, A.S.: A variational approach to optical flow estimation of unsteady incompressible flows. Int. J. Adv. Eng. Sci. Appl. Math. **7**(3), 149–167 (2015)
34. Serrin, J.: A symmetry problem in potential theory. Arch. Rational Mech. Anal. **43**, 304–318 (1971)
35. Skinner, S.N., Zare-Behtash, H.: State-of-the-art in aerodynamic shape optimisation methods. Appl. Soft Comput. **62**, 933–962 (2018)

Quasi-Monotonicity Formulas for Classical Obstacle Problems with Sobolev Coefficients and Applications

Matteo Focardi, Francesco Geraci, and Emanuele Spadaro

1 Introduction

The structure of free boundaries for classical obstacle problems has been described first by Caffarelli for quadratic energies having suitably regular matrix fields, and it is the resume of his long-term program on the subject (cf. for instance, [3–6] and the books [7, 17] and [21] for more details and references also on related problems). Similar results for smooth nonlinear operators can then be obtained via a freezing argument.

In the last years, such a topic has been investigated in the case in which the quadratic energy involved has a matrix of coefficients either Lipschitz continuous (cf. [11]) or belonging to a fractional Sobolev space (cf. [13]), with parameters suitably related. Let us also mention that obstacle problems for nondegenerate nonlinear variational energies have been studied in [12] through a linearization argument and the quoted results in the Lipschitz quadratic case.

The papers [11] and [13] follow the variational approach to free boundary analysis developed remarkably by Weiss [23] and Monneau [20], which is based on (quasi-)monotonicity formulas. The extensions of Weiss and Monneau's monotonicity formulas, obtained in [11] and [13], hinge upon a generalization of Rellich and Nečas' inequality due to Payne and Weinberger (cf. [18]). On a technical side, they involve the differentiation of the matrix field.

The aim of this short note is to extend the range of validity of Weiss and Monneau's type quasi-monotonicity formulas to classical obstacle problems, involving

---

M. Focardi (✉) · F. Geraci
DiMaI, Università degli Studi di Firenze, Firenze, Italy
e-mail: matteo.focardi@unifi.it

E. Spadaro
Dipartimento di Matematica, Università di Roma La Sapienza, Roma, Italy
e-mail: spadaro@mat.uniroma1.it

quadratic forms having a matrix of coefficients in a Sobolev space with summability exponent lager than the space dimension.

The main difference contained in the present note, with respect to the existing literature, concerns the (quasi-)monotone quantity itself. Indeed, rather than considering the natural quadratic energy associated with the obstacle problem under study, we establish quasi-monotonicity for a related constant coefficient quadratic form. The latter result is obtained thanks to a freezing argument inspired by some computations in a paper by Monneau (cf. [20, Section 6]) in combination with the well-known quadratic lower bound on the growth of solutions from free boundary points (see Sects. 4 and 5 for more details). Such an insight, though elementary, has been overlooked in the literature and enables us to obtain Weiss and Monneau's quasi-monotonicity formulas under mild assumptions (cf. (H1) and (H3) below, the latter having no role, if the obstacle function is null), since the matrix field is not differentiated along the derivation process of the quasi-monotonicity formulas. We stress again that the mentioned quasi-monotonicity formulas are instrumental to pursue the variational approach for the analysis of the corresponding free boundaries in classical obstacle problems.

Following the approach developed in [12], by means of Theorem 1, we generalize the results in [12, Theorem 3.8] there to infer similar results for nondegenerate, nonlinear classical obstacle problems (cf. Theorem 4 for more details).

To conclude this introduction, we briefly resume the structure of the chapter: Weiss and Monneau's quasi-monotonicity formulas, the main results of the chapter, together with their application to free boundaries, are stated in Sect. 2. Several preliminaries for the classical obstacle problem under study are collected in Sect. 3. The mentioned generalizations of Weiss and Monneau's quasi-monotonicity formulas are established in Sects. 4 and 5, respectively. Finally, Sect. 6 contains the proof of the quoted applications to the free boundary stratification for quadratic and nonlinear problems.

2 Statement of the Main Results

In this section, we state Weiss and Monneau's type quasi-monotonicity formulas for the quadratic problem and their application to the free boundary analysis.

We start off with introducing the variational problem related to free boundaries together with the necessary notations and assumptions in the next section.

2.1 Free Boundary Analysis: Statement

We consider the functional $\mathcal{E} : W^{1,2}(\Omega) \to \mathbb{R}$ given by

$$\mathcal{E}(v) := \int_\Omega \left(\langle \mathbb{A}(x) \nabla v(x), \nabla v(x) \rangle + 2h(x) v(x) \right) dx \qquad (1)$$

and study regularity issues related to its unique minimizer w on the set

$$\mathbb{K}_{\psi,g} := \left\{ v \in W^{1,2}(\Omega) : v \geq \psi \ \mathcal{L}^n\text{-a.e. on } \Omega, \ \text{Tr}(v) = g \text{ on } \partial\Omega \right\}.$$

Here, $\Omega \subset \mathbb{R}^n$ is a bounded Lipschitz open set, $n \geq 2$, $\psi \in C^{1,1}_{loc}(\Omega)$ and $g \in H^{1/2}(\partial\Omega)$ are such that $\psi \leq g$ $\mathcal{H}^{n-1}$-a.e. on $\partial\Omega$, $\mathbb{A} : \Omega \to \mathbb{R}^{n \times n}$ is a matrix-valued field and $f : \Omega \to \mathbb{R}$ is a function satisfying:

(H1) $\mathbb{A} \in W^{1,p}(\Omega; \mathbb{R}^{n \times n})$ with $p > n$.

(H2) $\mathbb{A}(x) = \left(a_{ij}(x)\right)_{i,j=1,\ldots,n}$ is symmetric, continuous and coercive, that is, $a_{ij}(x) = a_{ji}(x)$ for all $x \in \Omega$ and for all $i, j \in \{1, \ldots, n\}$, and for some $\Lambda \geq 1$,

$$\Lambda^{-1}|\xi|^2 \leq \langle \mathbb{A}(x)\xi, \xi \rangle \leq \Lambda|\xi|^2 \quad (2)$$

for all $x \in \Omega$, $\xi \in \mathbb{R}^n$.

(H3) $f := h - \text{div}\,(\mathbb{A}\nabla\psi) > c_0 \ \mathcal{L}^n$-a.e. on Ω, for some $c_0 > 0$, and f is Dini-continuous, namely

$$\int_0^1 \frac{\omega_f(t)}{t} \, dt < \infty, \quad (3)$$

where $\omega_f(t) := \sup_{x,y \in \Omega, \, |x-y| \leq t} |f(x) - f(y)|$.

In some instances in place of (H3), we will require the stronger condition.

(H4) $f > c_0 \ \mathcal{L}^n$-a.e. on Ω, for some $c_0 > 0$, and f is double Dini-continuous, that is,

$$\int_0^1 \frac{\omega_f(r)}{r} |\log r|^a \, dr < \infty, \quad (4)$$

for some $a \geq 1$.

Note that for the zero obstacle problem, i.e. $\psi = 0$, assumptions (H3) and (H4) involve only the lower order term h in the integrand and not the matrix field of coefficients $\mathbb{A}$. Moreover, the positivity condition on f corresponds to the concavity assumption on the obstacle function in the case of the Laplacian. Elementary examples show that it is needed to enforce free boundary regularity.

Given the assumptions introduced above, we provide a full free boundary stratification result.

Theorem 1 *Assume* (H1)–(H4) *to hold, and let w be the (unique) minimizer of $\mathcal{E}$ in* (1) *on $\mathbb{K}_{\psi,g}$.*

Then, w is $W^{2,p}_{loc} \cap C^{1,1-n/p}_{loc}(\Omega)$, and the free boundary can be decomposed as $\partial \{w = \psi\} \cap \Omega = \text{Reg}(w) \cup \text{Sing}(w)$, where $\text{Reg}(w)$ and $\text{Sing}(w)$ are called its regular and singular parts, respectively. Moreover, $\text{Reg}(w) \cap \text{Sing}(w) = \emptyset$ and

(i) *if $a > 2$ in* (H4), *then* $\mathrm{Reg}(w)$ *is relatively open in* $\partial\{w = \psi\}$ *and, for every point* $x_0 \in \mathrm{Reg}(w)$, *there exists a radius* $r = r(x_0) > 0$ *such that* $\partial\{w = \psi\} \cap B_r(x_0)$ *is a* C^1 $(n-1)$*-dimensional manifold with normal vector absolutely continuous.*
In particular, if f is Hölder continuous, there exists $r = r(x_0) > 0$ *such that* $\partial\{w = \psi\} \cap B_r(x_0)$ *is a* $C^{1,\beta}$ $(n-1)$*-dimensional manifold for some exponent* $\beta \in]0, 1[$.

(ii) *if $a \geq 1$ in* (H4), *then* $\mathrm{Sing}(w) = \cup_{k=0}^{n-1} S_k$, *with S_k contained in the union of at most countably many submanifolds of dimension k and class* C^1.

Remark 1 The fine structure of the set of singular points in the case of the Dirichlet energy has been the object of intense research in recent years. In particular, a logarithmic epiperimetric inequality has been established by Colombo, Spolaor and Velichkov [8], in turn implying a $C^{1,\log}$-regularity of the underlying manifolds in item (ii) above. Such a modulus of continuity is essentially sharp as shown by counterexamples due to Figalli and Serra [9]. In the latter paper, the stratification result is further detailed: singular points are locally contained in a C^2 curve in dimension $n = 2$ and in $C^{1,1}$ manifolds (or in countably many C^2 manifolds) in dimension $n \geq 3$ up to the presence of anomalous points of higher codimension.

Finally, Figalli et al. [10] have proven that, generically, the singular set has null $\mathcal{H}^{n-4}$ measure. In particular, it has codimension 3 inside the free boundary, and in dimension $n \leq 4$ the free boundary is generically a C^∞ manifold (in dimension $n = 2$, this had been proven by Monneau [20]).

2.2 Quasi-Monotonicity Formulas: Statements

Theorem 1 is a consequence of Weiss and Monneau's type quasi-monotonicity formulas that will be stated in this section (cf. Sect. 6 for the proofs). With this aim, we introduce some notation.

We first reduce ourselves to the zero obstacle problem. Let w be the unique minimizer of $\mathcal{E}$ in (1) over $\mathbb{K}_{\psi,g}$, and define $u := w - \psi$. Then, u is the unique minimizer of

$$\mathscr{E}(v) := \int_\Omega \left(\langle \mathbb{A}(x)\nabla v(x), \nabla v(x)\rangle + 2f(x)v(x) \right) dx, \tag{5}$$

over

$$\mathbb{K}_{\psi,g} := \left\{ v \in W^{1,2}(\Omega) : v \geq 0 \ \mathcal{L}^n\text{-a.e. on } \Omega, \ \mathrm{Tr}(v) = g - \psi \text{ on } \partial\Omega \right\},$$

where $f = h - \mathrm{div}\,(\mathbb{A}\nabla\psi)$. Clearly, $\partial\{w = \psi\} \cap \Omega = \partial\{u = 0\} \cap \Omega =: \Gamma_u$, and therefore we shall establish all the results in Theorem 1 for u (notice that assumptions (H3) and (H4) are formulated exactly in terms of f).

Let $x_0 \in \Gamma_u$ be any point of the free boundary; then, the affine change of variables

$$x \mapsto x_0 + f^{-1/2}(x_0)\mathbb{A}^{1/2}(x_0)x =: x_0 + \mathbb{L}(x_0)\,x$$

leads to

$$\mathscr{E}(u) = f^{1-\frac{n}{2}}(x_0) \det(\mathbb{A}^{1/2}(x_0))\,\mathscr{E}_{\mathbb{L}(x_0)}(u_{\mathbb{L}(x_0)}), \tag{6}$$

where $\Omega_{\mathbb{L}(x_0)} := \mathbb{L}^{-1}(x_0)\,(\Omega - x_0)$, and we have set

$$\mathscr{E}_{\mathbb{L}(x_0)}(v) := \int_{\Omega_{\mathbb{L}(x_0)}} \left(\langle \mathbb{C}_{x_0} \nabla v, \nabla v \rangle + 2\frac{f_{\mathbb{L}(x_0)}}{f(x_0)} v \right) dx, \tag{7}$$

with

$$u_{\mathbb{L}(x_0)}(x) := u\big(x_0 + \mathbb{L}(x_0)x\big), \tag{8}$$
$$f_{\mathbb{L}(x_0)}(x) := f\big(x_0 + \mathbb{L}(x_0)x\big),$$
$$\mathbb{C}_{x_0}(x) := \mathbb{A}^{-1/2}(x_0)\mathbb{A}(x_0 + \mathbb{L}(x_0)x)\mathbb{A}^{-1/2}(x_0).$$

Note that $f_{\mathbb{L}(x_0)}(0) = f(x_0)$ and $\mathbb{C}_{x_0}(0) = \mathrm{Id}$. Moreover, the free boundary is transformed under this map into

$$\Gamma_{u_{\mathbb{L}(x_0)}} := \mathbb{L}^{-1}(x_0)(\Gamma_u - x_0),$$

and the energy $\mathscr{E}$ in (5) is minimized by u, if and only if $\mathscr{E}_{\mathbb{L}(x_0)}$ in (7) is minimized by the function $u_{\mathbb{L}(x_0)}$ in (8).

In addition, writing the Euler–Lagrange equation for $u_{\mathbb{L}(x_0)}$ in non-divergence form, we get $\mathscr{L}^n$-a.e. on $\Omega_{\mathbb{L}(x_0)}$:

$$c_{ij}(x)\frac{\partial^2 u_{\mathbb{L}(x_0)}}{\partial x_i \partial x_j} + \mathrm{div}\,\mathbb{C}^i_{x_0}(x)\frac{\partial u_{\mathbb{L}(x_0)}}{\partial x_i} = \frac{f_{\mathbb{L}(x_0)}(x)}{f(x_0)} \chi_{\{u_{\mathbb{L}(x_0)}>0\}}$$

(using Einstein's convention) with $\mathbb{C}_{x_0} = (c_{ij})_{i,j=1,\ldots,n}$. Moreover, we may further rewrite the latter equation $\mathscr{L}^n$-a.e. on $\Omega_{\mathbb{L}(x_0)}$ as

$$\Delta u_{\mathbb{L}(x_0)} = 1 + \left(\frac{f_{\mathbb{L}(x_0)}(x)}{f(x_0)}\chi_{\{u_{\mathbb{L}(x_0)}>0\}} - 1\right)$$
$$- \big(c_{ij}(x) - \delta_{ij}\big)\frac{\partial^2 u_{\mathbb{L}(x_0)}}{\partial x_i \partial x_j} - \mathrm{div}\,\mathbb{C}^i_{x_0}(x)\frac{\partial u_{\mathbb{L}(x_0)}}{\partial x_i}\right) =: 1 + f_{x_0}(x). \tag{9}$$

Consider next Weiss' type boundary adjusted energy

$$\Phi_u(x_0, r) := \frac{1}{r^{n+2}} \int_{B_r} \left(|\nabla u_{\mathbb{L}(x_0)}|^2 + 2 u_{\mathbb{L}(x_0)} \right) dx - \frac{2}{r^{n+3}} \int_{\partial B_r} u_{\mathbb{L}(x_0)}^2 \, d\mathcal{H}^{n-1}, \tag{10}$$

for $x_0 \in \Gamma_u$. We claim its quasi-monotonicity.

Theorem 2 (Weiss' Quasi-Monotonicity Formula) *Under assumptions* (H1)–(H3), *for every compact set* $K \subset \Omega$, *there exists a positive constant* $C = C(n, p, \Lambda, c_0, K, \|f\|_{L^\infty}, \|\mathbb{A}\|_{W^{1,p}}) > 0$ *such that for all* $x_0 \in K \cap \Gamma_u$

$$\frac{d}{dr}\left(\Phi_u(x_0, r) + C \int_0^r \frac{\omega(t)}{t} dt\right) \geq \frac{2}{r^{n+4}} \int_{\partial B_r} (\langle \nabla u_{\mathbb{L}(x_0)}, x\rangle - 2 u_{\mathbb{L}(x_0)})^2 d\mathcal{H}^{n-1}, \tag{11}$$

for $\mathcal{L}^1$-*a.e.* $r \in]0, \frac{1}{2}\mathrm{dist}(K, \partial\Omega)[$, *where* $\omega(r) := \omega_f(r) + r^{1-\frac{n}{p}}$.

In particular, $\Phi_u(x_0, \cdot)$ *has finite right limit* $\Phi_u(x_0, 0^+)$ *in zero, and for all* $r \in]0, \frac{1}{2}\mathrm{dist}(K, \partial\Omega)[$,

$$\Phi_u(x_0, r) - \Phi_u(x_0, 0^+) \geq -C \int_0^r \frac{\omega(t)}{t} dt. \tag{12}$$

We recall that Weiss' original monotonicity formula for the Dirichlet energy provides an explicit expression for the derivative of $\Phi_u(x_0, \cdot)$. Namely, formula (11) is actually an equality for u, rather than for $u_{\mathbb{L}(x_0)}$, and ω is null.

Next, we introduce a second quasi-monotonicity formula to analyze a distinguished subset of points of the free boundary, that of singular points Sing(u). Namely, we assume that $x_0 \in \Gamma_u$ satisfies

$$\Phi_u(x_0, 0^+) = \Phi_v(\underline{0}, 1) \tag{13}$$

for some 2-homogeneous solution v of

$$\Delta v = 1 \quad \text{on } \mathbb{R}^n. \tag{14}$$

Note that, by 2-homogeneity, elementary calculations lead to

$$\Phi_v(\underline{0}, r) = \Phi_v(\underline{0}, 1) = \int_{B_1} v \, dy, \tag{15}$$

for all $r > 0$.

Theorem 3 (Monneau's Quasi-Monotonicity Formula) *Under hypotheses* (H1), (H2) *and* (H4) *with* $a = 1$, *if* $K \subset \Omega$ *is a compact set and* (15) *holds for* $x_0 \in K \cap \Gamma_u$, then a constant $C = C(n, p, \Lambda, c_0, K, \|f\|_{L^\infty}, \|\mathbb{A}\|_{W^{1,p}}) > 0$ exists such that the function

$$]0, \tfrac{1}{2}\mathrm{dist}(K, \partial\Omega)[\ni r \longmapsto \frac{1}{r^{n+3}} \int_{\partial B_r} (u_{L(x_0)} - v)^2 \, dx + C \int_0^r \frac{dt}{t} \int_0^t \frac{\omega(s)}{s} \, ds \quad (16)$$

is nondecreasing, where v is any 2-homogeneus polynomial solution of (14), and ω is the modulus of continuity provided by Theorem 2.

3 Preliminaries on the Classical Obstacle Problem

Throughout the section, we use the notation introduced in Sect. 2 and adopt Einstein' summation convention.

The next result has been established by Ural'tseva (cf., for instance, [22, Theorem 2.1]) for general variational inequalities with a penalization method. Our argument instead follows the approach in [11], inspired by the ideas of Weiss for the Laplacian in [23]. Let us briefly sketch our arguments. Consider the minimizer u of the energy $\mathcal{E}$ introduced in (5). It turns out that u satisfies a PDE both in the distributional sense and $\mathcal{L}^n$-a.e. on Ω, and elliptic regularity then applies to establish the smoothness of u itself.

Proposition 1 *Let u be the minimum of $\mathcal{E}$ on $\mathbb{K}_{\psi,g}$. Then,*

$$\mathrm{div}(\mathbb{A}\nabla u) = f \chi_{\{u>0\}} \quad (17)$$

$\mathcal{L}^n$-a.e. on Ω and in $\mathcal{D}'(\Omega)$. Moreover, $u \in W^{2,p}_{loc} \cap C^{1,1-\frac{n}{p}}_{loc}(\Omega)$.

Proof For the validity of (17), we refer to [12, Proposition 3.2], where the result is proven in the broader context of variational inequalities (see also [11, Proposition 2.2]).

From this, by taking into account that $\mathbb{A} \in C^{0,1-n/p}_{loc}(\Omega, \mathbb{R}^{n \times n})$ in view of Morrey embedding theorem, Schauder estimates yield $u \in C^{1,1-n/p}_{loc}(\Omega)$ (cf. [16, Theorem 3.13]).

Next, consider the equation

$$a_{ij} \frac{\partial^2 v}{\partial x_i \partial x_j} = f \chi_{\{u>0\}} - \mathrm{div}\mathbb{A}^j \frac{\partial u}{\partial x_j} =: \varphi, \quad (18)$$

where $\mathbb{A}^j$ denotes the j-column of $\mathbb{A}$. Being $\nabla u \in L^\infty_{loc}(\Omega, \mathbb{R}^n)$ and being $\mathrm{div}\,\mathbb{A}^j \in L^p(\Omega)$ for all $j \in \{1, \ldots, n\}$, then $\varphi \in L^p_{loc}(\Omega)$. The work [14, Corollary 9.18] implies the uniqueness of a solution $v \in W^{2,p}_{loc}(\Omega)$ to (18). By taking into account the identity $\text{Tr}(\mathbb{A}\nabla^2 v) = \text{div}(\mathbb{A}\nabla v) - \text{div}\mathbb{A}^j \frac{\partial v}{\partial x_j}$, (18) can be rewritten as

$$\text{div}(\mathbb{A}\nabla v) - \text{div}\mathbb{A}^j \frac{\partial v}{\partial x_j} = \varphi,$$

and then we have that u and v are two solutions. Then, by Miranda [19, Theorem 1.I] and (17), we deduce that $u = v$.

We recall next the standard notations for the coincidence set and for the corresponding free boundary

$$\Lambda_u := \{x \in \Omega : u(x) = 0\}, \qquad \Gamma_u := \partial \Lambda_u \cap \Omega.$$

For any point $x_0 \in \Gamma_u$, we introduce the family of rescaled functions

$$u_{x_0,r}(x) := \frac{u(x_0 + rx)}{r^2}$$

for $x \in \frac{1}{r}(\Omega - \{x_0\})$. The existence of $C^{1,\gamma}$-limits as $r \downarrow 0$ of the latter family is standard by noting that the rescaled functions satisfy an appropriate PDE and then uniform $W^{2,p}$ estimates.

Proposition 2 ([13, Proposition 4.1]) *Let u be the unique minimizer of $\mathcal{E}$ over $\mathbb{K}_{\psi,g}$ and $K \subset \Omega$ a compact set. Then, for every $x_0 \in K \cap \Gamma_u$, and for every $R > 0$, there exists a constant $C = C(n, p, \Lambda, R, K, \|f\|_{L^\infty}, \|\mathbb{A}\|_{W^{1,p}}) > 0$ such that, for every $r \in]0, \frac{1}{4R}\text{dist}(K, \partial\Omega)[$,*

$$\|u_{x_0,r}\|_{W^{2,p}(B_R)} \leq C. \tag{19}$$

In particular, $(u_{x_0,r})_r$ is equibounded in $C^{1,\gamma}_{\text{loc}}$ for $\gamma \in]0, 1 - n/p]$.

Then, up to extracting a subsequence, the rescaled functions have limits in the $C^{1,\gamma}$ topology. The functions arising in this process are called *blow-up limits*.

Corollary 1 (Existence of Blow-Ups) *Let u be the unique minimizer of $\mathcal{E}$ over $\mathbb{K}_{\psi,g}$, and let $x_0 \in \Gamma_u$. Then, for every sequence $r_k \downarrow 0$, there exists a subsequence $(r_{k_j})_j \subset (r_k)_k$ such that the rescaled functions $(u_{x_0,r_{k_j}})_j$ converge in $C^{1,\gamma}_{\text{loc}}$, $\gamma \in]0, 1 - n/p[$, to some function belonging to $C^{1,1-n/p}_{\text{loc}}$.*

Elementary growth conditions of the solution from free boundary points are easily deduced from Proposition 2 and the condition $p > n$. In turn, such properties will be crucial in the derivation of the quasi-monotonicity formulas.

Proposition 3 *Let u be the unique minimizer of $\mathcal{E}$ over $\mathbb{K}_{\psi,g}$. Then, for all compact sets $K \subset \Omega$, a constant $C = C(n, p, \Lambda, K, \|f\|_{L^\infty}, \|\mathbb{A}\|_{W^{1,p}}) > 0$ exists, such that* *for all points* $x_0 \in \Gamma_u \cap K$, *and for all* $r \in \,]0, \frac{1}{2}\mathrm{dist}(K, \partial\Omega)[$, *it holds*

$$\|u\|_{L^\infty(B_r(x_0))} \leq C\, r^2, \qquad \|\nabla u\|_{L^\infty(B_r(x_0), \mathbb{R}^n)} \leq C\, r \qquad (20)$$

and

$$\|\nabla^2 u\|_{L^p(B_r(x_0), \mathbb{R}^{n\times n})} \leq C\, r^{n/p}. \qquad (21)$$

Finally, we recall the fundamental quadratic detachment property from free boundary points that entails non-triviality of blow-up limits. It has been established by Blank and Hao in [1, Theorem 3.9] under the sole boundedness and measurability assumptions on the matrix field $\mathbb{A}$, hypotheses clearly weaker than (H1).

Lemma 1 ([1, Theorem 3.9]) *There exists a positive constant* ϑ, *with* $\vartheta = \vartheta(n, \Lambda, c_0, \|f\|_{L^\infty})$, *such that for every* $x_0 \in \Gamma_u$ *and* $r \in \,]0, \frac{1}{2}\mathrm{dist}(x_0, \partial\Omega)[$, *it holds*

$$\sup_{x \in \partial B_r(x_0)} u(x) \geq \vartheta\, r^2.$$

4 Weiss' Quasi-Monotonicity Formula: Proof of Theorem 2

In this section, we prove the quasi-monotonicity of Weiss' energy $\Phi_u(x_0, \cdot)$ defined in (10). The proof is based on equality (9) and Proposition 3.

Proof of Theorem 2 We analyse separately the volume and the boundary terms appearing in the definition of the Weiss energy in (10). For the sake of notational simplicity, we write u_{x_0} in place of $u_{\mathbb{L}(x_0)}$. In what follows, with C we denote a constant $C = C(n, p, \Lambda, c_0, K, \|f\|_{L^\infty}, \|\mathbb{A}\|_{W^{1,p}}) > 0$ that may vary from line to line.

We start off with the bulk term. The Coarea formula implies for $\mathcal{L}^1$-a.e. $r \in \,]0, \mathrm{dist}(K, \partial\Omega)[$

$$\frac{d}{dr}\left(\frac{1}{r^{n+2}} \int_{B_r} \left(|\nabla u_{x_0}|^2 + 2 u_{x_0}\right) dx\right) =$$

$$- \frac{n+2}{r^{n+3}} \int_{B_r} \left(|\nabla u_{x_0}|^2 + 2 u_{x_0}\right) dx + \frac{1}{r^{n+2}} \int_{\partial B_r} \left(|\nabla u_{x_0}|^2 + 2 u_{x_0}\right) dx. \qquad (22)$$

We use the divergence theorem together with the following Identities:

$$|\nabla u_{x_0}|^2 = \frac{1}{2}\operatorname{div}(\nabla(u_{x_0}^2)) - u_{x_0}\Delta u_{x_0},$$

$$\operatorname{div}\left(|\nabla u_{x_0}|^2 \frac{x}{r}\right) = \frac{n-2}{r}|\nabla u_{x_0}|^2 - 2\Delta u_{x_0}\langle\nabla u_{x_0}, \frac{x}{r}\rangle + 2\operatorname{div}\left(\langle\nabla u_{x_0}, \frac{x}{r}\rangle\nabla u_{x_0}\right),$$

$$\operatorname{div}\left(u_{x_0}\frac{x}{r}\right) = u_{x_0}\frac{n}{r} + \langle\nabla u_{x_0}, \frac{x}{r}\rangle,$$

to deal with the first, third and fourth addends in (22), respectively. Hence, we can rewrite the right-hand side of equality (22) as follows:

$$\frac{d}{dr}\left(\frac{1}{r^{n+2}}\int_{B_r}\left(|\nabla u_{x_0}|^2 + 2u_{x_0}\right)dx\right)$$

$$= \frac{2}{r^{n+2}}\int_{B_r}(\Delta u_{x_0} - 1)\left(2\frac{u_{x_0}}{r} - \langle\nabla u_{x_0}, \frac{x}{r}\rangle\right)dx$$

$$+ \frac{2}{r^{n+2}}\int_{\partial B_r}\langle\nabla u_{x_0}, \frac{x}{r}\rangle^2 d\mathcal{H}^{n-1} - \frac{4}{r^{n+2}}\int_{\partial B_r}\frac{u_{x_0}}{r}\langle\nabla u_{x_0}, \frac{x}{r}\rangle d\mathcal{H}^{n-1}. \qquad (23)$$

We consider next the boundary term in the expression of Φ_u. By scaling and a direct calculation, we get

$$\frac{d}{dr}\left(\frac{2}{r^{n+3}}\int_{\partial B_r}u_{x_0}^2 d\mathcal{H}^{n-1}\right) \stackrel{x=ry}{=} 2\int_{\partial B_1}\frac{d}{dr}\left(\frac{u_{x_0}(ry)}{r^2}\right)^2 d\mathcal{H}^{n-1}$$

$$= 4\int_{\partial B_1}\frac{u_{x_0}(ry)}{r^4}\left(\langle\nabla u_{x_0}(ry), y\rangle - 2\frac{u_{x_0}(ry)}{r}\right)d\mathcal{H}^{n-1}$$

$$\stackrel{x=ry}{=} \frac{4}{r^{n+2}}\int_{\partial B_r}\frac{u_{x_0}}{r}\langle\nabla u_{x_0}, \frac{x}{r}\rangle d\mathcal{H}^{n-1} - \frac{8}{r^{n+2}}\int_{\partial B_r}\frac{u_{x_0}^2}{r^2}d\mathcal{H}^{n-1}. \qquad (24)$$

Then, by combining together equations (23) and (24) and recalling equation (9), we obtain

$$\Phi_u'(x_0, r) = \frac{2}{r^{n+2}}\int_{B_r}f_{x_0}\left(2\frac{u_{x_0}}{r} - \langle\nabla u_{x_0}, \frac{x}{r}\rangle\right)dx$$

$$+ \frac{2}{r^{n+2}}\int_{\partial B_r}\left(\langle\nabla u_{x_0}, \frac{x}{r}\rangle - 2\frac{u_{x_0}}{r}\right)^2 d\mathcal{H}^{n-1}$$

$$= \frac{2}{r^{n+2}}\int_{B_r\setminus\Lambda_{u_{x_0}}}f_{x_0}\left(2\frac{u_{x_0}}{r} - \langle\nabla u_{x_0}, \frac{x}{r}\rangle\right)dx$$

$$+ \frac{2}{r^{n+2}}\int_{\partial B_r}\left(\langle\nabla u_{x_0}, \frac{x}{r}\rangle - 2\frac{u_{x_0}}{r}\right)^2 d\mathcal{H}^{n-1},$$

where in the last equality we used the unilateral obstacle condition to deduce that $\Lambda_{u_{x_0}} \subseteq \{\nabla u_{x_0} = \underline{0}\}$. Therefore, by the growth of u and ∇u from x_0 in (20), we obtain

$$\Phi'_u(x_0, r) \geq -\frac{C}{r^{n+1}} \int_{B_r \setminus \Lambda_{u_{x_0}}} |f_{x_0}| \, dx + \frac{2}{r^{n+2}} \int_{\partial B_r} \left(\langle \nabla u_{x_0}, \frac{x}{r}\rangle - 2\frac{u_{x_0}}{r}\right)^2 d\mathcal{H}^{n-1}. \tag{25}$$

Next, note that by (H1) and (H3), and by the very definition of f_{x_0} in (9), it follows that

$$\frac{1}{r^{n+1}} \int_{B_r \setminus \Lambda_{u_{x_0}}} |f_{x_0}| \, dx \leq \frac{\omega_f(r)}{c_0 r} + \frac{C}{r^{n(1+\frac{1}{p})}} \int_{B_r} |\nabla^2 u_{x_0}| \, dx + \frac{C}{r^n} \int_{B_r} |\text{div }\mathbb{C}_{x_0}| \, dx. \tag{26}$$

By (21), we estimate the second addend on the right-hand side of the last inequality as follows:

$$\frac{1}{r^{n(1+\frac{1}{p})}} \int_{B_r} |\nabla^2 u_{x_0}| \, dx \leq \frac{C}{r^{n(1+\frac{1}{p})}} \|\nabla^2 u_{x_0}\|_{L^p(B_r, \mathbb{R}^{n\times n})} (\omega_n r^n)^{1-\frac{1}{p}} \leq C r^{-\frac{n}{p}}, \tag{27}$$

and by Hölder inequality, we get for the third addend

$$\frac{1}{r^n} \int_{B_r} |\text{div }\mathbb{C}_{x_0}| \, dx \leq \frac{1}{r^n} \|\text{div }\mathbb{C}_{x_0}\|_{L^p(B_r, \mathbb{R}^n)} (\omega_n r^n)^{1-\frac{1}{p}} \leq C r^{-\frac{n}{p}}. \tag{28}$$

Therefore, we conclude from (25)–(28)

$$\Phi'_u(x_0, r) \geq -C\frac{\omega(r)}{r} + \frac{2}{r^{n+2}} \int_{\partial B_r} \left(\langle \nabla u_{x_0}, \frac{x}{r}\rangle - 2\frac{u_{x_0}}{r}\right)^2 d\mathcal{H}^{n-1},$$

where $\omega(r) := \omega_f(r) + r^{1-\frac{n}{p}}$.

Remark 2 Recalling that f is Dini-continuous by (H3), the modulus of continuity ω provided by Theorem 2 is in turn Dini-continuous as $p > n$.

Remark 3 More generally, the argument in Theorem 2 works for solutions to second-order elliptic PDEs in non-divergence form of the type

$$a_{ij}(x) u_{ij} + b_i(x) u_i + c(x) u = f(x)\chi_{\{u>0\}},$$

the only difference with the statement of Theorem 2 being that in this framework $\omega(r) := \omega_f(r) + r^{1-\frac{n}{p}} + r^2 \sup_{B_r} c$ (cf. [20, Appendix]).

5 Monneau's Quasi-Monotonicity Formula: Proof of Theorem 3

In this section, we prove Monneau's quasi-monotonicity formula for the L^2 distance on the boundary of $u_{\mathbb{L}(x_0)}$ from any 2-homogeneous solution to Eq. (14). As for Theorem 2, the Proof of Theorem 3 uses equality (9) and Proposition 3.

Proof of Theorem 3 For the sake of notational simplicity, we write u_{x_0} rather than $u_{\mathbb{L}(x_0)}$ (as in the Proof of Theorem 2).

Set $w := u_{x_0} - v$, and then arguing as in (24) and by applying the divergence theorem, we get

$$\frac{d}{dr}\left(\frac{1}{r^{n+3}}\int_{\partial B_r} w^2 \, d\mathcal{H}^{n-1}\right) = \frac{2}{r^{n+3}}\int_{\partial B_r} w\left(\langle\nabla w, \frac{x}{r}\rangle - 2\frac{w}{r}\right) d\mathcal{H}^{n-1}$$

$$= \frac{2}{r^{n+3}}\int_{B_r} \operatorname{div}(w\nabla w) \, dx - \frac{4}{r^{n+4}}\int_{\partial B_r} w^2 \, d\mathcal{H}^{n-1}$$

$$= \frac{2}{r^{n+3}}\int_{B_r} w\Delta w \, dx + \frac{2}{r^{n+3}}\int_{B_r} |\nabla w|^2 \, dx - \frac{4}{r^{n+4}}\int_{\partial B_r} w^2 \, d\mathcal{H}^{n-1}. \quad (29)$$

For what the first term on the right-hand side of (29) is concerned, recall that $u \in W^{2,p}_{loc}(\Omega)$; thus by locality of the weak derivatives, we have that $\mathcal{L}^n(\{\nabla u_{x_0} = \underline{0}\} \setminus \{\nabla^2 u_{x_0} = \underline{0}\}) = 0$. Being $\Lambda_{u_{x_0}} \subseteq \{\nabla u_{x_0} = \underline{0}\}$, we conclude that $\Delta u_{x_0} = 0$ $\mathcal{L}^n$-a.e. in $\Lambda_{u_{x_0}}$, and therefore in view of (9), we infer

$$w\Delta w = (u_{x_0} - v)(\Delta u_{x_0} - 1) = \begin{cases} (u_{x_0} - v) f_{x_0} & \mathcal{L}^n\text{-a.e. } \Omega \setminus \Lambda_{u_{x_0}} \\ v & \mathcal{L}^n\text{-a.e. } \Lambda_{u_{x_0}}. \end{cases}$$

Instead, estimating the second and third terms on the right-hand side of (29) thanks to (14) yields

$$\frac{1}{r^{n+3}}\int_{B_r} |\nabla w|^2 \, dx - \frac{2}{r^{n+4}}\int_{\partial B_r} w^2 \, d\mathcal{H}^{n-1} = \frac{1}{r^{n+3}}\int_{B_r} \left(|\nabla u_{x_0}|^2 + |\nabla v|^2\right) dx$$

$$- \frac{2}{r^{n+3}}\int_{B_r} \operatorname{div}(u_{x_0}\nabla v) \, dx + \frac{2}{r^{n+3}}\int_{B_r} u_{x_0} \, dx - \frac{2}{r^{n+4}}\int_{\partial B_r} w^2 \, d\mathcal{H}^{n-1}$$

$$\stackrel{(15)}{=} \frac{1}{r}\left(\Phi_{u_{x_0}}(x_0, r) - \Phi_v(x_0, r)\right) - \frac{2}{r^{n+4}}\int_{\partial B_r} u_{x_0}\left(\langle\nabla v, \frac{x}{r}\rangle - 2v\right) dx$$

$$\stackrel{(13)}{=} \frac{1}{r}\left(\Phi_{u_{x_0}}(x_0, r) - \Phi_{u_{x_0}}(x_0, 0^+)\right).$$

Then, (29) can be rewritten as

$$\frac{d}{dr}\left(\frac{1}{r^{n+3}}\int_{\partial B_r} w^2 \, d\mathcal{H}^{n-1}\right) = \frac{2}{r}\left(\Phi_u(x_0, r) - \Phi_u(x_0, 0^+)\right)$$
$$+ \frac{2}{r^{n+3}}\int_{B_r \setminus \Lambda_{u_{x_0}}} (u_{x_0} - v) f_{x_0} \, dx + \frac{2}{r^{n+3}} \int_{B_r \cap \Lambda_{u_{x_0}}} v \, dx.$$

Inequality (12) in Theorem 2, the growth of the solution u from free boundary points in (20), the 2-homogeneity and positivity of v yield the conclusion (cf. (26)–(28)):

$$\frac{d}{dr}\left(\frac{1}{r^{n+3}}\int_{\partial B_r} w^2 \, d\mathcal{H}^{n-1}\right)$$
$$\geq -\frac{C}{r}\int_0^r \frac{\omega(t)}{t} \, dt - \frac{C}{r^{n+1}} \int_{B_r \setminus \Lambda_{u_{x_0}}} |f_{x_0}| \, dx = -\frac{C}{r}\int_0^r \frac{\omega(t)}{t} \, dt$$

for some $C = C(n, p, \Lambda, c_0, K, \|f\|_{L^\infty}, \|\mathbb{A}\|_{W^{1,p}}) > 0$.

6 Free Boundary Analysis: Proof of Theorem 1

Weiss and Monneau's quasi-monotonicity formulas proved in Sects. 4 and 5, respectively, are important tools to deduce regularity of free boundaries for classical obstacle problems for variational energies, both in the quadratic and in the nonlinear setting (see [11–13, 20, 23] and [21]).

In this section, we improve upon [11, Theorems 4.12 and 4.14] in the quadratic case weakening the regularity of the coefficients of the relevant energies. This is possible thanks to the abovementioned new quasi-monotonicity formulas.

In the ensuing proof, we will highlight only the substantial changes, since the arguments are essentially those given in [11, 13]. In particular, we remark again that in the quadratic case, the main differences concern the quasi-monotonicity formulas established for the quantity Φ_u rather than for the natural candidate related to $\mathcal{E}$.

We follow the variational approach by Weiss [23] and Monneau [20] for the free boundary analysis in Theorem 1.

Proof of Theorem 1 First, recall that we may establish the conclusions for the function $u = w - \psi$ introduced in Sect. 3. Given this, the only minor change to be done to the arguments in [11, Section 4] is related to the freezing of the energy, where the regularity of the coefficients plays a substantial role. More precisely, in the current framework for all $v \in W^{1,2}(B_1)$, we have

$$\left|\int_{B_1} \left(\mathbb{A}(rx)\nabla v, \nabla v\right) + 2f(rx)v\right) dx - \int_{B_1} \left(|\nabla v|^2 + 2v\right) dx\right|$$
$$\leq (r^{1-\frac{n}{p}} + \omega_f(r)) \int_{B_1} \left(|\nabla v|^2 + 2v\right) dx.$$

We then describe shortly the route to the conclusion. To begin with, recall that the quasi-monotonicity formulas established in [11, Section 3] are to be substituted by those in Sects. 4 and 5. Then, the 2-homogeneity of blow-up limits in [11, Proposition 4.2] now follows from Theorem 2. The quadratic growth of solutions from free boundary points contained in [11, Lemma 4.3], which implies non-degeneracy of blow-up limits, is contained in Lemma 1. The classification of blow-up limits is performed exactly as in [11, Proposition 4.5]. The conclusions of [11, Lemma 4.8], a result instrumental for the uniqueness of blow-up limits at regular points, can be obtained with essentially no difference. The proofs of [11, Propositions 4.10, 4.11, Theorems 4.12, 4.14] remain unchanged. The theses then follow at once.

7 Free Boundary Regularity for Nonlinear Obstacle Problems

We are now ready to apply the main result of the chapter to a nonlinear setting slightly improving the results in [12, Theorem 3.8] for what the regularity of the obstacle function ψ is concerned. More precisely, we consider for $p > n$ the nonlinear classical obstacle problem

$$\inf_{v \in \mathbb{K}^p_{\psi,g}} \int_\Omega F(x, v(x), \nabla v(x))\, dx, \tag{30}$$

where

$$\mathbb{K}^p_{\psi,g} := \{v \in W^{1,p}(\Omega) : v \geq \psi \, \mathcal{L}^n\text{-a.e. on } \Omega,\ \mathrm{Tr}(v) = g \text{ on } \partial\Omega\} \tag{31}$$

and $g \in W^{1-1/p,p}(\partial\Omega)$, $\psi \leq g\ \mathcal{H}^{n-1}$-a.e on $\partial\Omega$ (note that $\mathbb{K}^2_{\psi,g} = \mathbb{K}_{\psi,g}$). Furthermore, we assume that

(H5) $F \in C^{2,1}_{loc}(\Omega \times \mathbb{R} \times \mathbb{R}^n)$ satisfies

(i) there are $c_1, c_2 > 0$, $c_3 \geq 0$, $q \geq 0$ and $\phi \in L^1(\Omega)$ such that for all $z \in \mathbb{R}, \xi \in \mathbb{R}^n$ and for $\mathcal{L}^n$ a.e. $x \in \Omega$,

$$c_1|\xi|^p - \phi(x) \leq F(x, z, \xi) \leq c_2|\xi|^p + c_3|z|^q + \phi(x); \tag{32}$$

(ii) there are $\Lambda > 0$ and $\phi_2 \in L^{\frac{p}{p-1}}(\Omega)$ such that for $\mathcal{L}^n$ a.e. $x \in \Omega$ and for all $(z, \xi) \in \mathbb{R} \times \mathbb{R}^n$

$$|\partial_z F(x, z, \xi)| \vee |\nabla_\xi F(x, z, \xi)| \leq \Lambda(|z|^{p-1} + |\xi|^{p-1}) + \phi_2(x);$$

(iii) there is $\Theta > 0$ such that for all $x, y \in \Omega$, $z, \zeta \in \mathbb{R}$ and $\xi \in \mathbb{R}^n$

$$|\nabla_\xi F(x, z, \xi) - \nabla_\xi F(y, \zeta, \xi)| \leq \Theta(|x-y| + |z-\zeta|)(1 + |\xi|^{p-1});$$

(H6) $F(x, z, \cdot)$ is convex w.r.t. to ξ (uniformly in (x, z)), i.e. there exists $\nu > 1$ such that for all $x \in \Omega$, $z \in \mathbb{R}$ and $\xi, \eta \in \mathbb{R}^n$,

$$\nu^{-1}(1 + |\eta|)^{p-2}|\xi|^2 \leq \nabla_\xi^2 F(x, z, \eta)\xi \cdot \xi \leq \nu(1 + |\eta|)^{p-2}|\xi|^2. \tag{33}$$

(H7) $\psi \in W^{2,p}_{loc}(\Omega)$.

Thanks to (H5) and (H7), then

$$h := -\mathrm{div}\big(\nabla_\xi F(x, \psi, \nabla\psi)\big) + \partial_z F(x, \psi, \nabla\psi) \in L^p_{loc}(\Omega). \tag{34}$$

Remark 4 In case $F = F(x, \xi)$, the structural conditions imposed on F, i.e. convexity and (32), imply item (ii) in (H5) (cf. [15, Lemma 5.2]). Therefore, besides uniform convexity, the only nontrivial assumption on F is (iii) in (H5). In turn, the latter is clearly satisfied in the autonomous case $F = F(\xi)$.

We refer to [12, Remarks 3.10 and 3.11] for further comments on all the assumptions above and the subsequent (H8)–(H9).

Theorem 1 and a linearization argument introduced in [12], which we resume briefly in what follows for the readers' convenience, imply the ensuing result.

Theorem 4 *Let $\Omega \subset \mathbb{R}^n$ be smooth, bounded and open, and $p \in (n, \infty)$. Assume F and ψ satisfy (H5)–(H7) above.*

Then, the minimum problem in (31) has (at least) a solution in $\mathbb{K}^p_{\psi,g}$, and every solution belongs to $C^{1,\gamma}_{loc}(\Omega)$ for some $\gamma \in (0, 1)$.

Let $u \in \mathbb{K}^p_{\psi,g}$ be a solution. If, moreover, ψ satisfies

(H8) *for some constant $c_0 > 0$, we have for $\mathcal{L}^n$ a.e. on Ω*

$$h = -\mathrm{div}\big(\nabla_\xi F(x, \psi, \nabla\psi)\big) + \partial_z F(x, \psi, \nabla\psi) \geq c_0 > 0;$$

(H9) *for some $\alpha \in (0, 1)$,*

$$\mathrm{div}\big(\nabla_\xi F(\cdot, u, \nabla\psi)\big) \in C^{0,\alpha}_{loc}(\Omega),$$

then the free boundary decomposes as $\partial\{u = \psi\} \cap \Omega = \mathrm{Reg}(u) \cup \mathrm{Sing}(u)$, where $\mathrm{Reg}(u)$ and $\mathrm{Sing}(u)$ are called its regular and singular parts, respectively. Moreover, $\mathrm{Reg}(u) \cap \mathrm{Sing}(u) = \emptyset$ and

(i) *$\mathrm{Reg}(u)$ is relatively open in $\partial\{u = \psi\}$ and, for every point $x_0 \in \mathrm{Reg}(u)$, there exist $r = r(x_0) > 0$ and $\beta = \beta(x_0) \in (0, 1)$ such that $\mathrm{Reg}(u) \cap B_r(x_0)$ is a $C^{1,\beta}$ submanifold of dimension $n-1$;*

(ii) $\text{Sing}(u) = \cup_{k=0}^{n-1} S_k$, with S_k contained in the union of at most countably many submanifolds of dimension k and class C^1.

To establish Theorem 4, we use the linearization technique proposed in [12, Lemma 3.12] to prove that a solution to (30) is also the solution to a classical obstacle problem (locally in Ω) of the type discussed in Theorem 1. We highlight the changes needed with respect to the original proof in [12, Lemma 3.12].

Lemma 2 *Let (H5)–(H8) hold true, and let $u \in W_{loc}^{2,p}(\Omega)$ be a solution of (30). Then, there exists a symmetric matrix field $\mathbb{A} : \Omega \to \mathbb{R}^{n \times n}$ such that*

$$\text{div}(\mathbb{A}(x)\nabla(u - \psi)) = \big(-\text{div}(\nabla_\xi F(x, u, \nabla\psi)) + \partial_z F(x, u, \nabla u)\big)\chi_{\{u > \psi\}} \quad (35)$$

$\mathcal{L}^n$ *a.e. in Ω and in $\mathcal{D}'(\Omega)$; with $\mathbb{A}$ satisfying*

(i) $\mathbb{A} \in W_{loc}^{1,p}(\Omega, \mathbb{R}^{n \times n})$,
(ii) *for all $K \subset\subset \Omega$, there is $\lambda_K \geq 1$ for which*

$$\lambda_K^{-1}|\xi|^2 \leq \mathbb{A}(x)\xi \cdot \xi \leq \lambda_K|\xi|^2 \quad \text{for all } x \in K \text{ and for all } \xi \in \mathbb{R}^n. \quad (36)$$

Proof We first note the inclusion $\{u = \psi\} \subseteq \{\nabla u = \nabla \psi\}$, which follows from the obstacle condition $u \geq \psi$ on Ω and from the regularity of u and ψ. Then, we use [12, Proposition 3.2, Theorem 3.4 and Corollary 3.5] together with assumption (H8) to rewrite the Euler–Lagrange equation satisfied by minimizers of (30) as

$$\text{div}\big(\nabla_\xi F(x, u, \nabla u) - \nabla_\xi F(x, u, \nabla\psi)\big)$$
$$= \big(-\text{div}(\nabla_\xi F(x, u, \nabla\psi)) + \partial_z F(x, u, \nabla u)\big)\chi_{\{u > \psi\}}. \quad (37)$$

On setting $w := u - \psi$, for all x in Ω, we have

$$\nabla_\xi F(x, u(x), \nabla u(x)) - \nabla_\xi F(x, u(x), \nabla\psi(x))$$
$$= \Big(\int_0^1 \nabla_\xi^2 F\big(x, u(x), \nabla\psi(x) + t\nabla w(x)\big)dt\Big)\nabla w(x) =: \mathbb{A}(x)\nabla w(x). \quad (38)$$

Hence, w satisfies (35) by taking into account (37) and (38).

For what item (i) is concerned, first note that being $F \in C_{loc}^{2,1}(\Omega \times \mathbb{R} \times \mathbb{R}^n)$ and $u, \psi \in W_{loc}^{2,p}(\Omega)$, $p > n$, then $\mathbb{A} \in L_{loc}^\infty(\Omega, \mathbb{M}^{n \times n})$. Furthermore, by the difference quotient characterization of Sobolev spaces (cf. [2, Proposition 9.3]), $\mathbb{A}$ turns out to be weakly differentiable with $|\nabla\mathbb{A}| \in L_{loc}^p(\Omega)$.

Let now $K \subset \mathbb{R}^n$ be a compact set; then in view of (33) in (H6), we have for all $x \in \Omega$ and for all $\xi \in K$

$$\nu^{-1}(2^{p-2} \wedge 1)|\xi|^2 \int_0^1 \left(1 + |\nabla \psi(x) + t\nabla w(x)|\right)^{p-2} dt$$

$$\leq \mathbb{A}(x)\xi \cdot \xi = \int_0^1 \nabla_\xi^2 F\bigl(x, u(x), \nabla\psi(x) + t\nabla w(x)\bigr) \xi \cdot \xi \, dt$$

$$\leq \|\nabla_\xi^2 F\|_{L^\infty(K \times B_{r_K} \times B_{r_K}, \mathbb{R}^{n \times n})} |\xi|^2,$$

with $r_K := \sup_K(|u| + |\nabla\psi| + |\nabla w|)$. The upper bound in (36) is then established. The lower bound in (36) is immediate if $p \geq 2$ (which is always the case for $n \geq 2$). Instead, if $n = 1$, we use that $u, \psi \in C_{loc}^{1,\gamma}(\Omega)$, for some $\gamma \in (0, 1)$, to conclude. □

We are ready to prove Theorem 4 as a direct consequence of Theorem 1 and Lemma 2.

Proof The existence of solutions to (30) follows from the direct method of the calculus of variations thanks to the convexity of $\xi \mapsto F(x, z, \xi)$ and to the growth conditions (32) (cf. [15, Theorem 4.5]).

The assertion that any minimizer u is $W_{loc}^{2,p}(\Omega)$ is a consequence of [12, Proposition 3.2, Theorem 3.4 and Corollary 3.5] and the standing assumptions (H5)–(H7) on F. Therefore, we use Lemma 2 to conclude that $w = u - \psi$ is the minimizer of the quadratic classical obstacle problem

$$\mathcal{E}[v] = \int_\Omega \left(\mathbb{A}(x)\nabla v(x) \cdot \nabla v(x) + 2f(x)v(x)\right) dx$$

over $\mathbb{K}_{g-\psi,0}$, with the matrix field $\mathbb{A} \in W_{loc}^{1,p}(\Omega, \mathbb{R}^{n \times n})$ provided there and with

$$f := -\text{div}(\nabla_\xi F(x, u, \nabla\psi)) + \partial_z F(x, u, \nabla u)$$

(cf. (35)). In addition, note that $\partial\{w = 0\} \cap \Omega = \partial\{u = \psi\} \cap \Omega$.

We claim that we can apply Theorem 1 locally in Ω. Indeed, recall first that $\{u = \psi\} \subseteq \{\nabla u = \nabla\psi\}$, being $u \geq \psi$ on Ω. Thus, given $\Omega' \subset\subset \Omega$ and any $\varepsilon > 0$, the set $\Omega'_\varepsilon := \{0 \leq u - \psi < \varepsilon\} \cap \{|\nabla(u - \psi)| < \varepsilon\} \cap \Omega'$ is open and such that $\{u = \psi\} \cap \Omega' \subset \Omega'_\varepsilon$ in view of the remark above. Moreover, as $h = -\text{div}(\nabla_\xi F(x, \psi, \nabla\psi)) + \partial_z F(x, \psi, \nabla\psi) \geq c_0 > 0$ (cf. (H8)), using that $F \in C_{loc}^{2,1}(\Omega \times \mathbb{R} \times \mathbb{R}^n)$, it is easy to infer that

$$f \geq h - \|h - f\|_{L^\infty(\Omega'_\varepsilon)} \geq \frac{c_0}{2} > 0$$

holds on Ω'_ε for ε sufficiently small (cf. [12, Theorem 3.8]). Furthermore, $f \in C_{loc}^{0,\alpha\wedge\gamma}(\Omega)$ by hypotheses (H7) and (H9) and being $u \in C_{loc}^{1,\gamma}(\Omega)$, for some $\gamma \in$

(0, 1). Therefore, all the assumptions of Theorem 1 are satisfied on the open set Ω'_ε. All the conclusions then follow at once on Ω'_ε for every $\Omega' \subset\subset \Omega$ and $\varepsilon > 0$. $\square$

8 Conclusions

We have established quasi-monotonicity formulas of Weiss and Monneau's type for quadratic energies having a matrix of coefficients in $W^{1,p}$, $p > n$, and we have given an application to the corresponding free boundary analysis for the related classical obstacle problem both in a quadratic and in a nonlinear setting improving upon the results first established in [12].

As pointed out in Sect. 1, concerning the quasi-monotonicity formulas, the main difference with the existing literature is related to the monotone quantity itself. Indeed, rather than considering the natural quadratic energy $\mathscr{E}$ associated with the obstacle problem under study, we may consider the classical Dirichlet energy thanks to a normalization. In doing this, we have been inspired by Monneau [20, Section 6]. The advantage of this formulation is that the matrix field $\mathbb{A}$ is not differentiated in deriving the quasi-monotonicity formulas contrary to [11] and [13]. Our additional insight is elementary but crucial: we further exploit the quadratic growth of solutions from free boundary points in Proposition 3 to establish quasi-monotonicity. In view of all of this, we are able to weaken the required regularity assumptions on the matrix field $\mathbb{A}$ (cf. (H1)).

As a consequence, we are able to improve upon [12, Theorem 3.8] and relax slightly the regularity assumptions on the obstacle function ψ involved in nondegenerate, nonlinear classical obstacle problems. In particular, the optimal $C^{1,1}_{loc}(\Omega)$ regularity of solutions, which was crucial there in order to apply the results of [11], is no longer needed.

Acknowledgments E. S. has been supported by the ERC-STG Grant no. 759229 HiCoS "Higher Co-dimension Singularities: Minimal Surfaces and the Thin Obstacle Problem." M. F., F. G. and E. S. are members of the Gruppo Nazionale per l'Analisi Matematica, la Probabilità e le loro Applicazioni (GNAMPA) of the Istituto Nazionale di Alta Matematica (INdAM).

References

1. Blank, I., Hao Z.: The mean value theorem and basic properties of the obstacle problem for divergence form elliptic operators. Commun. Analy. Geometry **23**, 129–158 (2015)
2. Brezis, H.: Functional Analysis, Sobolev Spaces and Partial Differential Equations. Universitext. Springer, New York (2011)
3. Caffarelli, L.A.: The regularity of free boundaries in higher dimensions. Acta Math. **139**, 155–184 (1977)
4. Caffarelli, L.A.: Compactness methods in free boundary problems. Comm. Partial Differ. Equ. **5**, 427–448 (1980)

5. Caffarelli, L.A.: The obstacle problem revisited. Lezioni Fermiane. Accademia Nazionale dei Lincei, Rome; Scuola Normale Superiore, Pisa (1998)
6. Caffarelli, L.A.: The obstacle problem revisited. J. Fourier Anal. Appl. **4**, 383–402 (1998)
7. Caffarelli, L.A., Salsa, S.: A Geometric Approach to Free Boundary Problems. Graduate Studies in Mathematics, vol. 68. American Mathematical Society, Providence, (2005)
8. Colombo, M., Spolaor, L., Velichkov, B.: A logarithmic epiperimetric inequality for the obstacle problem. Geom. Funct. Anal. **28**(4), 1029–1061 (2018)
9. Figalli, A., Serra, J.: On the fine structure of the free boundary for the classical obstacle problem. Invent. Math. **215**(1), 311–366 (2019)
10. Figalli, A., Ros-Oton, X., Serra, J.: Generic regularity of free boundaries for the obstacle problem. Publ. Math. Inst. Hautes Études Sci. **132**, 181–292 (2020)
11. Focardi, M., Gelli, M.S., Spadaro, E.: Monotonicity formulas for obstacle problems with Lipschitz coefficients. Calc. Var. Partial Differ. Equ. **54**, 1547–1573 (2015)
12. Focardi, M., Geraci, F., Spadaro, E.: The classical obstacle problem for nonlinear variational energies. Nonlinear Anal. **154**, 71–87 (2017)
13. Geraci, F.: The classical obstacle problem with coefficients in fractional Sobolev spaces. Ann. Mat. Pura Appl. **197**, 549–581 (2018)
14. Gilbarg, D., Trudinger, N.S.: Elliptic Partial Differential Equations of Second Order. Reprint of the 1998 edition. Classics in Mathematics. Springer, Berlin (2001)
15. Giusti, E.: Direct Methods in the Calculus of Variations. World Scientific Publishing, River Edge (2003)
16. Han, Q., Lin, F.: Elliptic Partial Differential Equations. Courant Lecture Notes in Mathematics, , 2nd edn. American Mathematical Society, Providence (2011)
17. Kinderlehrer, D., Stampacchia, G.: An Introduction to Variational Inequalities and Their Applications. Pure and Applied Mathematics, vol. 88. Academic, New York (1980)
18. Kukavica, I.: Quantitative uniqueness for second-order elliptic operators. Duke Math. J. **91**, 225–240 (1998)
19. Miranda, C.: Sulle equazioni ellittiche del secondo ordine di tipo non variazionale, a coefficienti discontinui. (Italian). Ann. Mat. Pura Appl. **63**, 353–386 (1963)
20. Monneau, R.: On the number of singularities for the obstacle problem in two dimensions. J. Geom. Anal. **13**, 359–389 (2003)
21. Petrosyan, A., Shahgholian, H., Uraltseva, N.: Regularity of Free Boundaries in Obstacle-Type Problems. Graduate Studies in Mathematics, vol. 136. American Mathematical Society, Providence, (2012)
22. Uraltseva, N.: Regularity of solutions of variational inequalities. Russian Math. Surveys **42**(6), 191–219 (1987)
23. Weiss, G.S.: A homogeneity improvement approach to the obstacle problem. Invent. Math. **138**, 23–50 (1999)

Optimal Feedback for Structures Controlled by Hydraulic Semi-active Dampers

Ido Halperin, Grigory Agranovich, and Yuri Ribakov

1 Introduction

Guaranteeing safety of civil structures and their occupants against earthquakes has been a major concern of researchers and engineers for many years. A contemporary approach for obtaining satisfactory level of safety is to use structural control in order to grant to buildings the ability to bear such unwanted dynamic phenomena [9]. Generally speaking, physical realization of structural control is done by actuators that apply forces to the vibrating structure in real time. There are many types of such actuators, and each one imposes different design constraints on the control law [1, 13, 15, 18]. A famous type of devices, known to be effective in many applications, is the controlled hydraulic damper. It is a type of semi-active device [9, 11] whose operation principle resembles that of viscous fluid damper [3]. The difference though is the presence of a valves system that dictates the flow of the fluid through the hydraulic damper's orifices [13]. The valves are adjusted electromechanically, leading to different damping's properties. Closing the valve increases the damping and vice versa when it opens. Incorporation of such dampers into a controller allows it to adjust the damping to a preferred value during the structure's dynamic response in real time. This kind of devices has been implemented in several full-scale structures [10, 13, 14].

Many control devices manifest highly nonlinear behavior. The problem is that taking into account such nonlinear complexities during the controller design can establish a significant hurdle for the control designer. A work-around solution is to separate between the system's and the damper's dynamics [21]. This allows for the nonlinear properties of the device to be considered separately from the system's

---

I. Halperin (✉) · G. Agranovich · Y. Ribakov
Ariel University, Ariel, Israel
e-mail: idoh@ariel.ac.il; agr@ariel.ac.il; ribakov@ariel.ac.il

controller. The latter is designed to generate control signals, mostly optimal by some sense, that are tracked by the device's controller. A dilemma that naturally emerges in such situation is what to do when the system's controller instructs a signal which is not feasible by the device's limitations. A simple and popular approach in such a case is to use arbitrary clipping [14, 16, 17, 22]. However, the arbitrary clipping of the control trajectory distorts it and therefore raises a theoretical question on its contribution to the controlled plant. This issue spurs formulation of optimal control designs that can account for semi-active devices' limitations and reduce the need of arbitrary clipping [4–8]. The present study suggests a new method for the computation of optimal feedback for a plant controlled by multiple semi-active controlled hydraulic dampers and subjected to external, a priori known deterministic excitation input.

2 Background

2.1 The Plant Model

The characteristics of civil structures, in conjunction with common engineering assumptions, allow to model them by linear approaches, such as dynamic linear models. Consider a model of an excited structure with lumped masses, linear damping, linear stiffness, and a controller comprised of multiple actuators. The equations of motion in the structure's degrees of freedom (DOFs) are given by the following second-order initial value problem [19]:

$$\mathbf{M}\ddot{\mathbf{z}}(t) + \mathbf{C}_d\dot{\mathbf{z}}(t) + \mathbf{K}\mathbf{z}(t) = \mathbf{\Psi}\mathbf{w}(t) + \mathbf{e}(t); \quad \mathbf{z}(0), \dot{\mathbf{z}}(0), \forall t \in]0, t_f[\qquad (1)$$

This is a *linear time invariant* (LTI) model, in which $\mathbf{M} > 0$, $\mathbf{C}_d \geq 0$, and $\mathbf{K} > 0$ are symmetric mass, damping, and stiffness matrices, respectively,[1] $\mathbf{z} : \mathbb{R} \to \mathbb{R}^{n_z}$ is a smooth vector function, which represents the DOF displacements, $\mathbf{w} : \mathbb{R} \to \mathbb{R}^{n_w}$ is a vector function of the control forces that are generated by the actuators, $\mathbf{\Psi} \in \mathbb{R}^{n_z \times n_w}$ is an input matrix that describes how the control force inputs affect the structure's DOF, and $\mathbf{e} : \mathbb{R} \to \mathbb{R}^{n_z}$ is a vector function that describes the external excitation force inputs. Here, $\mathbf{z}(t)$ is the intersection of $\mathbf{z}$ at t. That is, here, $\mathbf{z}(t)$ is used to signify a specific vector in $\mathbb{R}^{n_z}$, obtained at a given t, whereas $\mathbf{z}$ refers to the entire trajectory over $]0, t_f[$.

When dealing control theory, state-space representation is much more convenient than (1). Hence, transforming it to the state-space form yields

$$\dot{\mathbf{x}}(t) = \mathbf{A}\mathbf{x}(t) + \mathbf{B}\mathbf{w}(t) + \mathbf{g}(t); \quad \mathbf{x}(0), \forall t \in]0, t_f[\qquad (2)$$

---

[1] Recall that $\mathbf{M} > 0$, $\mathbf{K} > 0$ iff $\mathbf{z}^T\mathbf{M}\mathbf{z} > 0$, $\mathbf{z}^T\mathbf{K}\mathbf{z} > 0$ for all $\mathbf{z} \in \mathbb{R}^{n_z}$, $\mathbf{z} \neq \mathbf{0}$ and $\mathbf{C}_d \geq 0$ iff $\mathbf{z}^T\mathbf{C}_d\mathbf{z} \geq 0$ for all $\mathbf{z} \in \mathbb{R}^{n_z}$.

where

$$\mathbf{x}(t) \triangleq \begin{bmatrix} \mathbf{z}(t) \\ \dot{\mathbf{z}}(t) \end{bmatrix}; \qquad \mathbf{A} \triangleq \begin{bmatrix} \mathbf{0} & \mathbf{I} \\ -\mathbf{M}^{-1}\mathbf{K} & -\mathbf{M}^{-1}\mathbf{C}_d \end{bmatrix} \in \mathbb{R}^{n \times n}$$

$$\mathbf{B} \triangleq \begin{bmatrix} \mathbf{0} \\ \mathbf{M}^{-1}\mathbf{\Psi} \end{bmatrix} \in \mathbb{R}^{n \times n_w}; \qquad \mathbf{g}(t) \triangleq \begin{bmatrix} \mathbf{0} \\ \mathbf{e}(t) \end{bmatrix}$$

$n = 2n_z$.

Let the control devices, embedded in the structure, be semi-active. Even though such actuators have many advantages, which is why they garner attention from many researchers, they impose some constraints on the synthesized control. Basically, the semi-active dampers set limits on the control force—w_i, as follows:

1. w_i is always opposed to the relative velocity of the damper's anchors. This assures that the damper only consumes mechanical energy from the structure.
2. Physical considerations inhibit the device from generating a control force when the relative velocity in the damper is zero. In other words, w_i must vanish when there is no motion in the damper.
3. For some semi-active dampers, there is some minimal amount of damping that the device provides during its motion, even in *off-state*, i.e., when no damping effort is exerted.

In addition to these three semi-active constraints, which are related to the traits of semi-active dampers, many practical implementations require the control force to be bounded.

In order to include these constraints in a control problem, they should be quantified. Assume that the relative velocity of the damper's anchors can be represented as linear combination of the state variables, i.e., as $\mathbf{c}_i \mathbf{x}$ for some $\mathbf{c}_i^T \in \mathbb{R}^n$, and that a linear viscous damping is valid to the given problem [20]. Then, the above limitations are expressed by the following constraints:

C1: $w_i(t)\mathbf{c}_i\mathbf{x}(t) \leq 0$
C2: $\mathbf{c}_i\mathbf{x}(t) = 0 \rightarrow w_i(t) = 0$
C3: $|w_i(t)| \geq w_{i,min}(t, \mathbf{x}(t)) \geq 0$
C4: $w_{i,max} \geq |w_i(t)|$

for all $t \in [0, t_f]$ and for some $w_{i,min} : \mathbb{R} \times \mathbb{R}^n \rightarrow [0, w_{i,max}]$. Note that the lower bound must satisfy $w_{i,min}(t, \mathbf{x}(t)) = 0$ whenever $\mathbf{c}_i\mathbf{x}(t) = 0$; otherwise C2 and C3 might contradict.

In this work, constraints C1–C4 are adapted to the traits of a certain type of controlled hydraulic dampers. A control design must account for these constraints, especially when the optimal control design is sought. Otherwise, the design's relevancy to a constrained problem is dubious. The problem is that the inclusion of such constraints into optimal control design problem can turn it into a very nontrivial problem. A method that can be used to tackle such a problem is explained below.

2.2 Krotov's Method: A Global Method of Successive Improvements of Control

Krotov's method is aimed at numerically solving optimal control problems. However, its utilization depends on the successful formulation of a function's sequence with special properties. If such a sequence can be found, it allows to compute a candidate optimum of the addressed optimal control problem. This subsection describes elements from Krotov's theory, relevant to the addressed problem.

Let

$$\dot{\mathbf{x}}(t) = \mathbf{f}(t, \mathbf{x}(t), \mathbf{u}(t)); \quad \mathbf{x}(0), \forall t \in]0, t_f[\tag{3}$$

be a state equation, $\mathcal{U} \subseteq \{\mathbb{R} \to \mathbb{R}^{n_u}\}$ be the set of admissible control trajectories, and $\mathcal{X} \subseteq \{\mathbb{R} \to \mathbb{R}^n\}$ be the set of state trajectories that are reachable from $\mathcal{U}$ and $\mathbf{x}(0)$. The term *admissible process* refers to the state and control trajectories $(\mathbf{x} \in \mathcal{X}, \mathbf{u} \in \mathcal{U})$ which satisfy (3). The goal is to find an admissible process that minimizes the following performance index:

$$J(\mathbf{x}, \mathbf{u}) = \int_0^{t_f} l(t, \mathbf{x}(t), \mathbf{u}(t)) \mathrm{d}t + l_f(\mathbf{x}(t)) \tag{4}$$

Definition 1 (Improving Sequence) Let $\{(\mathbf{x}_k, \mathbf{u}_k)\}$ be a sequence of admissible processes, and assume that $\inf_{\substack{\mathbf{x} \in \mathcal{X} \\ \mathbf{u} \in \mathcal{U}}} J(\mathbf{x}, \mathbf{u})$ exists. If

$$J(\mathbf{x}_k, \mathbf{u}_k) \geq J(\mathbf{x}_{k+1}, \mathbf{u}_{k+1}) \tag{5}$$

for all $k = 1, 2, \ldots$ and

$$\lim_{k \to \infty} J(\mathbf{x}_k, \mathbf{u}_k) = \inf_{\substack{\mathbf{x} \in \mathcal{X} \\ \mathbf{u} \in \mathcal{U}}} J(\mathbf{x}, \mathbf{u}) \tag{6}$$

then $\{(\mathbf{x}_k, \mathbf{u}_k)\}$ is said to be an improving sequence.

Such a sequence is the outcome of Krotov's method. In order to obtain the improving sequence, the method successively improves admissible processes, as follows [12].

Theorem 1 *Let $(\mathbf{x}_k, \mathbf{u}_k)$ be a given admissible process and q be some smooth function, and define s and s_f as*

$$s(t, \xi, \mathbf{v}) \triangleq q_t(t, \xi) + q_\mathbf{x}(t, \xi)\mathbf{f}(t, \xi, \mathbf{v}) + l(t, \xi, \mathbf{v}) \tag{7}$$

$$s_f(\xi) \triangleq l_f(\xi) - q(t_f, \xi) \tag{8}$$

where $\xi \in \mathbb{R}^n$ and $\mathbf{v} \in \mathbb{R}^{n_u}$ are some vectors.

If q grants to s and s_f the next property:

$$s(t, \mathbf{x}_k(t), \mathbf{u}_k(t)) = \max_{\boldsymbol{\xi} \in \mathcal{X}(t)} s(t, \boldsymbol{\xi}, \mathbf{u}_k(t))$$
$$s_f(\mathbf{x}_k(t_f)) = \max_{\boldsymbol{\xi} \in \mathcal{X}(t_f)} s_f(\boldsymbol{\xi})$$
(9)

and if $\hat{\mathbf{u}}$ is a control feedback which satisfies

$$\hat{\mathbf{u}}(t, \boldsymbol{\xi}) = \arg \min_{\mathbf{v} \in \mathcal{U}(t)} s(t, \boldsymbol{\xi}, \mathbf{v}); \quad \forall t \in [0, t_f]$$
(10)

then $\mathbf{x}_{k+1}$, which solves

$$\dot{\mathbf{x}}_{k+1}(t) = \mathbf{f}(t, \mathbf{x}_{k+1}(t), \hat{\mathbf{u}}(t, \mathbf{x}_{k+1}(t))); \quad \mathbf{x}_{k+1}(0) = \mathbf{x}(0), \forall t \in]0, t_f[$$
(11)

and the control trajectory $\mathbf{u}_{k+1}(t) = \hat{\mathbf{u}}(t, \mathbf{x}_{k+1}(t))$ satisfy (5).

It follows from this theorem that if for a prescribed $(\mathbf{x}_k, \mathbf{u}_k)$, one can find q such that (9) holds, then it is possible to find an improved admissible process— $(\mathbf{x}_{k+1}, \mathbf{u}_{k+1})$. Such q is denoted as *improving function*. Solving this problem over and over yields an improving sequence and hence leads to the solution of the optimization problem. In his work, Krotov showed that if, at some point, the processes stop changing, then the obtained process satisfies Pontryagin's minimum principle.

Generally speaking, the iterative procedure is summarized in the following algorithm. Its initialization requires to compute some initial admissible process— $(\mathbf{x}_0, \mathbf{u}_0)$. Afterward, the following steps are iterated for $k = \{0, 1, 2, \ldots\}$ until convergence is attained:

1. Find q_k that grants s_k and $s_{f,k}$ the next property:

$$s_k(t, \mathbf{x}_k(t), \mathbf{u}_k(t)) = \max_{\boldsymbol{\xi} \in \mathcal{X}(t)} s_k(t, \boldsymbol{\xi}, \mathbf{u}_k(t))$$
$$s_{f,k}(\mathbf{x}_k(t_f)) = \max_{\boldsymbol{\xi} \in \mathcal{X}(t_f)} s_{f,k}(\boldsymbol{\xi})$$

at a given $(\mathbf{x}_k, \mathbf{u}_k)$ and for all t in $[0, t_f]$. Here, s_k and $s_{f,k}$ are the functions obtained by substituting q_k into (7) and (8).

2. Find a minimizing feedback

$$\hat{\mathbf{u}}_{k+1}(t, \mathbf{x}(t)) = \arg \min_{\mathbf{v} \in \mathcal{U}(t)} s_k(t, \mathbf{x}(t), \mathbf{v})$$

for all t in $[0, t_f]$

3. Propagate into the next improved state and control processes, by solving

$$\dot{\mathbf{x}}_{k+1}(t) = \mathbf{f}\big(t, \mathbf{x}_{k+1}(t), \hat{\mathbf{u}}_{k+1}(t, \mathbf{x}_{k+1}(t))\big)$$

and setting

$$\mathbf{u}_{k+1}(t) = \hat{\mathbf{u}}_{k+1}(t, \mathbf{x}_{k+1}(t))$$

As it can be seen from the above, the use of Krotov's method requires to formulate a sequence of improving functions—$\{q_k\}$. In general, the search for these improving functions can be a significant challenge. As of this writing, there is no known unified method for their formulation, and they usually differ from one optimal control problem to another.

3 Main Results

Consider a structure equipped with a set of controlled hydraulic dampers and subjected to an a priori known external excitation—$\mathbf{g} : \mathbb{R} \to \mathbb{R}^n$. It is assumed that the control forces follow a linear viscous damping law and that each device features merely two control phases—*on* or *off*. In many works, (2) is used for modeling such a system in conjunction with a set of limitations, reflecting the constraints induced by the nature of the semi-active dampers. In this study, however, a bilinear representation is used, allowing to account for the system's dynamics and constraints C1–C4. It will be shown that the alternative representation is equivalent to that based on (2).

Consider the bilinear state-space equation:

$$\dot{\mathbf{x}}(t) = \left(\mathbf{A} - \sum_{i=1}^{n_u} \mathbf{b}_i u_i(t) \mathbf{c}_i\right) \mathbf{x}(t) + \mathbf{g}(t); \quad \mathbf{x}(0), \forall t \in]0, t_f[\quad (12)$$

where $n_u = n_w$; $\mathbf{c}_i^T \in \mathbb{R}^n$ is constructed such that $\mathbf{c}_i \mathbf{x}$ is the relative velocity of the damper's anchors, positive when the damper elongates, and u_i is a control trajectory that satisfies $u_i(t) \in \mathscr{U}_i(t, \mathbf{x})$, where $\mathscr{U}_i(\mathbf{x})$ is the set of control trajectories, which are admissible in the i-th device. $\mathscr{U}_i(t, \mathbf{x})$ is the set of admissible values at some time—t. It is defined by

$$\mathscr{U}_i(t, \mathbf{x}) = \begin{cases} \{d_i, D_i\}, & D_i|\mathbf{c}_i \mathbf{x}(t)| \leq w_{i,max} \\ d_i, & \text{otherwise} \end{cases} \quad (13)$$

Here, $D_i \geq d_i \geq 0$ are the damper's on/off gains. Physically, they are the maximal and minimal viscous damping coefficients of the i-th control device, respectively. When the valve is opened, the device provides a minimal damping force—$w_{i,min}(t, \mathbf{x}(t)) = d_i|\mathbf{c}_i\mathbf{x}(t)|$.

The following proposition shows that the suggested representation accords to a model, governed by (2) with constraints C1–C4.

Proposition 1 *If* $(\mathbf{x}, \mathbf{u})$ *is an admissible process by means of* (12) *and* (13), *and if* $d_i|\mathbf{c}_i\mathbf{x}(t)| < w_{i,max}$ *is met, then* $(\mathbf{x}, (-u_i\mathbf{c}_i\mathbf{x})_{i=1}^{n_u})$ *is an admissible process by means of* (2) *with constraints C1–C4.*

Proof Let $(\mathbf{x}, \mathbf{u})$ be an admissible process by means of (12) and (13), and let

$$\hat{w}_i(\mathbf{x}(t), t) \triangleq -u_i(t)\mathbf{c}_i\mathbf{x}(t) \tag{14}$$

It follows that

$$\hat{w}_i(\mathbf{x}(t), t)\mathbf{c}_i\mathbf{x}(t) = -u_i(t)(\mathbf{c}_i\mathbf{x}(t))^2 \leq 0 \tag{15}$$

i.e., C1 is satisfied. The compliance of $\hat{w}_i$ with C2 is straightforward from its definition. C3 is satisfied because

$$|\hat{w}_i(\mathbf{x}(t), t)| = u_i(t)|\mathbf{c}_i\mathbf{x}(t)| \geq d_i|\mathbf{c}_i\mathbf{x}(t)| = w_{i,min}(t, \mathbf{x}(t))$$

C4 is satisfied by the hypothesis. Hence, $(\mathbf{x}, \hat{\mathbf{w}}(\mathbf{x}))$ is admissible by means of (2) with constraints C1–C4. □

Therefore, assuming that the problem is defined with large enough $w_{i,max}$, representations (2) and (12) are interchangeable.

The next definition formally states the addressed optimal control problem.

Definition 2 (CBQR) The continuous-time bilinear quadratic regulator (CBQR) control problem is a search for an optimal and admissible process $(\mathbf{x}^*, \mathbf{u}^*)$ that minimizes the quadratic performance index:

$$J(\mathbf{x}, \mathbf{u}) = \frac{1}{2}\int_0^{t_f} \mathbf{x}(t)^T \mathbf{Q}\mathbf{x}(t) + \sum_{i=1}^{n_u} u_i(t)^2 r_i \, dt + \frac{1}{2}\mathbf{x}(t)^T \mathbf{H}\mathbf{x}(t) \tag{16}$$

where $0 \leq \mathbf{Q}, \mathbf{H} \in \mathbb{R}^{n \times n}$, and $r_i > 0$ for $i = 1, \ldots, n_u$. An admissible process is a pair $(\mathbf{x}, \mathbf{u})$ which satisfies (12) and $u_i \in \mathcal{U}_i(\mathbf{x})$ for $i = 1, \ldots, n_u$.

From physical viewpoint, the performance index weighs the states' response against the time-varying damping gains. Smaller values of $(r_i)_{i=1}^{n_u}$ will produce a control law which tends to produce more frequent closed-valve pulses.

The CBQR problem will be solved here by Krotov's method. To this end, a class of improving functions and minimizing feedback, which suit to the CBQR problem are formulated in the next lemmas.

Lemma 1 Let $q(t, \pmb{\xi}) = \frac{1}{2}\pmb{\xi}^T \mathbf{P}(t)\pmb{\xi} + \mathbf{p}(t)^T \pmb{\xi}$, where $\pmb{\xi} \in \mathbb{R}^n$, $\mathbf{P} : \mathbb{R} \to \mathbb{R}^{n \times n}$ is a continuous, piecewise smooth and symmetric, matrix function, and $\mathbf{p} : \mathbb{R} \to \mathbb{R}^n$ is a continuous and piecewise smooth, vector function.

Let $v_i(t, \pmb{\xi}) \triangleq \frac{\mathbf{b}_i^T (\mathbf{P}(t)\pmb{\xi} + \mathbf{p}(t))\mathbf{c}_i \pmb{\xi}}{r_i}$. The vector of control laws, $(\hat{u}_i)_{i=1}^{n_u}$, which minimizes $s(t, \mathbf{x}(t), \mathbf{u}(t))$ over $\{\mathbf{u}(t) \in \mathcal{U}(t, \mathbf{x})\}$, is given by

$$\hat{u}_i(t, \mathbf{x}(t)) = \begin{cases} d_i, & D_i|\mathbf{c}_i\mathbf{x}(t)| > w_{i,max} \\ \arg \min_{v_i \in \{d_i, D_i\}} (v_i - v_i(t, \mathbf{x}(t)))^2, & \text{otherwise} \end{cases} \quad (17)$$

Proof The partial derivatives of q are

$$q_t(t, \pmb{\xi}) = \frac{1}{2}\pmb{\xi}^T \dot{\mathbf{P}}(t)\pmb{\xi} + \dot{\mathbf{p}}(t)^T \pmb{\xi}; \quad q_{\mathbf{x}}(t, \pmb{\xi}) = \pmb{\xi}^T \mathbf{P}(t) + \mathbf{p}(t)^T \quad (18)$$

Let $\mathbf{v} \in \mathbb{R}^{n_u}$. By explicitly writing (7) and rearranging, we obtain

$$\begin{aligned} s(t, \mathbf{x}(t), \mathbf{v}) =& q_t(t, \mathbf{x}(t)) + q_{\mathbf{x}}(t, \mathbf{x}(t))\mathbf{f}(t, \mathbf{x}(t), \mathbf{v}) \\ &+ \frac{1}{2}\left(\mathbf{x}(t)^T \mathbf{Q}\mathbf{x}(t) + \sum_{i=1}^{n_u} v_i^2 r_i\right) \end{aligned} \quad (19)$$

$$\begin{aligned} =& \frac{1}{2}\mathbf{x}(t)^T \left(\dot{\mathbf{P}}(t) + \mathbf{P}(t)\mathbf{A} + \mathbf{A}^T \mathbf{P}(t) + \mathbf{Q}\right)\mathbf{x}(t) \\ &+ \mathbf{x}(t)^T \left(\dot{\mathbf{p}}(t) + \mathbf{A}^T \mathbf{p}(t) + \mathbf{P}(t)\mathbf{g}(t)\right) + \mathbf{p}(t)^T \mathbf{g}(t) \\ &+ \frac{1}{2}\sum_{i=1}^{n_u} r_i v_i^2 - 2r_i v_i v_i(t, \mathbf{x}(t)) \end{aligned} \quad (20)$$

where v_i was defined in the lemma. Completing the squares leads to

$$\begin{aligned} s(t, \mathbf{x}(t), \mathbf{v}) =& \frac{1}{2}\mathbf{x}(t)^T \left(\dot{\mathbf{P}}(t) + \mathbf{P}(t)\mathbf{A} + \mathbf{A}^T \mathbf{P}(t) + \mathbf{Q}\right)\mathbf{x}(t) \\ &+ \mathbf{x}(t)^T \left(\dot{\mathbf{p}}(t) + \mathbf{A}^T \mathbf{p}(t) + \mathbf{P}(t)\mathbf{g}(t)\right) + \mathbf{p}(t)^T \mathbf{g}(t) \\ &+ \frac{1}{2}\sum_{i=1}^{n_u} r_i (v_i - v_i(t, \mathbf{x}(t)))^2 - r_i v_i(t, \mathbf{x}(t))^2 \\ =& f_2(t, \mathbf{x}(t)) + \frac{1}{2}\sum_{i=1}^{n_u} r_i (v_i - v_i(t, \mathbf{x}(t)))^2 \end{aligned}$$

where $f_2 : \mathbb{R} \times \mathbb{R}^n \to \mathbb{R}$ is some function that is independent of v_i. It follows that a minimum of $s(t, \mathbf{x}(t), \mathbf{v})$ over $\{\mathbf{v}|\mathbf{v} \in \mathcal{U}(t, \mathbf{x})\}$ is the minimum of the quadratic sum with relation to each $\{v_i|v_i \in \mathcal{U}_i(t, \mathbf{x})\}$, independently. Thereby, the admissible minimum is attained at $v_i = \arg\min_{v_i \in \mathcal{U}_i(t,\mathbf{x})} (v_i - v_i(t, \mathbf{x}(t)))^2$ for each device. This fact is reflected by (17).

Lemma 2 *Let* $(\mathbf{x}_k, \mathbf{u}_k)$ *be a given admissible process, and let* $\mathbf{P}_k$ *and* $\mathbf{p}_k$ *be the solutions of*

$$\dot{\mathbf{P}}_k(t) = -\mathbf{P}_k(t)(\mathbf{A} - \mathbf{B}\,\mathrm{diag}(\mathbf{u}_k(t))\mathbf{C}) \\ - (\mathbf{A} - \mathbf{B}\,\mathrm{diag}(\mathbf{u}_k(t))\mathbf{C})^T \mathbf{P}_k(t) - \mathbf{Q}; \qquad \mathbf{P}_k(t_f) = \mathbf{H} \qquad (21)$$

and

$$\dot{\mathbf{p}}_k(t) = -(\mathbf{A} - \mathbf{B}\,\mathrm{diag}(\mathbf{u}_k(t))\mathbf{C})^T \mathbf{p}_k(t) - \mathbf{P}_k(t)\mathbf{g}(t); \qquad \mathbf{p}_k(t_f) = \mathbf{0} \qquad (22)$$

then

$$q_k(t, \boldsymbol{\xi}) = \frac{1}{2}\boldsymbol{\xi}^T \mathbf{P}_k(t)\boldsymbol{\xi} + \mathbf{p}_k(t)^T \boldsymbol{\xi}$$

grants s_k *and* $s_{f,k}$ *the property:*

$$s_k(t, \mathbf{x}_k(t), \mathbf{u}_k(t)) = \max_{\boldsymbol{\xi} \in \mathcal{X}(t)} s_k(t, \boldsymbol{\xi}, \mathbf{u}_k(t))$$

$$s_{f,k}(\mathbf{x}_k(t_f)) = \max_{\boldsymbol{\xi} \in \mathcal{X}(t_f)} s_{f,k}(\boldsymbol{\xi})$$

and thus is an improving function.

Proof Substituting q_k into (8) yields

$$s_{f,k}(\mathbf{x}(t_f)) = \frac{1}{2}\mathbf{x}(t_f)^T \mathbf{H}\mathbf{x}(t_f) - \left(\frac{1}{2}\mathbf{x}(t_f)^T \mathbf{P}_k(t_f)\mathbf{x}(t_f) + \mathbf{p}_k(t_f)^T \mathbf{x}(t_f)\right) = 0$$

for all $\mathbf{x}(t_f) \in \mathcal{X}(t_f)$. Hence, $s_{f,k}(\mathbf{x}(t_f)) \le s_{f,k}(\mathbf{x}_k(t_f))$.

By substituting v_i into (20) and then reordering terms, $s_k(t, \mathbf{x}(t), \mathbf{u}_k(t))$ becomes

$$s_k(t, \mathbf{x}(t), \mathbf{u}_k(t)) = \frac{1}{2}\mathbf{x}(t)^T \left(\dot{\mathbf{P}}_k(t) + \mathbf{P}_k(t)(\mathbf{A} - \mathbf{B}\,\mathrm{diag}(\mathbf{u}_k(t))\mathbf{C})\right.$$

$$\left. + (\mathbf{A} - \mathbf{B}\,\mathrm{diag}(\mathbf{u}_k(t))\mathbf{C})^T \mathbf{P}_k(t) + \mathbf{Q}\right) \mathbf{x}(t)$$

$$+ \mathbf{x}(t)^T \left(\dot{\mathbf{p}}_k(t) + (\mathbf{A} - \mathbf{B}\,\mathrm{diag}(\mathbf{u}_k(t))\mathbf{C})^T \mathbf{p}_k(t) + \mathbf{P}_k(t)\mathbf{g}(t)\right)$$

$$+ \mathbf{p}_k(t)^T \mathbf{g}(t) + \frac{1}{2}\sum_{i=1}^{n_u} u_{i,k}(t)^2 r_i$$

As $\dot{\mathbf{P}}_k(t)$ and $\dot{\mathbf{p}}_k(t)$ satisfy (21) and (22), we have

$$s_k(t, \mathbf{x}(t), \mathbf{u}_k(t)) = \frac{1}{2}\mathbf{x}(t)^T \mathbf{0}\mathbf{x}(t) + \mathbf{x}(t)^T \mathbf{0} + \mathbf{p}_k(t)^T \mathbf{g}(t)$$

$$= \mathbf{p}_k(t)^T \mathbf{g}(t) + \frac{1}{2}\sum_{i=1}^{n_u} u_{i,k}(t)^2 r_i$$

Since $s_k(t, \mathbf{x}(t), \mathbf{u}_k(t)) = s_k(t, \mathbf{x}_k(t), \mathbf{u}_k(t))$, it is obvious that

$$s_k(t, \mathbf{x}(t), \mathbf{u}_k(t))) \leq s_k(t, \mathbf{x}_k(t), \mathbf{u}_k(t))$$

for all $\mathbf{x}(t)$.

It can be seen that, if $\mathbf{g} = \mathbf{0}$, then $\mathbf{p}_k(t) = \mathbf{0}$, and the problem reduces to the free vibrations case that is described in [2].

By putting together Sect. 2.2 and the above two lemmas, the sequences $\{q_k\}$ and $\{(\mathbf{x}_k, \mathbf{u}_k)\}$ can be computed where the second one is an improving sequence. As J is nonnegative, it has an infimum and $\{(\mathbf{x}_k, \mathbf{u}_k)\}$ gets arbitrarily close to a candidate optimum.

The resulting algorithm is summarized in Algorithm 1. Its output is an arbitrary approximation for $\mathbf{P}^*$ and $\mathbf{p}^*$, which define the optimal control law. It should be noted that, seemingly, the use of absolute value in step (8) of the iterations stage is theoretically unnecessary. However, it is needed due to practical considerations. Sometimes, numerical computation errors may cause the algorithm to lose its monotonicity when J starts converging.

4 Numerical Example

This section demonstrates the seismic response of a controlled structure whose control trajectories are calculated by the suggested method. The simulations were carried out numerically by MATLAB computational framework.

The model that is used here is the same one suggested by Spencer et al. [19] as a control benchmark problem for seismically excited buildings, except for slight modifications. Here, its response was simulated to El-Centro horizontal ground acceleration input [3]. Peak ground acceleration was set to 0.3 g.

Nine controlled on/off hydraulic dampers are assumed to be embedded in the structure. The model's and the control devices' configuration are shown in Fig. 1. In this figure, z_i is the i-th DOF and w_i is the control force in the adjacent device. The devices are numbered from 1 to 9 in an increasing order, starting from the device mounted in the first floor.

Algorithm 1 CBQR: Algorithm for successive improvement of control process

1: Input: $\mathbf{A}$, $\mathbf{B} = \begin{bmatrix} \mathbf{b}_1 & \mathbf{b}_2 & \ldots \end{bmatrix}$, $\mathbf{C} = \begin{bmatrix} \mathbf{c}_1 & \mathbf{c}_2 & \ldots \end{bmatrix}$, $\mathbf{g}$, $(d_i)_{i=1}^{n_u}$, $(D_i)_{i=1}^{n_u}$, $(w_{i,max})_{i=1}^{n_u}$, $\mathbf{x}(0)$, $\mathbf{Q} \geq 0$, $(r_i | r_i > 0)_{i=1}^{n_u}$, $\mathbf{H} \geq 0$.
2: Initialization:

   (1) Select a convergence tolerance - $\epsilon > 0$.
   (2) Solve:

$$\dot{\mathbf{x}}_0(t) = (\mathbf{A} - \mathbf{B}\,\text{diag}((d_i)_{i=1}^{n_u})\mathbf{C})\mathbf{x}_0(t) + \mathbf{g}(t); \quad \mathbf{x}(0)$$

and set $\mathbf{u}_0(t) \equiv (d_i)_{i=1}^{n_u}$. Solve

$$\dot{\mathbf{P}}_0(t) = -\mathbf{P}_0(t)(\mathbf{A} - \mathbf{B}\,\text{diag}(\mathbf{u}_0(t)))\mathbf{C}) - (\mathbf{A} - \mathbf{B}\,\text{diag}(\mathbf{u}_0(t)))\mathbf{C})^T \mathbf{P}_0(t) - \mathbf{Q}; \quad \mathbf{P}_0(t_f) = \mathbf{H}$$

$$\dot{\mathbf{p}}_0(t) = -(\mathbf{A} - \mathbf{B}\,\text{diag}(\mathbf{u}_0(t)))\mathbf{C})^T \mathbf{p}_0(t) - \mathbf{P}_0(t)\mathbf{g}(t); \quad \mathbf{p}_0(t_f) = \mathbf{0}$$

   (3) Compute: $J_0(\mathbf{x}_0, \mathbf{u}_0) = \frac{1}{2} \int_0^{t_f} \mathbf{x}_0(t)^T \mathbf{Q} \mathbf{x}_0(t) + \sum_{i=1}^{n_u} u_{i,0}(t)^2 r_i \, dt$

3: **for** $k = \{0, 1, 2, \ldots\}$ **do**
4:     Propagate to the improved process by solving:

$$\dot{\mathbf{x}}_{k+1}(t) = (\mathbf{A} - \mathbf{B}\,\text{diag}(\hat{\mathbf{u}}_{k+1}(t, \mathbf{x}_{k+1}(t)))\mathbf{C})\mathbf{x}_{k+1}(t) + \mathbf{g}(t); \quad \mathbf{x}_{k+1}(0) = \mathbf{x}(0)$$

where $v_{i,k}(t, \mathbf{x}(t)) \triangleq \mathbf{b}_i^T (\mathbf{P}_k(t)\mathbf{x}(t) + \mathbf{p}_k(t))\mathbf{c}_i \mathbf{x}(t)/r_i$ and

$$\hat{u}_{i,k+1}(t, \mathbf{x}(t)) = \begin{cases} d_i, & D_i |\mathbf{c}_i \mathbf{x}(t)| > w_{i,max} \\ \arg\min_{v_i \in \{d_i, D_i\}} (v_i - v_{i,k}(t, \mathbf{x}(t)))^2, & \text{otherwise} \end{cases}$$

5:     Set $\mathbf{u}_{k+1}(t) = \hat{\mathbf{u}}_{k+1}(t, \mathbf{x}_{k+1}(t))$.
6:     Solve:

$$\dot{\mathbf{P}}_{k+1}(t) = -\mathbf{P}_{k+1}(t)(\mathbf{A} - \mathbf{B}\,\text{diag}(\mathbf{u}_{k+1}(t))\mathbf{C}) - (\mathbf{A} - \mathbf{B}\,\text{diag}(\mathbf{u}_{k+1}(t))\mathbf{C})^T \mathbf{P}_{k+1}(t) - \mathbf{Q}$$

$$\dot{\mathbf{p}}_{k+1}(t) = -(\mathbf{A} - \mathbf{B}\,\text{diag}(\mathbf{u}_{k+1}(t))\mathbf{C})^T \mathbf{p}_{k+1}(t) - \mathbf{P}_{k+1}(t)\mathbf{g}(t)$$

for $\mathbf{P}_{k+1}(t_f) = \mathbf{H}$ and $\mathbf{p}_{k+1}(t_f) = \mathbf{0}$.
7:     Compute:

$$J(\mathbf{x}_{k+1}, \mathbf{u}_{k+1}) = \frac{1}{2} \int_0^{t_f} \mathbf{x}_{k+1}(t)^T \mathbf{Q} \mathbf{x}_{k+1}(t) + \sum_{i=1}^{n_u} u_{i,k+1}(t)^2 r_i \, dt$$

8:     If $|J(\mathbf{x}_k, \mathbf{u}_k) - J(\mathbf{x}_{k+1}, \mathbf{u}_{k+1})| < \epsilon$, stop iterating, otherwise—continue.
9: **end for**
10: return $\mathbf{P}_{k+1}, \mathbf{p}_{k+1}$.

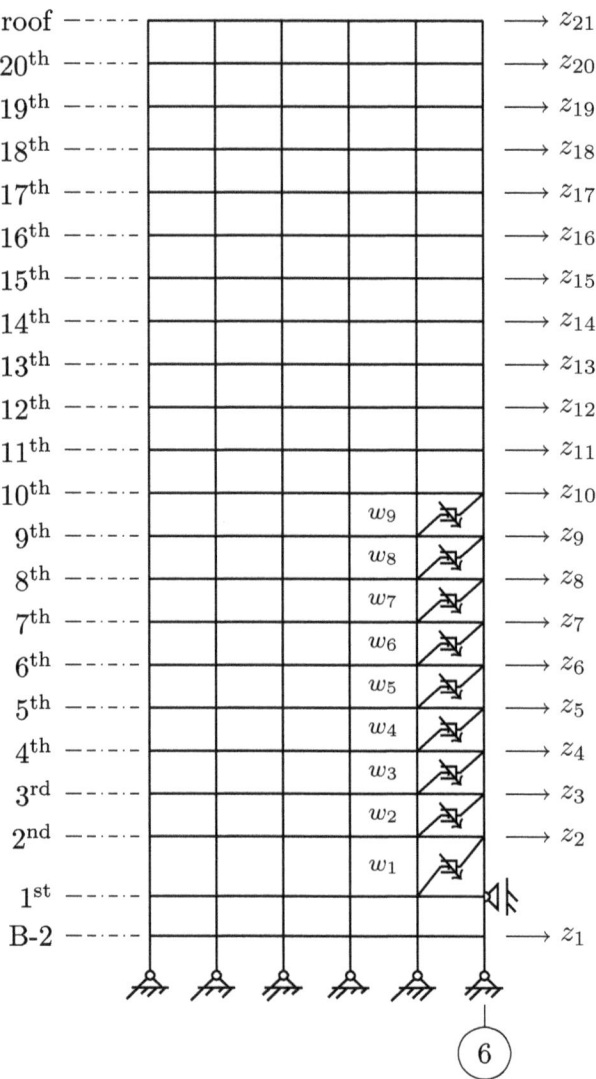

Fig. 1 Evaluation model and dampers configuration

State-space model was formulated according to (2). The external excitation is $\mathbf{e} = \boldsymbol{\gamma}^T \ddot{z}_g$, where $\boldsymbol{\gamma} = \begin{bmatrix} 1 & 1 & \ldots & 1 \end{bmatrix}^T \in \mathbb{R}^{21}$ and $\ddot{z}_g$ is the earthquake input.

The response of three cases was analyzed:

Case 1: There are no control devices.
Case 2: The control law is the clipped optimal control law [13].
Case 3: The control law is a CBQR one.

The clipped optimal control logic, which was used in case 1, is based on the prevalent LQR control law. The clipping logic is described in previous studies [13].

In accordance with (14), each control force w_i is associated with an equivalent damping gain—u_i. Identical properties were set for all the devices. The on/off gains were defined as $D = 5 \times 10^6$ kg/s and $d = 2 \times 10^5$ kg/s. The maximal allowable control force was set to $w_{max} = 20 \times 10^3$ kN.

The i-th rows in the observation matrix $\mathbf{C} \in \mathbb{R}^{9 \times 42}$ are $\mathbf{c}_i$. The state weighting matrix $\mathbf{Q}$ for cases 1 and 2 was chosen such that

$$\mathbf{x}(t)^T \mathbf{Q} \mathbf{x}(t) = 5 \times 10^{18} \left(z_1(t)^2 + z_2(t)^2 + \sum_{i=3}^{21} (z_i(t) - z_{i-1}(t))^2 \right)$$

Such a weighting accounts for the inter-story drifts in the structure, which is a common evaluation quantity in seismic practice [19]. It can be obtained here by letting $\mathbf{Q} = 5 \times 10^{18} \mathbf{N}^T \mathbf{N}$, where $\mathbf{N} \in \mathbb{R}^{n \times n}$ is defined by $(\mathbf{N})_{i,i} = 1$ for $1 \leq i \leq 21$, $(\mathbf{N})_{i+1,i} = -1$ for $2 \leq i \leq 20$, and $(\mathbf{N})_{i,j} = 0$ in the other elements. Unlike the states' weighting, which has the same meaning in cases 1 and 2, the control weighting for case 1 has different interpretation than that of case 2. In the LQR method, which underpins case 1, the control weighting relates to the control forces, whereas in case 2 the CBQR control weighting relates to the equivalent damping gains. It means that case 1 and case 2 have completely different design goals. Hence, in order to create a common comparison basis, case 1 control weighting was chosen such that the Euclidean norm $\|(u_i^{c1})_{i=1}^{n_u}\|$ will be approximately the same as $\|(u_i^{c2})_{i=1}^{n_u}\|$. To this end, $(r_i^{c1})_{i=1}^{9} = (1, 1, \ldots, 1) \times 4.7 \times 10^5$ and $(r_i^{c2})_{i=1}^{9} = (1, 1, \ldots, 1) \times 10^{-4}$ were set for cases 1 and 2, respectively.

The initial state vector was set to zero.

Figure 2 shows the progress of the performance index during 9 design iterations. A dramatic improvement can be seen after the first iteration. Practically, the algorithm converged after the second one. Figure 3 shows the inter-story drifts of the 10th floor during the first 10 s of the response. It can be seen that cases 2 and 3 present pretty close performance with a slight advantage in favor of case 3. The peak inter-story drifts throughout the building are given in Fig. 4. It can be seen that

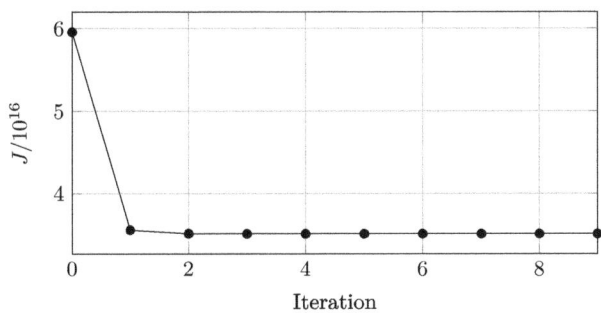

Fig. 2 Performance index values in each iteration

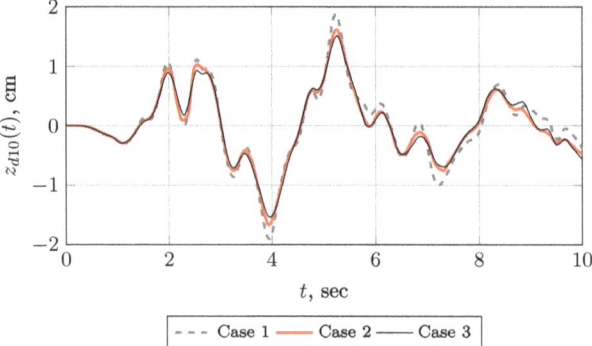

Fig. 3 Inter-story drifts in the 10th floor

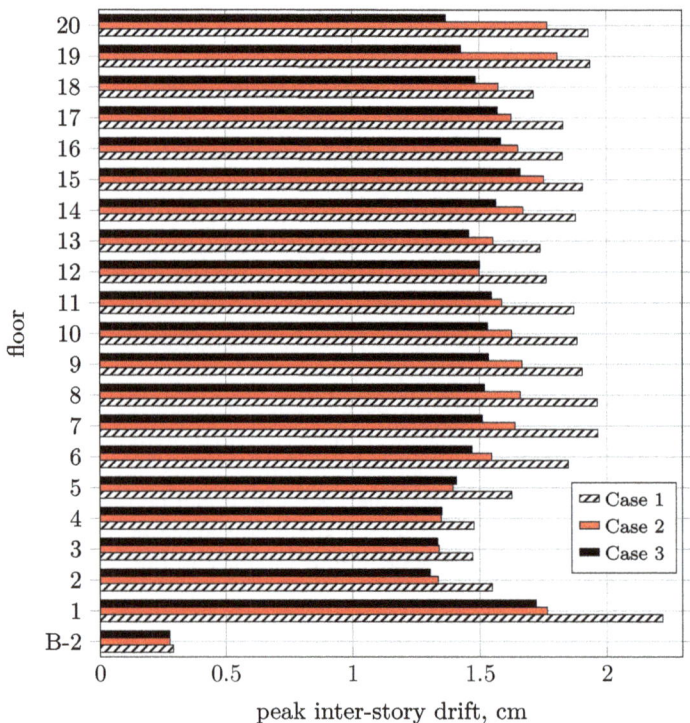

Fig. 4 Peak inter-story drifts throughout the structure

case 3 attained additional improvement, compared to case 2. The control policies of cases 2 and 3 resulted with different control signals. Figure 5 shows the first 10 s of the control signal u_1, synthesized by each control law for the first device, located in the first floor. The variations in the signal express the valve's open/close commands in this device, generated in effort to regulate the vibrations. Figure 6 shows the form of control force w_1, generated during the first 10 s in the same device. The sharp changes reflect moments when the valve's state was switched in the device.

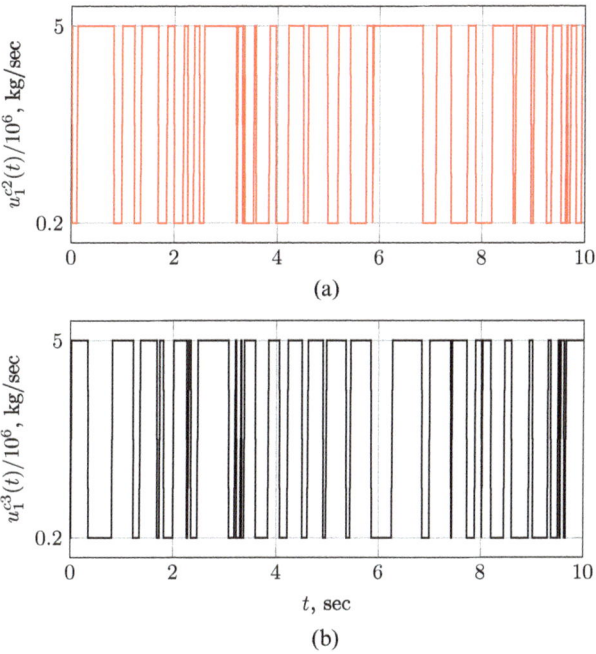

Fig. 5 Control signal u_1 in (**a**) case 2 and (**b**) case 3

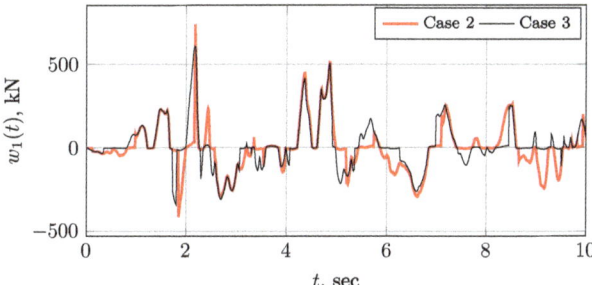

Fig. 6 Control force w_1

References

1. Agrawal, A., Yang, J.: A semi-active electromagnetic friction damper for response control of structures. In: Advanced Technology in Structural Engineering, chap. 5, pp. 1–8. American Society of Civil Engineers, Reston (2000)
2. Halperin, I., Agranovich, G., Ribakov, Y.: a method for computation of realizable optimal feedback for semi-active controlled structures. In: EACS 2016 6th European Conference on Structural Control, pp. 1–11 (2016)
3. Halperin, I., Ribakov, Y., Agranovich, G.: Optimal viscous dampers gains for structures subjected to earthquakes. Struct. Control Health Monit. **23**(3), 458–469 (2016). https://doi.org/10.1002/stc.1779
4. Halperin, I., Agranovich, G., Ribakov, Y.: Optimal control of a constrained bilinear dynamic system. J. Optim. Theory Appl. **174**, 1–15 (2017). https://doi.org/10.1007/s10957-017-1095-2
5. Halperin, I., Agranovich, G., Ribakov, Y.: Using constrained bilinear quadratic regulator for the optimal semi-active control problem. J. Dyn. Syst. Meas. Control **139**(11), 111011 (2017). https://doi.org/10.1115/1.4037168
6. Halperin, I., Agranovich, G., Ribakov, Y.: Optimal control synthesis for the constrained bilinear biquadratic regulator problem. Optim. Lett. **12**(8), 1855–1870 (2018)
7. Halperin, I., Agranovich, G., Ribakov, Y.: Extension of the constrained bilinear quadratic regulator to the excited multi-input case. J. Optim. Theory Appl. (2019). https://doi.org/10.1007/s10957-019-01479-x
8. Halperin, I., Agranovich, G., Ribakov, Y.: Multi-input control design for a constrained bilinear biquadratic regulator with external excitation. Optim. Control Appl. Methods **40**(6), 1045–1053 (2019). https://doi.org/10.1002/oca.2533. https://onlinelibrary.wiley.com/doi/abs/10.1002/oca.2533
9. Housner, G., Bergman, L., Caughey, T., Chassiakos, A., Claus, R., Masri, S., Skelton, R., Soong, T., Jr., B.S., Yao, J.: Structural control: Past, present and future. J. Eng. Mech. **123**(9), 897–971 (1997)
10. Ikeda, Y.: Active and semi-active control of buildings in Japan. J. Jpn Assoc. Earthquake Eng. **4**(3), 278–282 (2004)
11. Karnopp, D.: Design principles for vibration control systems using semi-active dampers. J. Dyn. Syst. Meas. Control **112**(3), 448–455 (1990). https://doi.org/10.1115/1.2896163
12. Krotov, V.F.: Global Methods in Optimal Control Theory. Chapman & Hall/CRC Pure and Applied Mathematics. Taylor & Francis, Milton Park (1995)
13. Luca, S.G., Pastia, C.: Case study of variable orifice damper for seismic protection of structures. Buletinul Institutului Politehnic din Iasi. Sectia Constructii, Arhitectura **55**(1), 39 (2009)
14. Patten, W.N., Kuo, C.C., He, Q., Liu, L., Sack, R.L.: Seismic structural control via hydraulic semi-active vibration dampers (SAVD). In: Proceedings of the 1st World Conference on Structural Control (1994)
15. Ribakov, Y., Gluck, J.: Active control of MDOF structures with supplemental electrorheological fluid dampers. Earthquake Eng. Struct. Dyn. **28**(2), 143–156 (1999)
16. Robinson, W.D.: A pneumatic semi-active control methodology for vibration control of air spring based suspension systems. Ph.D. Thesis, Iowa State University (2012)
17. Sadek, F., Mohraz, B.: Semi-active control algorithms for structures with variable dampers. J. Eng. Mech. **124**(9), 981–990 (1998)
18. Spencer Jr., B.F., Nagarajaiah, S.: State of the art of structural control. J. Struct. Eng. **127**(7), 845–856 (2003)
19. Spencer Jr., B.F., Christenson, R.E., Dyke, S.J.: Next generation benchmark control problems for seismically excited buildings. In: Proceedings of the 2nd World Conference on Structural Control, Japan, vol. 2, pp. 1335–1360 (1998)

20. Symans, M.D., Constantinou, M.C., Taylor, D.P., Garnjost, K.D.: Semi-active fluid viscous dampers for seismic response control. In: First World Conference on Structural Control (vol. 3) (1994)
21. Wang, D.H., Liao, W.H.: Semiactive controllers for magnetorheological fluid dampers. J. Intell. Mat. Syst. Struct. **16**(11–12), 983–993 (2005)
22. Yuen, K.V., Shi, Y., Beck, J.L., Lam, H.F.: Structural protection using MR dampers with clipped robust reliability-based control. Structural and multidisciplinary optimization **34**(5), 431–443 (2007). https://doi.org/10.1007/s00158-007-0097-3

Multi-Displacement Requirement in a Topology Optimization Algorithm Based on Non-uniform Rational Basis Spline Hyper-Surfaces

Marco Montemurro, Thibaut Rodriguez, Paul Le Texier, and Jérôme Pailhès

1 Introduction

During the last 30 years, Topology Optimization (TO) has obtained an increasing attention to such an extent that, today, it constitutes an important field of research in both academic and industrial communities. Generally speaking, TO for structural applications aims at determining the best distribution of the material in a given domain to satisfy the design requirements (DRs) of the problem at hand. To this end, a significant amount of studies has been devoted to the development of suitable algorithms for TO. Since the first pioneering works, using the so-called homogenisation method [1–3] for shape optimization problems in structural mechanics, important steps forwards have been done. For instance, the Evolutionary Structural Optimization (ESO) method, introduced in [4], is based on the combination of a metaheuristic algorithm and the Finite Element (FE) method. An extension of the ESO method is the well-known Bi-directional Evolutionary Structural Optimization (BESO) [5]. Later, the BESO approach has been reformulated in [6, 7], by adding features to obtain mesh-independent results, without checker-board pattern and by introducing a sensitivity number averaging method to make easy convergence.

M. Montemurro (✉) · J. Pailhès
Arts et Métiers Institute of Technology, Université de Bordeaux, CNRS, INRA, Bordeaux INP, HESAM Université, Talence, France
e-mail: marco.montemurro@ensam.eu; marco.montemurro@u-bordeaux.fr

T. Rodriguez
Arts et Métiers Institute of Technology, Université de Bordeaux, CNRS, INRA, Bordeaux INP, HESAM Université, Talence, France

French Atomic Energy Commission, Le Barp Cedex, France

P. Le Texier
French Atomic Energy Commission, Le Barp Cedex, France

Although new meta-heuristic-based methods are increasingly developed in the literature to deal with TO problems, the density-based approaches [1, 8, 9], and the Level-Set Method (LSM) [10, 11], are the most popular and well-established techniques.

In the framework of the LSM, the topology descriptor is represented by a level-set function (LSF), whose sign can be conventionally associated to solid or void zones, while the zero value represents the boundary of the optimized structure [12]. It is noteworthy that the LSM makes use of the mesh of the FE model only to assess the physical responses involved in the problem formulation. A detailed discussion of the LSM for TO is available in [10, 11]. Often, the LSF is parametrized on the design domain by using dedicated *basis functions*, which can be of different mathematical nature. The most commonly employed basis functions for TO problems are either Radial Basis Functions, Spectral Parametrization Functions, and Non-Uniform Rational Basis Spline (NURBS) entities. A wide discussion on this topic can be found in [13].

Due to their efficiency and robustness, pioneering density-based methods for TO [1, 8, 9] are still widely studied and employed in both scientific and industrial communities. In these methodologies, the topological variable is represented by a fictitious density function, taking values in the interval [0, 1], which is affected to each element of the mesh of the FE model to penalize the constitutive tensors used to describe the physical behaviour of the structure. Lower and upper bounds of the density function correspond to "void" and "solid" phases, respectively. In this background, to force convergence towards well-defined topology boundary, elements characterized by intermediate values of the density function are penalized during optimization. The Solid Isotropic Material with Penalization (SIMP) scheme is the most common penalty approach used for TO [8]. The success of the SIMP method is due to its efficiency and compactness [9]: several applications of this method can be found in the literature [8]. Moreover, commercial software for TO, like OptiStruct® [14] and TOSCA® [15], running in Altair-Hyperworks® and Abaqus® environments, respectively, make use of the SIMP approach. Several research works make use of the SIMP method to deal with various optimization problems: compliance minimization, mass minimization, maximization of the first buckling load or of the first natural frequency, etc. [8]. Nevertheless, two main drawbacks affect the SIMP method. Firstly, the element-wise description of the topology does not allow obtaining a smooth topology compatible with computer-aided design (CAD) software. Accordingly, a time-consuming CAD reconstruction/reassembly phase must be performed to post-process the results. Secondly, to overcome the well-known checker-board and mesh dependence effects characterizing the SIMP approach, problem-dependent distance-based filters [8] or projection methods [16, 17] must be introduced. Moreover, when the problem formulation includes several DRs (especially those involving local responses, like failure criteria, local displacements, damage phenomena, etc.), the optimization constraints on the CAD reassembled geometry are often not met.

To overcome the aforementioned issues, NURBS entities have been coupled to the SIMP method. To this end, a general approach has been developed at the I2M laboratory in Bordeaux [18, 19], which is referred to as *NURBS-based SIMP method* in the following. Unlike the classical SIMP approach, the NURBS-based SIMP method separates the pseudo-density field, describing the topology of the continuum, from the mesh of the FE model. More precisely, if the dimension of the TO problem is D, a NURBS hyper-surface of dimension $D + 1$ is needed as a topology descriptor. This entity is used to describe the pseudo-density field, which is projected onto the mesh of the FE model in order to penalize the element stiffness matrix according to the SIMP method. In this background, the optimisation variables are both the density at the control points (CPs) and the associated weights of the NURBS entity. The use of NURBS entities to describe the topology makes the CAD reconstruction phase a straightforward task for both 2D [20] and 3D [21] TO problems. More details on the NURBS-based SIMP method are available in [18, 19, 22–27]. Further research works make use of the isogeometric analysis (IGA) approach in the framework of TO in order to fully exploit the advantages related to NURBS entities. In particular, in [28] the IGA approach is coupled to trimmed spline surfaces to represent topology changes during the optimization process. Moreover, recently, the Moving Morphable Component (MMC) [29] and the Moving Morphable Void (MMV) [30] approaches have been proposed for TO problems.

Regarding the integration of a DR on structural displacements in TO problems, this DR is often implemented as a constraint in the volume (or mass) minimization problem. For example, the SIMP method associated with a design space adjustment technique to consider multi-displacement constraints is used in [31]. However, the resulting optimized topologies are strongly mesh-dependent and problem-dependent because the whole process is based on a sort of "repairing algorithm" (i.e. a heuristic procedure) whose tuning parameters must be changed according to the DRs of the problem at hand. Mesh-dependent optimized topologies occur also in the analyses presented in [32, 33], which make use of the SIMP method for solving TO problems involving requirements on mass/volume and structural displacement as well.

In [34], the element independent nodal density (EIND) method is applied to TO problems involving a *global formulation* of the structural displacement constraint. In this work, the maximum displacement occurring in the structure is approximated through the p-norm operator and its gradient is evaluated thanks to the adjoint method. The advantage of the formulation proposed in [34] is that the user does not need to know a-priori the location where the maximum displacement occurs. Conversely, since a local DR is transformed into a global one due to the introduction of the p-norm function, the main drawback of such an approach is that the overall topology is modified, at each iteration, instead of producing only local modifications in order to satisfy the constraint on the structural displacement. The independent continuous mapping (ICM) method is used in [35], where the constraint on the local displacement is integrated in the TO of multi-material structures with the aim of minimizing the structural weight. However, only 2D problems are considered in [35] and only the displacements of the loaded regions are included in the problem formulation. A Bi-level Programming Approach for truss TO problems subject to local displacement DR is presented in [36]. Of course, the DR on the structural displacement is often integrated in design problems dealing with compliant mechanisms, firstly introduced in [37]. Compliant mechanisms have also been analysed in the framework of either robust density-based algorithms for TO, by adding a suitable filtering technique [38], or within the LSM [39]. An alternative approach using meta-heuristics for designing compliant mechanisms is proposed in [31]. The integration of requirements on multiple displacements into the problem formulation is discussed in [40], as well.

In this work, the theoretical formulation and numerical framework proposed in [24] to include requirements on structural displacements in the NURBS-based SIMP method is here extended to the case of multi-displacement DRs. As discussed in [24], the analytic form of the generic structural displacement requirement (and of its gradient) for both loaded and non-loaded regions of the design domain is derived by exploiting the main properties of NURBS entities and the adjoint method. The multi-displacement DR can be integrated into the problem formulation as either objective function or constraint function. In particular, the formulation of such a requirement in the framework of NURBS hyper-surfaces takes advantage of the *local support property* of the NURBS blending functions [41], which establishes an implicit relationship among the pseudo-densities of adjacent elements. Thanks to this property there is no need of introducing problem-dependent filtering schemes, unlike the classical SIMP method. Moreover, the optimized topology boundary is expressed as a native CAD entity which can be directly imported into a CAD software and exploited for Additive Layer Manufacturing (ALM) production, via a STEP-NC (STandard for the Exchange of Product model data compliant Numerical Control) model [42]. The effectiveness of the proposed approach is tested on both 2D and 3D benchmark problems taken from the literature and by comparing the results to those provided by commercial software.

The chapter is organized as follows. In Sect. 2, the theoretical background of the NURBS hyper-surfaces is briefly recalled. Section 3 presents the problem formulation in the framework of the NURBS-based SIMP method for TO including the multi-displacement DRs, together with its gradient. The effectiveness of the proposed formulation is tested on both 2D and 3D benchmark problems in Sect. 4. Moreover, the influence of the NURBS blending functions discrete parameters on the optimized topology is also investigated. Finally, Sect. 5 ends the paper with some conclusions and prospects.

Notation

Upper-case bold letters and symbols are used to indicate tensors and matrices, while lower-case bold letters and symbols indicate column vectors. $\sharp(\cdots)$ denotes the cardinality of the generic quantity $(\cdots)$, i.e. the number of elements belonging to a generic set, vector, array, etc.

2 Fundamentals of NURBS Hyper-Surfaces

The fundamentals of NURBS hyper-surfaces are briefly recalled here below. Curves and surfaces formulæ, widely discussed in [41], can be easily deduced from the following relations. A NURBS hyper-surface is a polynomial-based function, defined over a *parametric space* (domain), taking values in the *NURBS space* (co-domain). Therefore, if N is the dimension of the *parametric space* and M is the dimension of the *NURBS space*, a NURBS entity is defined as $\mathbf{h} : \mathbb{R}^N \longrightarrow \mathbb{R}^M$. For example, one scalar parameter ($N = 1$) can describe both a plane curve ($M = 2$) and a 3D curve ($M = 3$). In the case of a surface, two scalar parameters are needed ($N = 2$) together with, of course, three physical coordinates $M = 3$. The mathematical formula of a generic NURBS hyper-surface is

$$\mathbf{h}(u_1, \ldots, u_N) = \sum_{i_1=0}^{n_1} \cdots \sum_{i_N=0}^{n_N} R_{i_1,\ldots,i_N}(\zeta_1, \ldots, \zeta_N) \mathbf{P}_{i_1,\ldots,i_N}, \qquad (1)$$

where $R_{i_1,\ldots,i_N}(\zeta_1, \ldots, \zeta_N)$ are the piece-wise rational basis functions, which are related to the standard NURBS blending functions $N_{i_k,p_k}(\zeta_k)$, $k = 1, \ldots, N$ by means of the relationship

$$R_{i_1,\ldots,i_N}(\zeta_1, \ldots, \zeta_N) = \frac{\omega_{i_1,\ldots,i_N} \prod_{k=1}^{N} N_{i_k,p_k}(\zeta_k)}{\sum_{j_1=0}^{n_1} \cdots \sum_{j_N=0}^{n_N} \left[\omega_{j_1,\ldots,j_N} \prod_{k=1}^{N} N_{j_k,p_k}(\zeta_k) \right]}. \qquad (2)$$

In Eqs. (1) and (2), $\mathbf{h}(\zeta_1, \ldots, \zeta_N)$ is a M-dimension vector-valued rational function, $(\zeta_1, \ldots, \zeta_N)$ are scalar dimensionless parameters defined in the interval $[0, 1]$, whilst $\mathbf{P}_{i_1,\ldots,i_N}$ are the CPs. The j-th CP coordinate ($X^{(j)}_{i_1,\ldots,i_N}$) is stored in the array $\mathbf{X}^{(j)} \in \mathbb{R}^{(n_1+1) \times \cdots \times (n_N+1)}$. The explicit expression of CPs coordinates in $\mathbb{R}^M$ is:

$$\mathbf{P}_{i_1,\ldots,i_N} = \{X^{(1)}_{i_1,\ldots,i_N}, \ldots, X^{(M)}_{i_1,\ldots,i_N}\},$$

$$\mathbf{X}^{(j)} \in \mathbb{R}^{(n_1+1) \times \cdots \times (n_N+1)}, \quad j = 1, \ldots, M. \qquad (3)$$

For NURBS surfaces, $\mathbf{P}^{\mathrm{T}}_{i_1,i_2} = \{X^{(1)}_{i_1,i_2}, X^{(2)}_{i_1,i_2}, X^{(3)}_{i_1,i_2}\}$ and each coordinate is arranged in a matrix defined in $\mathbb{R}^{(n_1+1) \times (n_2+1)}$. The CPs layout is referred as *control polygon* for NURBS curves, *control net* for surfaces and *control hyper-net* otherwise [41]. The overall number of CPs constituting the hyper-net is:

$$n_{\mathrm{CP}} := \prod_{i=1}^{N} (n_i + 1). \qquad (4)$$

The generic CP does not actually belong to the NURBS entity but it affects the NURBS shape by means of its coordinates. A suitable weight $\omega_{i_1,\ldots,i_N}$ is related to the respective CP $\mathbf{P}_{i_1,\ldots,i_N}$. The higher the weight, the more the NURBS entity is attracted towards the CP. For each parametric direction ζ_k, $k = 1, \ldots, N$, the NURBS blending functions are of degree p_k and can be defined in a recursive way as

$$N_{i_k,0}(\zeta_k) = \begin{cases} 1, & \text{if } v_{i_k}^{(k)} \leq \zeta_k < v_{i_k+1}^{(k)}, \\ 0, & \text{otherwise}, \end{cases} \tag{5}$$

$$N_{i_k,q}(\zeta_k) = \frac{\zeta_k - v_{i_k}^{(k)}}{v_{i_k+q}^{(k)} - v_{i_k}^{(k)}} N_{i_k,q-1}(\zeta_k) + \\ + \frac{v_{i_k+q+1}^{(k)} - \zeta_k}{v_{i_k+q+1}^{(k)} - v_{i_k+1}^{(k)}} N_{i_k+1,q-1}(\zeta_k), \ q = 1, \ldots, p_k, \tag{6}$$

where each constitutive blending function is defined on the knot vector

$$\mathbf{v}^{(k)\mathrm{T}} = \{\underbrace{0, \ldots, 0}_{p_k+1}, v_{p_k+1}^{(k)}, \ldots, v_{m_k-p_k-1}^{(k)}, \underbrace{1, \ldots, 1}_{p_k+1}\}, \tag{7}$$

whose dimension is $m_k + 1$, with

$$m_k = n_k + p_k + 1. \tag{8}$$

Each knot vector $\mathbf{v}^{(k)}$ is a non-decreasing sequence of real numbers that can be interpreted as a discrete collection of values of the related dimensionless parameter u_k. The NURBS blending functions are characterized by several interesting properties: the interested reader is addressed to [41] for a deeper insight into the matter. Here, only the *local support property* is recalled because it is of paramount importance for the NURBS-based SIMP method for TO [18, 19]:

$$\text{if } (u_1, \ldots, u_N) \in \left[v_{i_1}^{(1)}, v_{i_1+p_1+1}^{(1)}\right[\times \cdots \times \left[v_{i_N}^{(N)}, v_{i_N+p_N+1}^{(N)}\right[. \tag{9}$$

The above formula means that each CP (and the respective weight) affects only a precise zone of the *parametric space*, which is denoted as *local support* or *influence zone*.

3 The NURBS-Based SIMP Method for Topology Optimization

3.1 Generalities

The details of the formulation of the SIMP method in the NURBS hyper-surfaces framework are given in [18, 19]. The main features of the method are briefly recalled here only for 3D TO problems for a fruitful understanding of the study. Consider the compact Euclidean space $\mathcal{D} \subset \mathbb{R}^3$ in a Cartesian orthogonal frame $O(x_1, x_2, x_3)$:

$$\mathcal{D} := \{\mathbf{x}^\mathrm{T} = (x_1, x_2, x_3) \in \mathbb{R}^3 : x_1 \in [0, a_1], x_2 \in [0, a_2], x_3 \in [0, a_3]\}, \tag{10}$$

where a_j, $j = 1, 2, 3$ is the reference length of the domain along the j-th axis. Without loss of generality, the mathematical formulation is here limited, for the sake of clarity, to the problem of minimizing the compliance of a structure subject to an inequality constraint on the volume. This problem can be mathematically well-posed through several techniques, widely discussed in [8]. The aim of TO is to search for the best distribution of a given "heterogeneous material" (i.e. the definition of void and material zones) satisfying the requirements of the problem at hand.

Consider the equilibrium equation (static case) of the FE model in the case of zero Dirichlet's boundary conditions (BCs) and non-zero Neumann's BCs:

$$\mathbf{K}\mathbf{u} = \mathbf{f}, \ \mathbf{u}, \mathbf{f} \in \mathbb{R}^{N_\mathrm{DOF}}, \ \mathbf{K} \in \mathbb{R}^{N_\mathrm{DOF} \times N_\mathrm{DOF}}, \tag{11}$$

where N_DOF is the overall number of unknown degrees of freedom (DOFs), $\mathbf{K}$ is the stiffness matrix of the FE model, while $\mathbf{f}$ and $\mathbf{u}$ are the vectors of the external generalized nodal forces and displacements, respectively. It is noteworthy that, in the case of zero Dirichlet's BCs, the compliance of the structure is defined as:

$$c := \mathbf{f}^\mathrm{T}\mathbf{u}. \tag{12}$$

In this case, the compliance coincides with the work of external forces (which equals the work of internal forces of the FE model). In the SIMP approach, the *material domain* $\Omega \subseteq \mathcal{D}$ is identified by means of a pseudo-density function $\rho(\mathbf{x}) \in [0, 1]$ for $\mathbf{x} \in \mathcal{D}$: $\rho(\mathbf{x}) = 0$ denotes the void phase, whereas $\rho(\mathbf{x}) = 1$ identifies the solid phase. The density field affects the element stiffness matrix and, accordingly, the global stiffness matrix of the FE model as follows:

$$\mathbf{K} := \sum_{e=1}^{N_e} \rho_e^\alpha \mathbf{L}_e^\mathrm{T} \mathbf{K}_e^0 \mathbf{L}_e = \sum_{e=1}^{N_e} \mathbf{L}_e^\mathrm{T} \mathbf{K}_e \mathbf{L}_e, \tag{13}$$

$$\mathbf{K}_e^0, \mathbf{K}_e \in \mathbb{R}^{N_\mathrm{DOF}^e \times N_\mathrm{DOF}^e}, \ \mathbf{L}_e \in \mathbb{R}^{N_\mathrm{DOF}^e \times N_\mathrm{DOF}},$$

where ρ_e is the fictitious density computed at the centroid of the generic element e, whilst $\alpha \geq 1$ is a suitable parameter that aims at penalizing the intermediate densities between 0 and 1, in agreement with the classic SIMP approach ($\alpha = 3$ in this study). N_e is the total number of elements and N_{DOF}^e is the number of DOFs of the generic element. In Eq. (13), $\mathbf{K}_e^0$ and $\mathbf{K}_e$ are the non-penalized and the penalized stiffness matrices of element e, expressed in the global reference frame of the FE model, whilst $\mathbf{L}_e$ is the connectivity matrix of element e defined as:

$$\mathbf{u}_e = \mathbf{L}_e \mathbf{u}, \tag{14}$$

where $\mathbf{u}_e \in \mathbb{R}^{N_{\text{DOF}}^e}$ is the vector of nodal displacements for element e. In the context of the NURBS-based SIMP method, the pseudo-density field for a TO problem of dimension D is represented through a NURBS hyper-surface of dimension $D+1$. Therefore, for a 3D problem a 4D entity is needed and the pseudo-density field is defined as:

$$\rho(\zeta_1, \zeta_2, \zeta_3) = \sum_{i_1=0}^{n_1} \sum_{i_2=0}^{n_2} \sum_{i_3=0}^{n_3} R_{i_1,i_2,i_3}(\zeta_1, \zeta_2, \zeta_3) \rho_{i_1,i_2,i_3}. \tag{15}$$

In Eq. (15), $\rho(\zeta_1, \zeta_2, \zeta_3)$ constitutes the fourth coordinate of the array $\mathbf{h}$ of Eq. (1), while $R_{i_1,i_2,i_3}(\zeta_1, \zeta_2, \zeta_3)$ are the NURBS rational basis functions of Eq. (2). The dimensionless parameter ζ_j can be related to the Cartesian coordinates as follows:

$$\zeta_j = \frac{x_j}{a_j}, \quad j = 1, 2, 3. \tag{16}$$

As discussed in Sect. 2, different parameters affect the shape of a NURBS entity. Among them, the pseudo-density at CPs and the associated weights are referred to as *design variables* in the following and are collected in the vectors $\boldsymbol{\xi}_1$ and $\boldsymbol{\xi}_2$, respectively, defined as:

$$\boldsymbol{\xi}_1^T := (\rho_{0,0,0}, \cdots, \rho_{n_1,n_2,n_3}), \quad \boldsymbol{\xi}_2^T := (\omega_{0,0,0}, \cdots, \omega_{n_1,n_2,n_3}), \quad \boldsymbol{\xi}_1, \boldsymbol{\xi}_2 \in \mathbb{R}^{n_{\text{CP}}}. \tag{17}$$

According to the above formula, in the most general case, the overall number of design variables is $n_{\text{var}} = 2n_{\text{CP}}$. Thus, the classic TO problem of compliance minimization subject to an inequality constraint on the volume can be formulated as:

$$\min_{\boldsymbol{\xi}_1, \boldsymbol{\xi}_2} \frac{c}{c_{\text{ref}}}, \text{ s.t. } \begin{cases} \mathbf{K}\mathbf{u} = \mathbf{f}, \quad \dfrac{V}{V_{\text{ref}}} - \gamma \leq 0, \\ \xi_{1k} \in [\rho_{\min}, \rho_{\max}], \quad \xi_{2k} \in [\omega_{\min}, \omega_{\max}], \\ \forall k = 1, \ldots, n_{\text{CP}}. \end{cases} \tag{18}$$

In Eq. (18), V_{ref} is a reference volume, V is the volume of the material domain Ω, while γ is the volume fraction. $\rho_{\min}$ and $\rho_{\max}$ are the lower and upper bounds on the pseudo-density at each CP, while $\omega_{\min}$ and $\omega_{\max}$ are the bounds on the generic weight. It is noteworthy that the lower bound of the pseudo-density is strictly positive to prevent any singularity for the solution of the equilibrium problem. The objective function is divided by a reference compliance, c_{ref}, to obtain a dimensionless value.

The volume of the material domain appearing in Eq. (18) is defined as:

$$V := \sum_{e=1}^{N_e} \rho_e V_e, \tag{19}$$

where V_e is the volume of element e. Moreover, in Eq. (18), the linear index k has been introduced for the sake of compactness. The relation between k and i_j, ($j = 1, 2, 3$) is:

$$k := 1 + i_1 + i_2(n_1 + 1) + i_3(n_1 + 1)(n_2 + 1). \tag{20}$$

The other parameters involved in the definition of the NURBS entity (i.e. degrees, knot-vector components and number of CPs) are kept constant and their value is set a-priori at the beginning of the TO analysis.

The computation of the derivatives of both objective and constraint functions with respect to the design variables is needed to solve problem (18) through a deterministic algorithm. This task is achieved by exploiting the local support property of Eq. (9). For instance, the general expressions of the derivatives of both the compliance and the volume [18, 19] read

$$\frac{\partial c}{\partial \xi_{ik}} = -\alpha \sum_{e \in S_k} \frac{c_e}{\rho_e} \frac{\partial \rho_e}{\partial \xi_{ik}}, \quad i = 1, 2, \; k = 1, \cdots, n_{\text{CP}}, \tag{21}$$

$$\frac{\partial V}{\partial \xi_{ik}} = \sum_{e \in S_k} V_e \frac{\partial \rho_e}{\partial \xi_{ik}}, \quad i = 1, 2, \; k = 1, \cdots, n_{\text{CP}}, \tag{22}$$

where c_e is the compliance of the generic element, whilst S_k is the discretized version of the local support of Eq. (9), while $\dfrac{\partial \rho_e}{\partial \xi_{ik}}$ reads

$$\frac{\partial \rho_e}{\partial \xi_{ik}} = \begin{cases} R_k^e, & \text{if } i = 1, \\ \dfrac{R_k^e}{\xi_{2k}} (\xi_{1k} - \rho_e), & \text{if } i = 2. \end{cases} \tag{23}$$

The scalar quantity R_k^e appearing in Eq. (23) is the NURBS rational basis function of Eq. (2) evaluated at the element centroid.

The NURBS-based SIMP approach is characterized by the following advantages: (1) the number of design variables is unrelated to the number of elements; (2) the optimized topology is unrelated to the quality of the mesh of the FE model; (3) the local support property implicitly ensures a filtering effect, i.e. each CP (with the related weight) affects only those elements whose centroid falls in the local support $\mathcal{S}_k$. This fact is equivalent to the definition of an explicit filter in classic SIMP approaches, which is introduced to avoid numerical artefacts (such as the well-known "checker-board effect"). For a deeper insight in the NURBS-based SIMP method, the reader is addressed to [18, 19].

3.2 Formulation of the Structural Displacement Requirement

Let **u** be the solution of Eq. (11). Let $\mathcal{I}_d := \{i \in \mathbb{N} \mid 1 \leq i \leq N_{\text{DOF}}\}$, with $\sharp \mathcal{I}_d = N_d < N_{\text{DOF}}$, be the set collecting the N_d indices of the DOFs on which a constraint is imposed. Let index τ denotes the position, in **u**, of the component of the displacement along the x_j axis of the generic point P, which corresponds to the DOF numbered τ in the global frame of the FE model. It is convenient to introduce the vector $\mathbf{a}_\tau \in \mathbb{R}^{N_{\text{DOF}}}$, whose components are all equal to zero, except the one in position τ which takes a unit value, i.e.

$$a_r := \begin{cases} 0, & \text{if } r \neq \tau, \\ 1, & \text{if } r = \tau. \end{cases} \quad (24)$$

In this background, the DOF at position $1 \leq \tau \leq N_{\text{DOF}}$ can be expressed as:

$$u_\tau := \mathbf{a}_\tau^{\text{T}} \mathbf{u}, \quad \tau \in \mathcal{I}_d. \quad (25)$$

Let $u_{\tau,\text{ref}} > 0$ be a suitable reference value of the DOF of index τ. Therefore, the multi-displacement DR can be expressed as:

$$\begin{aligned} g_{1\tau} &:= \frac{u_\tau}{u_{\tau,\text{ref}}} - 1 \leq 0, \forall \tau \in \mathcal{I}_d, \\ g_{2\tau} &:= -\frac{u_\tau}{u_{\tau,\text{ref}}} - 1 \leq 0, \forall \tau \in \mathcal{I}_d. \end{aligned} \quad (26)$$

The TO problem can be formulated as a Constrained Non-Linear Programming Problem (CNLPP), where the multi-displacement DR of Eq. (26) is integrated as a constraint function. The following two CNLPPs are considered in this study:

$$\min_{\xi_1, \xi_2} \frac{V}{V_{\text{ref}}}, \text{ s.t. :} \begin{cases} \mathbf{Ku} = \mathbf{f}, \\ g_{j\tau} \leq 0, \ j = 1, 2, \ \forall \tau \in \mathcal{I}_d, \\ \xi_{1k} \in [\rho_{\min}, \rho_{\max}], \ \xi_{2k} \in [\omega_{\min}, \omega_{\max}], \\ \forall k = 1, \ldots, n_{\text{CP}}. \end{cases} \quad (27)$$

$$\min_{\boldsymbol{\xi}_1, \boldsymbol{\xi}_2} \frac{c}{c_{\text{ref}}}, \text{ s.t. :} \begin{cases} \mathbf{Ku} = \mathbf{f}, \\ \dfrac{V}{V_{\text{ref}}} - \gamma \leq 0, \\ g_{j\tau} \leq 0, \ j = 1, 2, \ \forall \tau \in \mathcal{I}_d, \\ \xi_{1k} \in [\rho_{\min}, \rho_{\max}], \ \xi_{2k} \in [\omega_{\min}, \omega_{\max}], \\ \forall k = 1, \ldots, n_{\text{CP}}. \end{cases} \quad (28)$$

Of course, problem (28) is well-posed if and only if the set $\mathcal{I}_d$ does not include all the DOFs where external forces are applied (otherwise it constitutes a measure of the compliance of the continuum, giving, thus, a redundant information because the compliance is already integrated as objective function in the problem formulation).

In order to solve problems (27) and (28) by means of a suitable deterministic algorithm, the gradient of the physical responses with respect to the design variables $\boldsymbol{\xi}_1$ and $\boldsymbol{\xi}_2$ must be computed. The gradient of the compliance and that of the volume are given in Eqs. (21) and (22), respectively. Conversely, the derivation of the analytical expression of the gradient of the multi-displacement DR needs the use of the NURBS local support property of Eq. (9) and the use of the adjoint method [43]. To this end, consider the following proposition.

Proposition 3.1 *Consider a deformable isotropic medium subject to given BCs. Under the hypothesis that the vector of the external forces* $\mathbf{f}$ *does not depend on the pseudo-density field, i.e.* $\dfrac{\partial \mathbf{f}}{\partial \xi_{ik}} = \mathbf{0}$, $(i = 1, 2, k = 1, \ldots, n_{\text{CP}})$ *the gradient of the multi-displacement constraint of Eq. (26) reads:*

$$\begin{cases} \dfrac{\partial g_{1\tau}}{\partial \xi_{ik}} = \sum_{e \in S_k} \dfrac{\alpha}{\rho_e} \dfrac{\partial \rho_e}{\partial \xi_{ik}} \boldsymbol{\eta}_{\tau e}^{\mathrm{T}} \mathbf{f}_e, \\ \mathbf{K} \boldsymbol{\eta}_\tau = -\dfrac{1}{u_{\tau, \text{ref}}} \mathbf{a}_\tau, \\ \dfrac{\partial g_{2\tau}}{\partial \xi_{ik}} = -\dfrac{\partial g_{1\tau}}{\partial \xi_{ik}}, \\ i = 1, 2, \ k = 1, \ldots, n_{\text{CP}}, \ \forall \tau \in \mathcal{I}_d, \end{cases} \quad (29)$$

where $\mathbf{f}_e$ *and* $\boldsymbol{\eta}_{\tau e}$ *are defined as:*

$$\mathbf{f}_e := \mathbf{K}_e \mathbf{u}_e, \quad (30)$$

$$\boldsymbol{\eta}_{\tau e} := \mathbf{L}_e \boldsymbol{\eta}_\tau. \quad (31)$$

Remark 3.1 The second formula in Eq. (29) represents the so-called adjoint system, whose solution is the adjoint vector $\boldsymbol{\eta}_\tau$.

The proof of Proposition 3.1 is provided here below.

Proof Inasmuch as Eq. (11) holds, the first multi-displacement constraint of Eq. (26) can be written as:

$$g_{1\tau} = \frac{u_\tau}{u_{\tau,\text{ref}}} - 1 + \boldsymbol{\eta}_\tau^T (\mathbf{Ku} - \mathbf{f}), \tag{32}$$

where $\boldsymbol{\eta}_\tau \neq \mathbf{0}$ is the arbitrary adjoint vector. Under the hypothesis that $\dfrac{\partial \mathbf{f}}{\partial \xi_{ik}} = \mathbf{0}$, and by considering Eq. (25), by differentiating Eq. (32) one gets:

$$\frac{\partial g_{1\tau}}{\partial \xi_{ik}} = \left(\frac{1}{u_{\tau,\text{ref}}} \mathbf{a}_\tau^T + \boldsymbol{\eta}_\tau^T \mathbf{K} \right) \frac{\partial \mathbf{u}}{\partial \xi_{ik}} + \boldsymbol{\eta}_\tau^T \frac{\partial \mathbf{K}}{\partial \xi_{ik}} \mathbf{u}. \tag{33}$$

The adjoint vector $\boldsymbol{\eta}_\tau$ is chosen in such a way that the term multiplying $\dfrac{\partial \mathbf{u}}{\partial \xi_{ik}}$ vanishes from Eq. (33), i.e. $\boldsymbol{\eta}_\tau$ is the solution of the following adjoint system

$$\mathbf{K}\boldsymbol{\eta}_\tau = -\frac{1}{u_{\tau,\text{ref}}} \mathbf{a}_\tau. \tag{34}$$

Accordingly, Eq. (33) simplifies to:

$$\frac{\partial g_{1\tau}}{\partial \xi_{ik}} = \boldsymbol{\eta}_\tau^T \frac{\partial \mathbf{K}}{\partial \xi_{ik}} \mathbf{u}. \tag{35}$$

The above formula can be further simplified by considering the expression of matrix $\mathbf{K}$ of Eq. (13) as well as the expressions of vectors $\mathbf{u}_e$, $\mathbf{f}_e$, and $\boldsymbol{\eta}_{\tau e}$ provided in Eqs. (14), (30), and (31), respectively:

$$\frac{\partial g_{1\tau}}{\partial \xi_{ik}} = \boldsymbol{\eta}_\tau^T \sum_{e=1}^{N_e} \frac{\alpha}{\rho_e} \frac{\partial \rho_e}{\partial \xi_{ik}} \rho_e^\alpha \mathbf{L}_e^T \mathbf{K}_e^0 \mathbf{L}_e \mathbf{u} = \sum_{e \in S_k} \frac{\alpha}{\rho_e} \frac{\partial \rho_e}{\partial \xi_{ik}} \boldsymbol{\eta}_{\tau e}^T \mathbf{f}_e. \tag{36}$$

Of course, the gradient of $g_{2\tau}$ is equal to the opposite of the gradient of $g_{1\tau}$. This last statement concludes the proof.

4 Numerical Results

In this section, the effectiveness of the proposed method is proven through 2D and 3D benchmark problems. For each case, the pseudo-density field and the optimum geometry are shown. For each CNLPP, lower and upper bounds of design variables are set as: $\rho_{\min} = 10^{-3}$, $\rho_{\max} = 1$; $\omega_{\min} = 0.5$, $\rho_{\max} = 10$. Moreover, the non-trivial knot-vector's components in Eq. (7) have been uniformly distributed in the interval [0, 1] for both 2D and 3D problems.

Table 1 GC-MMA algorithm parameters

Parameter	Value
move	0.1
albefa	0.1
Stop criterion	Value
Maximum n. of function evaluations	$100 \times n_{\text{var}}$
Maximum n. of iterations[a]	100 / 200
Tolerance on objective function	10^{-6}
Tolerance on constraints	10^{-6}
Tolerance on input variables change	10^{-6}
Tolerance on Karush–Kuhn–Tucker norm	10^{-6}

[a] The maximum number of iterations varies depending on the problem formulation.

The results presented in this section are obtained by means of the Python version of the code SANTO (SIMP And NUBRS for Topology Optimization) developed at the I2M laboratory in Bordeaux [18, 19]. This version exhibits an easily operable code, with a structure adapted to work with any FE code. In this study, the FE commercial code ANSYS is utilized to build the FE models and assess the mechanical responses of the structure, i.e. structural displacements and compliance. Moreover, the Globally-Convergent Method of Moving Asymptotes (GC-MMA) algorithm [44] has been used to perform the solution search for each CNLPP. The parameters tuning the behaviour of the GC-MMA algorithm as well as the user-defined convergence criteria are listed in Table 1.

Post-processing operations are performed in ParaView® environment for the visualization of the optimized geometry for 2D and 3D cases, and Catia V5® for obtaining the CAD model in 2D cases.

As far as numerical tests are concerned, the following aspects are considered: (1) the influence of the geometric entity, i.e. B-spline or NURBS, used to describe the pseudo-density field on the optimized topology is studied for 2D and 3D cases; (2) the sensitivity of the optimized topology to the integer parameters involved in the definition of the NURBS entity, i.e. blending functions degree and CPs number, is investigated; (3) different TO problems are carried out to show the versatility of the NURBS-based SIMP approach in 2D and 3D cases; (4) for some benchmarks, results are compared to those provided by the classic SIMP method implemented within the TOSCA module [15] of the commercial FE software ABAQUS/CAE®.

4.1 2D Benchmark Problems

4.1.1 Description of 2D Benchmark Problems

The proposed 2D benchmark problems have been chosen to give an idea of the capabilities of the NURBS-based SIMP approach when dealing with different

CNLPPs. Depending on the problem formulation, the reference volume V_{ref} is the overall volume of the considered structure. c_{ref} is the compliance evaluated for the starting solution and $u_{\tau,\text{ref}}$ is the reference value of the displacement at the point of interest P along the x_j axis.

The first benchmark problem (BK1-2D) is illustrated in Fig. 1a and deals with a 2D cantilever plate made of a material with a linear elastic isotropic behaviour ($E = 110000$ MPa and $\nu = 0.33$) with the following size: $a_1 = 240$ mm, $a_2 = 160$ mm and a thickness $t = 2$ mm. The structure is clamped at $x_1 = 0$ mm and a point force $F_A = 500$ N, is applied at point A (as shown in Fig. 1a) along the x_2 axis. The FE model is made of 60 × 40 PLANE182 elements (plane stress hypothesis with thickness, 4 nodes, 2 DOFs per node).

For BK1-2D the CNLPP formulation of Eq. (27) is enhanced by adding a symmetry constraint: the optimized topology must be symmetric with respect to the plane $x_2 = a_2/2$. Moreover, an extensive numerical campaign of tests has been performed on BK1-2D. The aim of this campaign is to study the sensitivity of the optimized topology to the integer parameters involved in the definition of B-spline and NURBS entities.

The second benchmark problem (BK2-2D), taken from [6], deals with the topology optimization of a 2D roller support, as shown in Fig. 1b. The geometry of BK2-2D is characterized by the following dimensions: $a_1 = 100$ mm, $a_2 = 50$ mm, $t = 1$ mm. The constitutive material has a linear elastic isotropic behaviour with the following properties: $E = 1000$ MPa and $\nu = 0.33$. As illustrated in Fig. 1b, the force is applied at point C along the x_2 axis and its value is $F_C = 100$ N. The structure is clamped at $x_1 = x_2 = 0$ mm, whilst $u_2 = 0$ at point $(x_1 = 100, x_2 = 0)$ mm. The FE model is made of 100 × 100 PLANE182 elements (plane stress hypothesis). In this case, three Non-Design Regions (NDRs) are considered in the neighbourhood of the zones where BCs are applied (grey colour in Fig. 1b). The three NDRs are defined as follows:

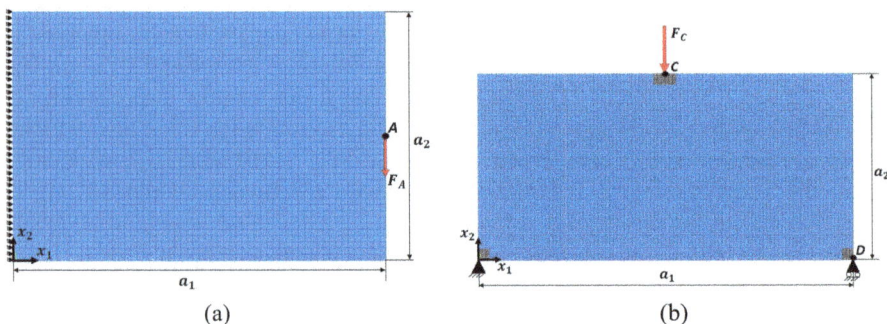

Fig. 1 Geometry and boundary conditions for benchmarks problems (**a**) BK1-2D and (**b**) BK2-2D

- $NDR_1 = \{(x_1, x_2) \mid x_1 \in [0, 3] \text{ mm}, x_2 \in [0, 3] \text{ mm}\}$;
- $NDR_2 = \{(x_1, x_2) \mid x_1 \in [47, 53] \text{ mm}, x_2 \in [47, 50] \text{ mm}\}$;
- $NDR_3 = \{(x_1, x_2) \mid x_1 \in [97, 100] \text{ mm}, x_2 \in [0, 3] \text{ mm}\}$.

CNLPPs of Eqs. (18) and (28) are considered for BK2-2D. The displacement is measured at point D along the x_1 axis and a symmetry constraint with respect to plane $x_1 = \dfrac{a_1}{2}$ is added to the problem formulation.

The third benchmark problem (BK3-2D) deals with the topology optimization of a 2D square domain, as shown in Fig. 2. The geometry of BK3-2D is characterized by the following dimensions: $a_1 = a_2 = 100$ mm, $t = 1$ mm. The constitutive material has a linear elastic isotropic behaviour with the following properties: $E = 72000$ MPa and $\nu = 0.33$. As shown in Fig. 2, the force is applied at point P along the x_2 axis and its value is $F_P = 1000$ N. The structure is subject to the following BCs: $u_1 = 0$ at $x_1 = x_2 = 0$ mm (point M) and at $x_1 = x_2 = 100$ mm (point O), $u_2 = 0$ at $(x_1 = 0, x_2 = 100)$ mm (point N). The FE model is made of 50 × 50 PLANE182 elements (plane stress hypothesis). In this case, four Non-Design Regions (NDRs) made of four elements are considered around points M, N, O, and P, as illustrated through grey colour in Fig. 2. CNLPPs of Eqs. (18) and (28) are considered for BK3-2D. The reference value of the displacement is measured at points N and O along x_1 and x_2 axes, respectively.

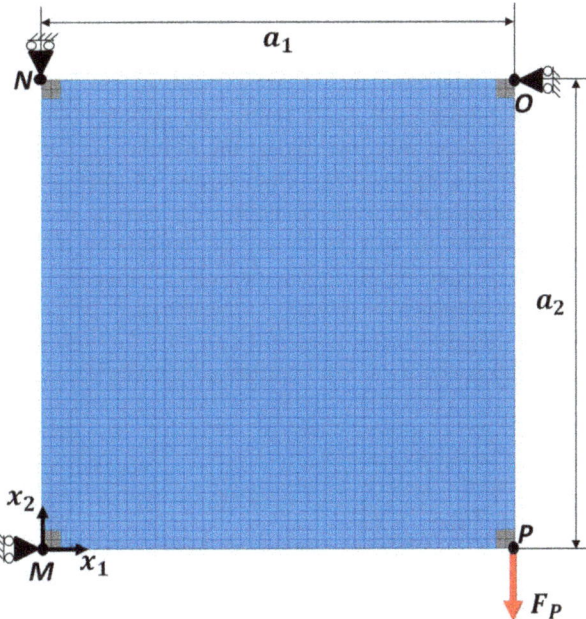

Fig. 2 Geometry and boundary conditions for benchmark problem BK3-2D

4.1.2 Sensitivity of the Optimized Topology to the B-Spline and NURBS Entities Integer Parameters

Problem (27) is solved for BK1-2D for several values of blending functions degrees p_j, $(j = 1, 2)$ and numbers of CPs n_{CP}. The maximum number of iterations is $N_{\text{iter}}^{\max} = 100$. The reference value of the displacement for problem (27) is $u_{A2,\text{ref}} = 0.2$ mm. Degrees and number of CPs are chosen as follows:

- $p_j = 2, 3, 4$;
- $n_{CP} = (n_1 + 1) \times (n_2 + 1) = 30 \times 20, 40 \times 30, 50 \times 36$.

The ratio of CPs to FE mesh elements number N_e is given in Table 2 along x_1 and x_2 axes. Results are provided in terms of volume fraction V/V_{ref} and number of iterations N_{iter} for B-spline and NURBS entities in Figs. 3 and 4, respectively. For each solution the requirement on the displacement is always satisfied. In each case, $V_{\text{ref}} = 76800 \text{ mm}^3$ which corresponds to the volume of the whole domain.

The following remarks can be inferred from the analysis of results.

1. The greater the number of CPs (for a given degree) or the smaller the degree (for a given number of CPs) the smaller the objective function value.
2. The CPs number and basis functions degree along each direction affect the size of the local support, see Eq. (9). As far as this point is concerned, the same remarks as in [18, 19] can be made: the higher the degree, the greater the local support, thus each CP affects a wider region of the mesh during optimization. The higher the degree, the worse the solution in terms of objective function because thinner topological branches disappear due to the local support size. Conversely, the higher the degree, the smoother the topology boundary after CAD reconstruction. In the same way, the higher the number of CPs, the smaller the local support. Therefore, as a general rule, a high number of CPs and a small degree should be considered if performances are of paramount importance. High degree and/or small number of CPs should be considered if the smoothness of the boundary is privileged and if small topological branches must be avoided (especially for manufacturing purposes). As discussed in [22], the local support of the NURBS blending functions constitutes an implicit filter zone which enforces a minimum length scale in the final optimized topologies.

Table 2 Sensitivity analysis for benchmark BK1-2D—ratio of the CPs number to the elements number

Mesh and control net	x_1-axis	x_2-axis	Total
N_e	60	40	2400
n_{CP}	30	20	600
n_{CP}/N_e	50%	50%	25%
n_{CP}	40	30	1200
n_{CP}/N_e	67%	75%	50%
n_{CP}	50	36	1800
n_{CP}/N_e	83%	90%	75%

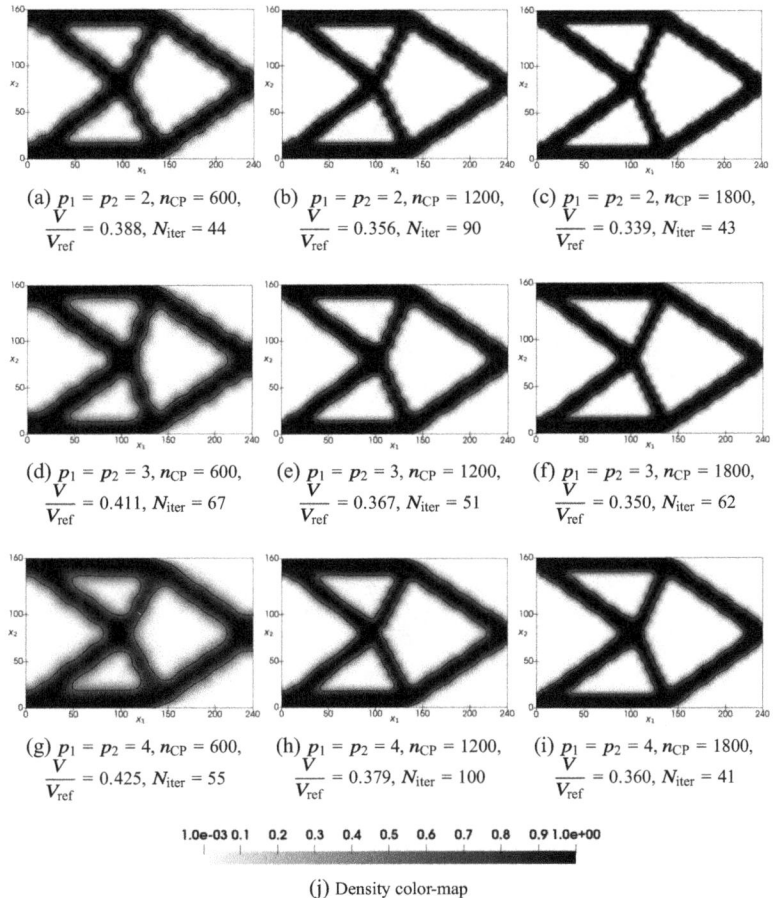

Fig. 3 Sensitivity analysis for benchmark BK1-2D—B-Spline solutions. (**a**) $p_1 = p_2 = 2, n_{CP} = 600, \frac{V}{V_{ref}} = 0.388, N_{iter} = 44$. (**b**) $p_1 = p_2 = 2, n_{CP} = 1200, \frac{V}{V_{ref}} = 0.356, N_{iter} = 90$. (**c**) $p_1 = p_2 = 2, n_{CP} = 1800, \frac{V}{V_{ref}} = 0.339, N_{iter} = 43$. (**d**) $p_1 = p_2 = 3, n_{CP} = 600, \frac{V}{V_{ref}} = 0.411, N_{iter} = 67$. (**e**) $p_1 = p_2 = 3, n_{CP} = 1200, \frac{V}{V_{ref}} = 0.367, N_{iter} = 51$. (**f**) $p_1 = p_2 = 3, n_{CP} = 1800, \frac{V}{V_{ref}} = 0.350, N_{iter} = 62$. (**g**) $p_1 = p_2 = 4, n_{CP} = 600, \frac{V}{V_{ref}} = 0.425, N_{iter} = 55$. (**h**) $p_1 = p_2 = 4, n_{CP} = 1200, \frac{V}{V_{ref}} = 0.379, N_{iter} = 100$. (**i**) $p_1 = p_2 = 4, n_{CP} = 1800, \frac{V}{V_{ref}} = 0.360, N_{iter} = 41$. (**j**) Density colour-map

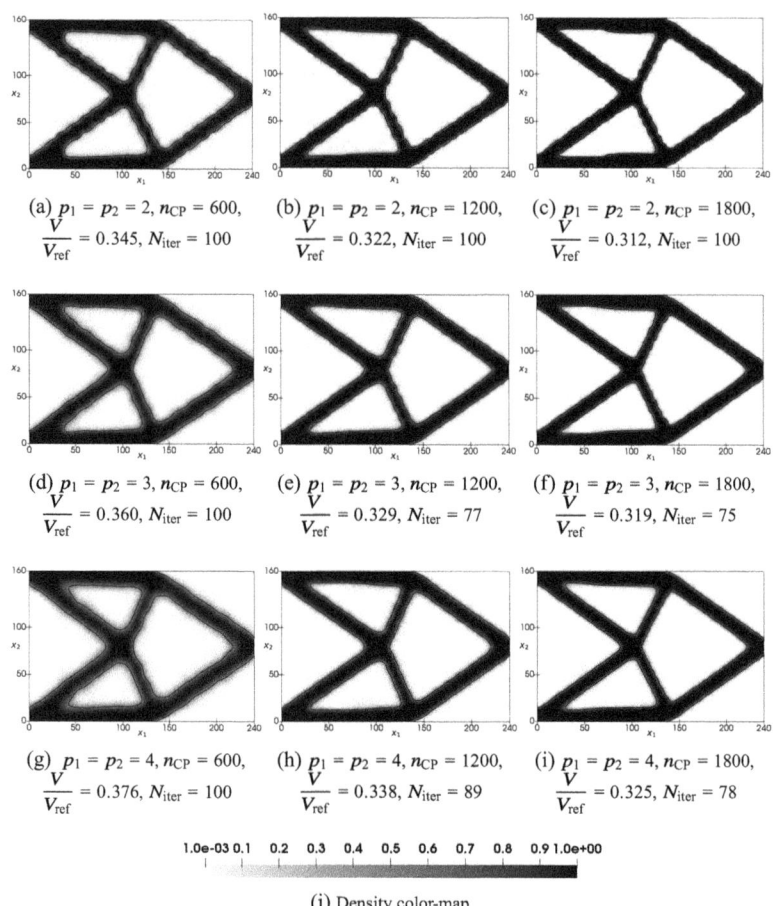

Fig. 4 Sensitivity analysis for benchmark BK1-2D—NURBS solutions. (a) $p_1 = p_2 = 2$, $n_{CP} = 600$, $\frac{V}{V_{ref}} = 0.345$, $N_{iter} = 100$. (b) $p_1 = p_2 = 2$, $n_{CP} = 1200$, $\frac{V}{V_{ref}} = 0.322$, $N_{iter} = 100$. (c) $p_1 = p_2 = 2$, $n_{CP} = 1800$, $\frac{V}{V_{ref}} = 0.312$, $N_{iter} = 100$. (d) $p_1 = p_2 = 3$, $n_{CP} = 600$, $\frac{V}{V_{ref}} = 0.360$, $N_{iter} = 100$. (e) $p_1 = p_2 = 3$, $n_{CP} = 1200$, $\frac{V}{V_{ref}} = 0.329$, $N_{iter} = 77$. (f) $p_1 = p_2 = 3$, $n_{CP} = 1800$, $\frac{V}{V_{ref}} = 0.319$, $N_{iter} = 75$. (g) $p_1 = p_2 = 4$, $n_{CP} = 600$, $\frac{V}{V_{ref}} = 0.376$, $N_{iter} = 100$. (h) $p_1 = p_2 = 4$, $n_{CP} = 1200$, $\frac{V}{V_{ref}} = 0.338$, $N_{iter} = 89$. (i) $p_1 = p_2 = 4$, $n_{CP} = 1800$, $\frac{V}{V_{ref}} = 0.325$, $N_{iter} = 78$. (j) Density color-map

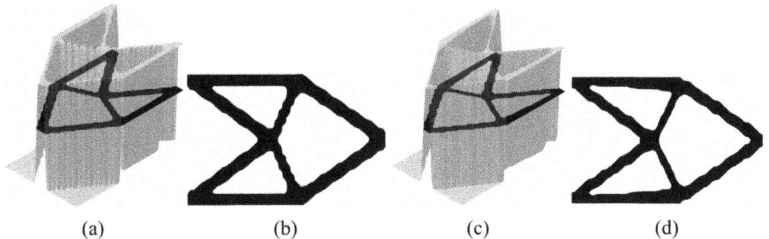

Fig. 5 CAD model of the optimized topology, for BK1-2D, obtained with a B-spline surface (**a**)–(**b**) and a NURBS surface (**c**)–(**d**), for the same integer parameters: $(n_1 + 1) \times (n_2 + 1) = 50 \times 36$ and $p_1 = p_2 = 2$

3. Fig. 5 illustrates the outstanding advantage provided by the NURBS-based SIMP method. It consists in the possibility to export a CAD-compatible entity in order to rebuild in a straightforward way the boundary of the optimized 2D structure. This task can be achieved by evaluating a threshold value for the density function meeting the optimization constraint, i.e. the displacement constraint (this operation is automatically done by the SANTO algorithm at the end of the optimization process).
4. Optimized topologies obtained using NURBS surfaces are characterized by values of the objective function lower than those resulting from B-spline surfaces when considering the same number of CPs and the same degrees, as shown in Fig. 6a. In particular, from the analysis of Figs. 3 and 4, it appears that NURBS topologies are smoother than B-spline ones for each case.
5. Fig. 6 shows the *projected* and the *reconstructed* volume fraction for both B-spline and NURBS solutions. The projected volume fraction $\dfrac{V}{V_{\text{ref}}}$ is that evaluated by means of the formula in Eq. (19), whilst the reconstructed volume fraction is the true volume fraction obtained after CAD reconstruction. As it can be easily inferred from Fig. 6a, b, the true volume fraction is always lower than the projected one since the threshold plane is evaluated in order to meet the requirement on the displacement, after the cutting operation. In fact, as a post-processing operation, the threshold value of the pseudo-density field ρ_{th} is evaluated by means of the secant method to satisfy the DR on the displacement: the value of this constraint before and after CAD reconstruction is shown in Fig. 6c, d, respectively.

4.1.3 Comparison Between SIMP and NURBS-Based SIMP Methods

The benchmark problem BK1-2D has been solved by using the TOSCA® module of the commercial FE software ABAQUS/CAE® in order to compare results provided by the classical SIMP method and those proposed in this work.

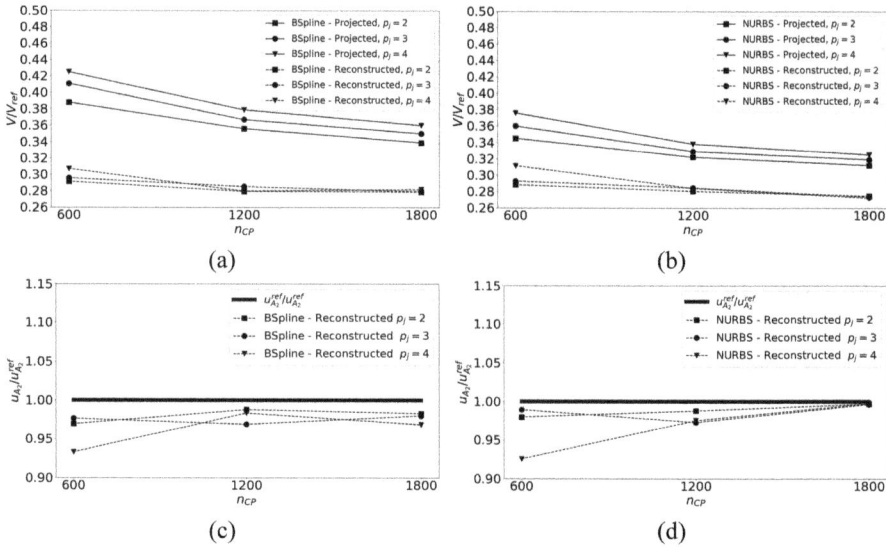

Fig. 6 BK1-2D—projected and reconstructed volume fraction of (**a**) B-spline and (**b**) NURBS solutions—displacement value after reconstruction for (**c**) B-spline and (**d**) NURBS solutions

Problem (27) has been run with SIMP method under identical BCs and load. The FE model is made of 60 × 40 CPS4 elements (plane stress hypothesis with thickness), with 4 nodes and 2 DOFs per node. A minimum dimension of the topological branches is set as additional constraint in order to satisfy the minimum length scale of the implicit TO filter behind the NURBS-based SIMP method (as detailed in [22]), when considering a NURBS surface with $(n_1 + 1) \times (n_2 + 1) = 50 \times 36$ CPs and $p_1 = p_2 = 2$. Following the procedure detailed in [22], the minimum length scale is 5 mm and this corresponds to the minimum member size requirement to be imposed in TOSCA®. A symmetry condition is added with respect to the plane located at $x_2 = a_2/2$, thus the design variables number is reduced to 1200.

The topology and the material distribution of the optimized solution are given in Fig. 7. At a first glance, the optimized topology of Fig. 7a is quite different from those illustrated in Figs. 3 and 4 for $(n_1 + 1) \times (n_2 + 1) = 50 \times 36$ configuration and $p_1 = p_2 = 2$. Instead of two thin branches, the TOSCA® solution is characterized by two coarse branches and four thinner branches connecting them to the outer boundary of the structure. As far as the pseudo-density field is concerned, the solution provided by TOSCA is characterized by the well-known checker-board effect in the region where the transition between the two thick branches and the four smaller ones occurs. Moreover, from Fig. 7, it is evident that the optimized topology provided by TOSCA does not satisfy the requirement on the minimum length scale (the four thinner branches have a thickness lower than 5 mm).

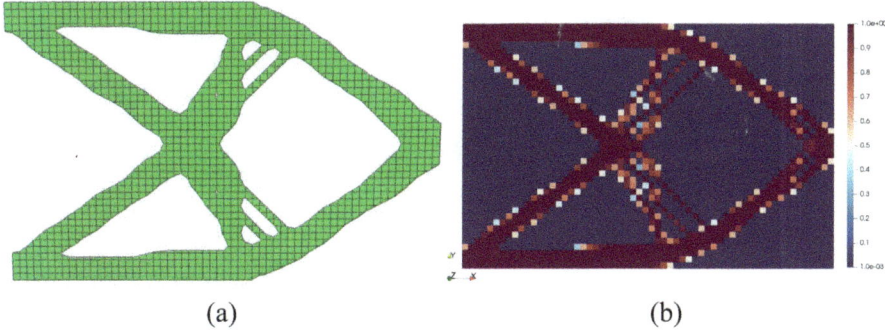

Fig. 7 Solution for Benchmark BK1-2D provided by SIMP method implemented in TOSCA®—(**a**) Optimized topology and (**b**) corresponding density distribution in FE model

The projected volume fraction provided by TOSCA® is $\frac{V}{V_{\text{ref}}} = 0.310$ at the final iteration $N_{\text{iter}} = 58$. When looking at the results of Fig. 6a, all B-spline and NURBS solutions produce higher values of the projected volume fraction. However, due to the checker-board pattern affecting the optimized solution provided by TOSCA, the reconstructed volume fraction is higher than the projected one. The import of the reconstructed topology in a CAD software (Catia V5® in this case) confirms this fact: the reconstructed optimized topology found by TOSCA is characterized by $\frac{V}{V_{\text{ref}}} = 0.386$ which is considerably higher than that characterizing both B-spline and NURBS optimized topologies (see Fig. 6). Finally, the whole optimization process requires a computational time (CT) of approximately 20 and 30 min for B-spline and NURBS solutions, respectively, when four cores of a machine with an Intel Xeon E5-2697v2 processor (2.70–3.50 GHz) are dedicated to the ANSYS solver. Conversely, about 15 min are required for the optimization performed by TOSCA.

4.1.4 Effect of the Structural Displacement Requirement on a Non-loaded Region

The benchmark problem BK2-2D is used to show the influence of the displacement requirement over a non-loaded region on the final optimized topology. To this purpose, both problems (18) and (28) are solved for BK2-2D. In both cases, the CNLPP is solved by using: (a) a B-spline surface characterized by $n_{\text{CP}} = 80 \times 80$ CPs and $p_j = 2, 4$, and (b) a NURBS surface characterized by $n_{\text{CP}} = 60 \times 60$ CPs and $p_j = 2, 4$. The maximum number of iterations is $N_{\text{iter}}^{\max} = 200$.

Firstly, problem (18) is solved by considering a volume fraction $\gamma = 0.3$ (with $V_{\text{ref}} = 5000$ mm^3). The optimized topology, for both B-spline and NURBS surfaces, is illustrated in Fig. 8: in the caption of each figure, the values of compliance, volume

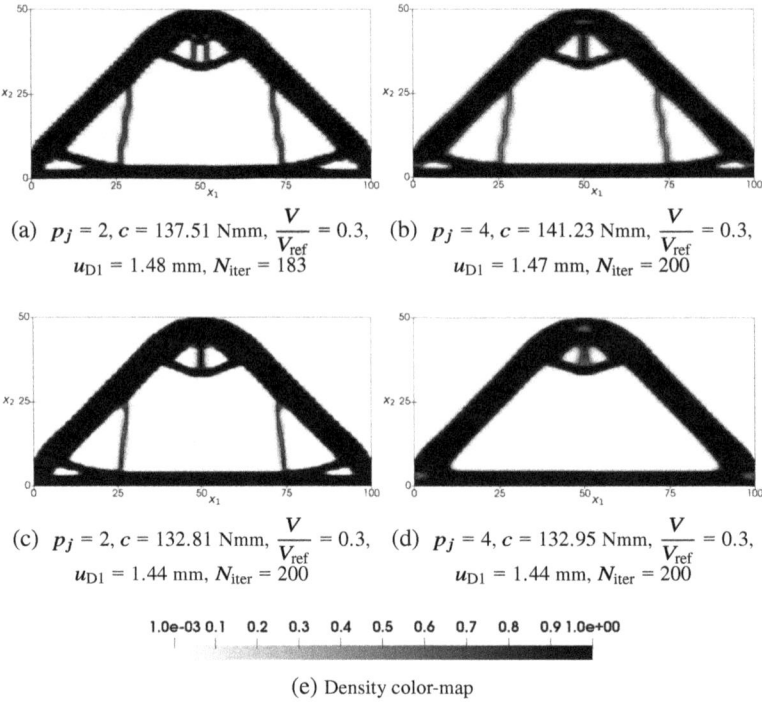

Fig. 8 BK2-2D—optimized solutions of problem (18) when using (**a**), (**b**) a B-spline surface and (**c**), (**d**) a NURBS surface

fraction, displacement at point D along x_1 axis and number of iterations to achieve convergence are also reported.

Secondly, problem (28) is solved by considering the same requirement on the volume fraction γ as in problem (18) and by setting a suitable reference value for displacement at point D along x_1 axis, i.e. $u_{\text{D1,ref}} = 1$ mm. The optimized topologies for both B-spline and NURBS solutions are shown in Fig. 9, wherein the mechanical responses (in terms of compliance, volume fraction and displacement at point D) of each optimized configuration are reported in the caption of each image.

Some interesting remarks can be inferred from the analysis of these results. As expected, the optimized topologies, solutions of problem (18), are characterized by a compliance lower than their counterparts' solutions of problem (28). The constraint on the volume fraction is met for both problems. Indeed, the requirement on the structural displacement at point D is the main cause at the basis of the differences observed in the optimized topologies.

Firstly, the presence of such a requirement enforces major modifications in the optimized topology. A quick glance to Figs. 8 and 9 suffices to understand this point. In particular, small horizontal/oblique topological branches appear in the solution of

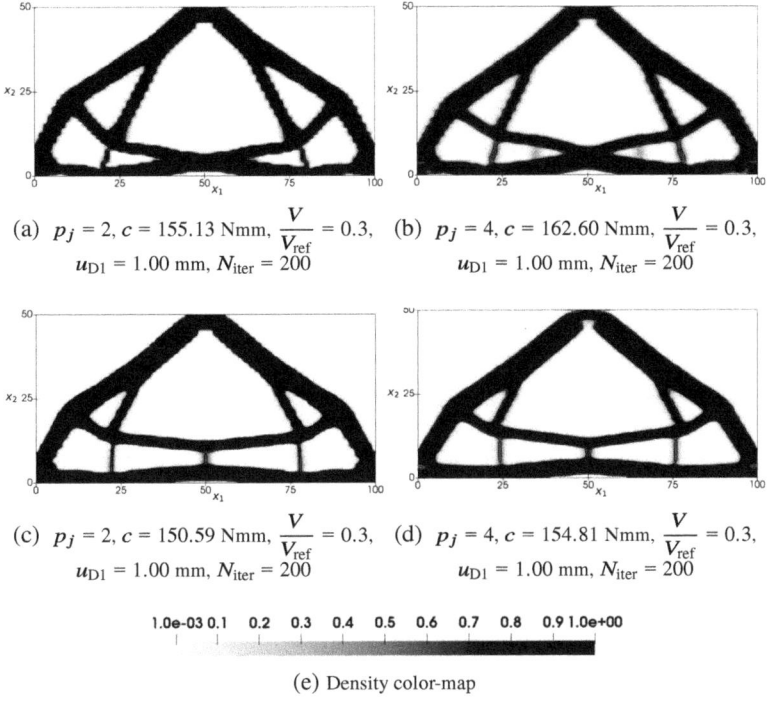

Fig. 9 BK2-2D—optimized solutions of problem (28) when using (**a**), (**b**) a B-spline surface and (**c**), (**d**) a NURBS surface

problem (28) to increase the overall stiffness along the x_1 axis in order to satisfy the requirement on the displacement at point D.

Secondly, NURBS solution provides better performances in term of compliance at the end of optimization process for problem (18) and (28) with less CPs but more optimization variables (because of the presence of NURBS weights).

Thirdly, it is noteworthy that for problem (28) the GC-MMA algorithm stops because the criterion on the maximum number of iterations is met for both B-spline and NURBS solutions, as it can be inferred from Fig. 9.

Concerning the threshold plane for the construction of the CAD model, it is automatically calculated in order to respect both constraints, i.e. volume fraction and displacement. The displacement field along x_1 axis for the B-spline solutions of problem (28), obtained as a result of a FE analysis on the topology resulting from the CAD reassembly operation, is illustrated in Fig. 10. The volume fraction and the displacement along x_1 axis at point D of the optimized topology are reported in the figure caption.

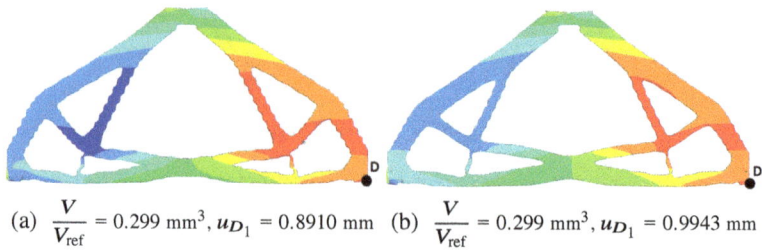

(a) $\frac{V}{V_{\text{ref}}} = 0.299$ mm^3, $u_{D_1} = 0.8910$ mm (b) $\frac{V}{V_{\text{ref}}} = 0.299$ mm^3, $u_{D_1} = 0.9943$ mm

Fig. 10 BK2-2D—displacement field along x_1 axis after CAD reassembly for the optimized solutions of problem (28) when using a B-spline surface with (**a**) $p_j = 2$ and (**b**) $p_j = 4$

4.1.5 Multi-Displacement Requirement on a Non-loaded Region

The benchmark problem BK3-2D aims at showing the influence of the displacement requirement, imposed on multiple non-loaded regions, on the final optimized topology. To this purpose, four different problems are considered:

- Problem (18) is solved by considering a volume fraction $\gamma = 0.4$ (with $V_{\text{ref}} = 10000$ mm^3);
- Problem (28) is solved by considering the same volume fraction of problem (18) and by introducing a constraint on the displacement component along x_2 axis at point O, $u_{\text{O2,ref}} = -0.5$ mm;
- Problem (28) is solved by considering the same volume fraction of problem (18) and by introducing a constraint on the displacement component along x_1 axis at point N, $u_{\text{N1,ref}} = 0.25$ mm;
- Problem (28) is solved by considering the same volume fraction of problem (18) and by introducing a constraint on both the displacement components at point N and O, i.e. $u_{\text{N1,ref}} = 0.25$ mm and $u_{\text{O2,ref}} = -0.5$ mm.

Each CNLPP is solved by using: (a) a B-spline surface characterized by $n_{\text{CP}} = 46 \times 46$ CPs and $p_j = 2$, and (b) a NURBS surface characterized by $n_{\text{CP}} = 32 \times 32$ CPs and $p_j = 2$. The maximum number of iterations is $N_{\text{iter}}^{\text{max}} = 200$.

The optimized topology, for both B-spline and NURBS surfaces, are illustrated in Figs. 11 and 12, respectively. In the caption of each figure, the problem type, the number of iterations as well as the values of compliance, volume, and displacement components at point N and O are also indicated.

From the analysis of the results, one can infer the following remarks. Firstly, unlike the results found for BK2-2D, the compliance of the optimized topologies solution of problem (28) is lower than the compliance of the optimal solution of problem (18) in the case of BK3-2D, for both B-spline and NURBS surfaces. This is an unexpected result that can be justified through the non-convexity of the objective function of problems (18) and (28). Probably, in the case of BK3-2D, adding constraints on the displacement of non-loaded regions allows the GC-MMA algorithm to better explore the design space in order to find a feasible local

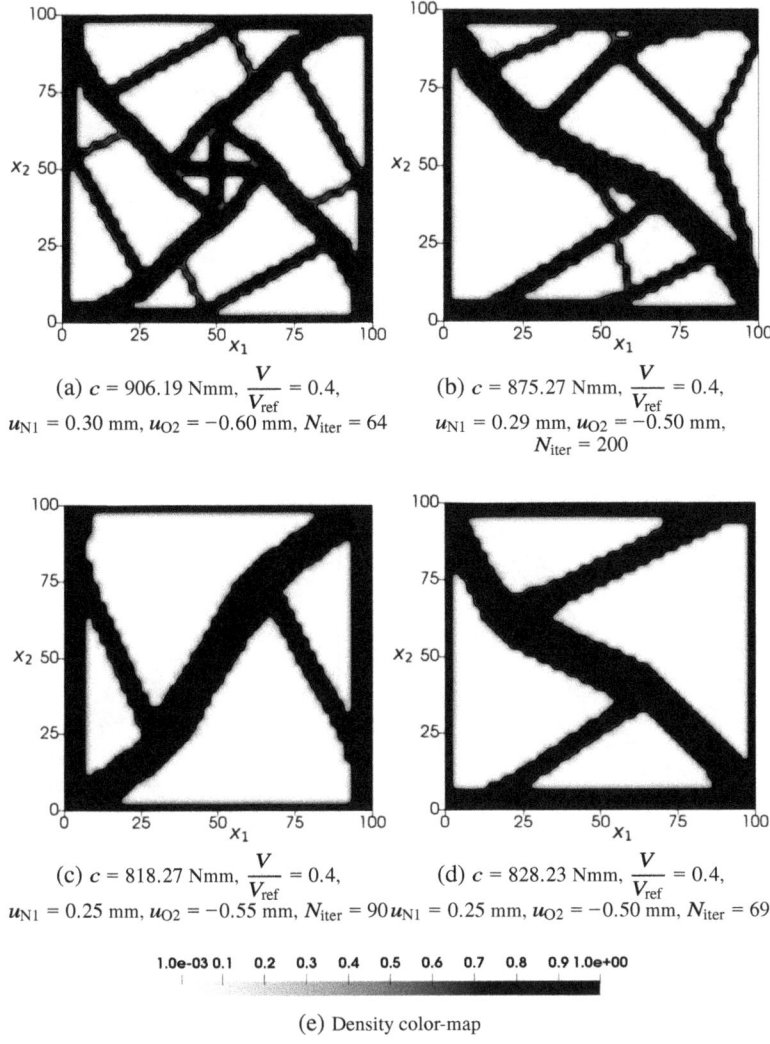

Fig. 11 BK3-2D—optimized B-spline solutions of (a) problem (18), and of problem (28) with a constraint on (b) u_{O2}, (c) u_{N1}, (d) u_{O2} and u_{N1}

minimizer. This is confirmed by the nature of the local feasible minimizer found for each problem formulation: in each case the optimized solution is located on the boundary of the feasible region because the constraints are active, i.e. the constraint functions on the volume fraction and on the multi-displacement DR are almost null (the residual is always in the interval $[\simeq -10^{-6}, 0]$).

Secondly, as shown in Figs. 11 and 12, B-spline solutions are characterized by topological branches thinner than those of the NURBS counterparts. This is due to the size of the local support of Eq. (9): since B-spline and NURBS entities used

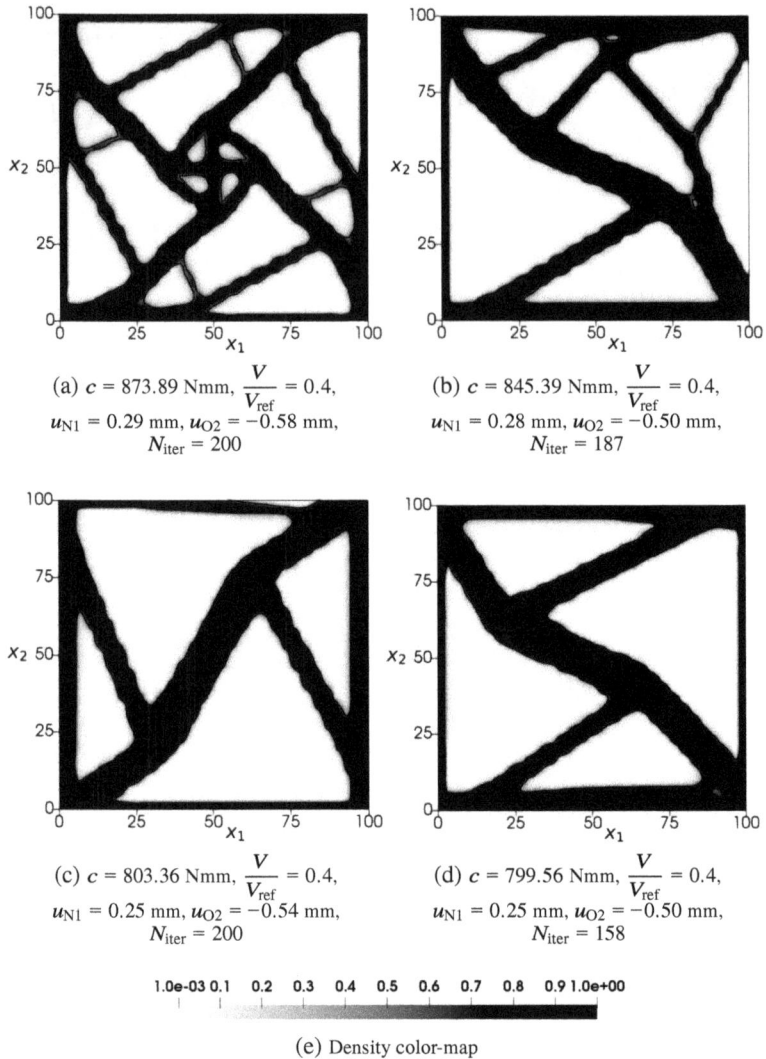

Fig. 12 BK3-2D—optimized NURBS solutions of (**a**) problem (18), and of problem (28) with a constraint on (**b**) u_{O2}, (**c**) u_{N1}, (**d**) u_{O2} and u_{N1}

in this example are characterized by the same degrees $p_j = 2$ ($j = 1, 2$), the local support of the B-spline solutions is smaller than the one of NURBS solutions because the CPs number used for B-spline entities is greater than the one used for NURBS entities.

4.2 3D Benchmark Problems

4.2.1 Description of 3D Benchmark Problems

In the general 3D case, two benchmarks problem are considered. The first one (BK1-3D) is a box, whose sizes are $a_1 = a_2 = a_3 = 40$ mm. The material is Ti6Al4V ($E = 110000$ MPa and $\nu = 0.33$). The load is applied along the x_3 axis at point E (Fig. 13a), with a magnitude $F_E = 20000N$. As shown in Fig. 13a, the load is applied at a master node, located at ($x_1 = 20$, $x_2 = 20$ $x_3 = 40$) mm, which is linked to a set of slave nodes in the upper zone of the box, i.e. $S_E = \{(x_1, x_2, x_3) \mid x_1 \in [18, 22]$ mm, $x_2 \in [18, 22]$ mm, $x_3 = 40$ mm$\}$. The link between the DOFs of the master node and those of the slave nodes is ensured through multi-point constraint elements (MPC184) with a rigid beam behaviour. In particular, MPC184 elements are used to define (locally) a rigid surface region. The box is clamped at the four corners placed at $x_3 = 0$ mm. NDRs (grey colour in Fig. 13a) are considered near to the regions where BCs and load are applied, i.e.

- $NDR_1 = \{(x_1, x_2, x_3) \mid x_1 \in [0, 4]$ mm, $x_2 \in [0, 4]$ mm, $x_3 \in [0, 4]$ mm$\}$;
- $NDR_2 = \{(x_1, x_2, x_3) \mid x_1 \in [36, 40]$ mm, $x_2 \in [0, 40]$ mm, $x_3 \in [0, 4]$ mm$\}$;
- $NDR_3 = \{(x_1, x_2, x_3) \mid x_1 \in [0, 4]$ mm, $x_2 \in [36, 40]$ mm, $x_3 \in [0, 4]$ mm$\}$;
- $NDR_4 = \{(x_1, x_2, x_3) \mid x_1 \in [36, 40]$ mm, $x_2 \in [36, 40]$ mm, $x_3 \in [0, 4]$ mm$\}$;
- $NDR_5 = \{(x_1, x_2, x_3) \mid x_1 \in [16, 24]$ mm, $x_2 \in [16, 24]$ mm, $x_3 \in [36, 40]$ mm$\}$.

Static FE analyses are carried out using $20 \times 20 \times 20$ SOLID185 elements (8 nodes, 3 DOFs per node).

The second benchmark problem (BK2-3D) focuses on the 3D version of a special cantilever beam. The material properties are the same as those characterizing BK2-2D. The geometry of the 3D model is illustrated in Fig. 13b: the dimensions are

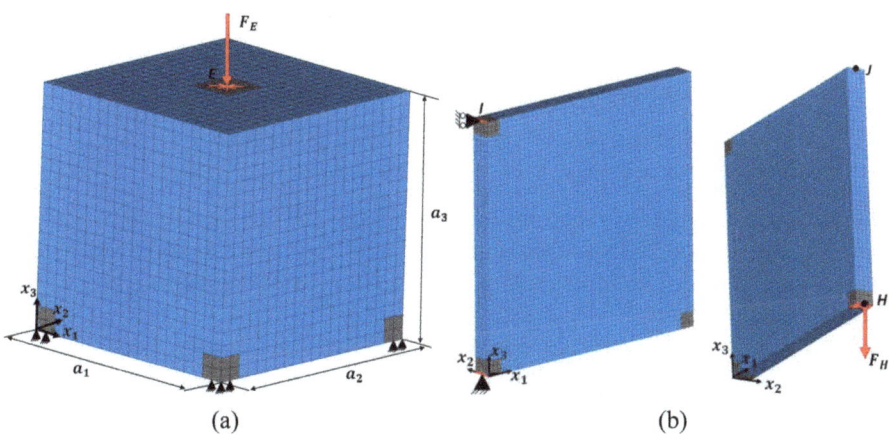

Fig. 13 Geometry and boundary conditions for benchmarks (**a**) BK1-3D and (**b**) BK2-3D

$a_1 = 500$ mm, $a_2 = 40$ mm, $a_3 = 500$ mm. As in the previous case, MPC184 elements are used for load application: a force $F_H = 3000$ N is applied at the master node H along the x_3 axis. Node H has coordinates $x_1 = 500$ mm, $x_2 = 20$ mm, $x_3 = 0$ mm and is linked to the set of slave nodes belonging to the following region: $S_H = \{(x_1, x_2, x_3) \mid x_1 \in [470, 500]$ mm, $x_2 \in [0, 40]$ mm, $x_3 = 0$ mm$\}$. Clamped BCs are applied on the nodes belonging to the following sets $S_{\text{clamp}_1} = \{(x_1, x_2, x_3) \mid x_1 = 0$ mm, $x_2 \in [0, 40]$ mm, $x_3 = 0$ mm$\}$, and roller support-like BCs (i.e. $u_1 = u_2 = 0$, whilst u_3 displacement is set as free) are applied on the following nodes : $S_{RS_3} = \{(x_1, x_2, x_3) \mid x_1 = 0$ mm, $x_2 \in [0, 40]$ mm, $x_3 = 500$ mm$\}$. As usual, NDRs are defined in the neighbourhood of the regions where BCs and loads are applied. For this example, three NDRs are defined as follows:

- NDR$_1$ = $\{(x_1, x_2, x_3) \mid x_1 \in [0, 30]$ mm, $x_2 \in [0, 40]$ mm, $x_3 \in [0, 30]$ mm$\}$;
- NDR$_2$ = $\{(x_1, x_2, x_3) \mid x_1 \in [470, 500]$ mm, $x_2 \in [0, 40]$ mm, $x_3 \in [0, 30]$ mm$\}$;
- NDR$_3$ = $\{(x_1, x_2, x_3) \mid x_1 \in [0, 30]$ mm, $x_2 \in [0, 40]$ mm, $x_3 \in [470, 500]$ mm$\}$.

Static FE analysis is carried out by using a mesh composed of $50 \times 4 \times 50$ SOLID185 elements.

The last benchmark problem (BK3-3D) is characterized by the same material properties, geometry, mesh, loads, and BCs of BK2-3D. The only difference between BK2-3D and BK3-3D consists in a further roller support-like BCs (i.e. $u_2 = u_3 = 0$, whilst u_1 is free), which is applied on the following set of nodes: $S_{RS_1} = \{(x_1, x_2, x_3) \mid x_1 = 500$ mm, $x_2 \in [0, 40]$ mm, $x_3 = 500$ mm$\}$. The node located in the middle of the segment indicated by S_{RS_1} is denoted J, as shown in Fig. 13b.

4.2.2 Displacement Requirement on Loaded Region

Problem (27) is solved for BK1-3D by considering both B-spline and NURBS hyper-surfaces. The reference value of the displacement for problem (27) is $u_{E3,\text{ref}} = 0.1$ mm. Degrees and number of CPs are chosen as follows:

- B-spline hyper-surface: $n_{CP} = 18 \times 18 \times 18$ and $p_j = 2, 4$;
- NURBS hyper-surface: $n_{CP} = 14 \times 14 \times 14$ and $p_j = 2, 4$.

The maximum number of iterations is $N_{\text{iter}}^{\text{max}} = 100$. The optimized topologies for both B-spline and NURBS solutions are shown in Fig. 14, wherein the mechanical responses (in terms of volume fraction and displacement at point E) are provided in the related sub-caption.

The optimized topology relative to the NURBS hyper-surface is characterized by a value of the objective function lower than the B-spline counterpart, even with half CPs in NURBS hyper-surface configuration. Both solutions satisfy the constraint on the displacement at point E. Moreover, the solution obtained when using a NURBS

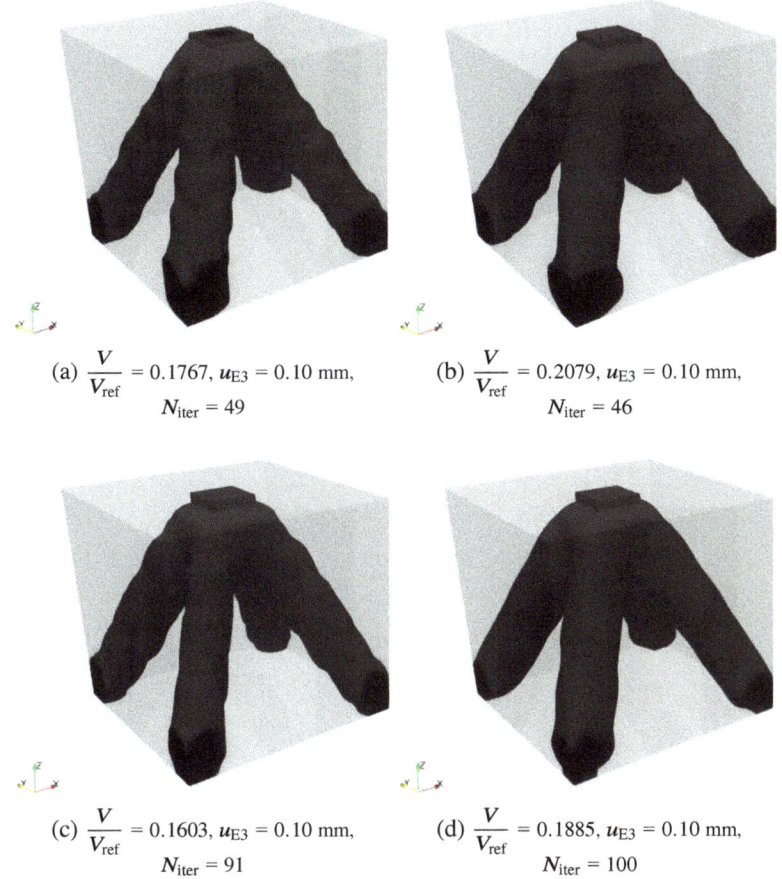

Fig. 14 Benchmark BK1-3D, optimized solutions of problem (27) for a B-spline hyper-surface with $n_{CP} = 18 \times 18 \times 18$ and (**a**) $p_j = 2$, (**b**) $p_j = 4$ and for a NURBS hyper-surface with $n_{CP} = 14 \times 14 \times 14$ and (**c**) $p_j = 2$, (**d**) $p_j = 4$

hyper-surface is characterized by a boundary surface smoother than the B-spline counterpart, as shown in Fig. 14.

4.2.3 Displacement Requirement on Non-loaded Region

The influence of the displacement constraint over a non-loaded region on the optimized topology is considered in the case of BK2-3D. The analysis strategy is the same as that used for BK2-2D: both problems (18) and (28) are solved, and the resulting topologies are compared. A further constraint on the symmetry of the topology with respect to plane $x_2 = \frac{a_2}{2}$ is added to both problem formulations. CPs and degrees are set as follows:

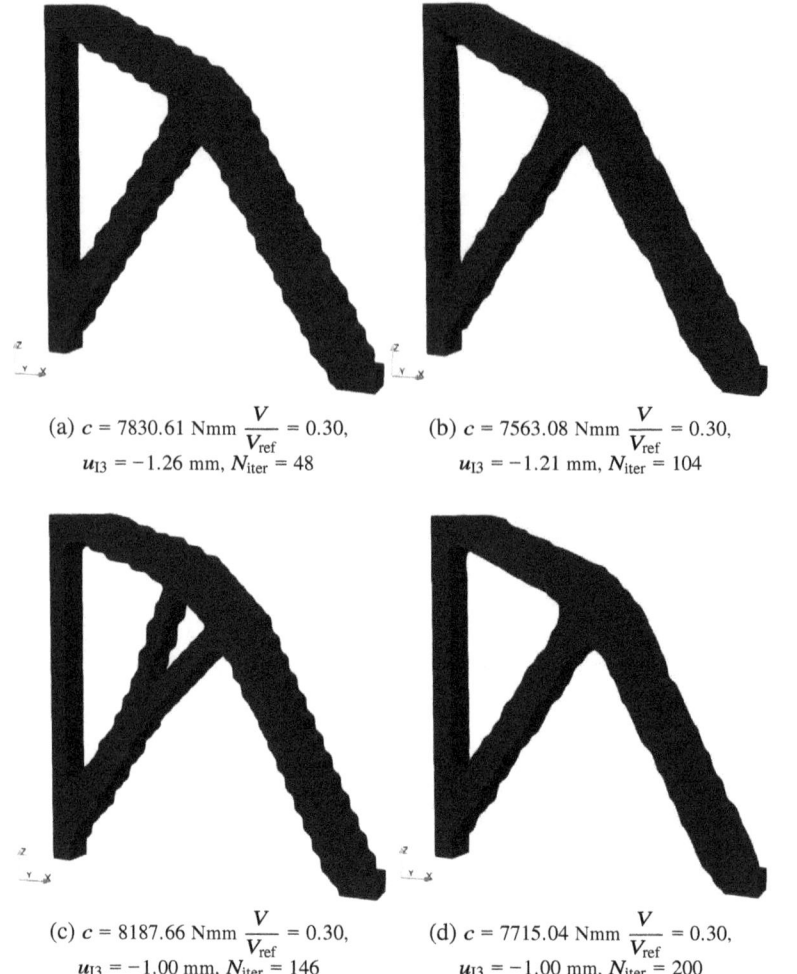

Fig. 15 Benchmark BK2-3D, optimized solutions of problem (18) for (**a**) B-spline and (**b**) NURBS hyper-surfaces and optimized solutions of problem (28) for (**c**) B-spline and (**d**) NURBS hyper-surfaces

- B-spline hyper-surface: $n_{CP} = 40 \times 4 \times 40$ and $p_j = 2$;
- NURBS hyper-surface: $n_{CP} = 30 \times 4 \times 30$ and $p_j = 2$.

Firstly, problem (18) is solved with a volume fraction $\gamma = 0.3$ ($V_{\text{ref}} = 10^7$ mm^3). Secondly, problem (28) is solved by introducing a requirement on the displacement component along x_3 axis at point I, i.e. $u_{I3,\text{ref}} = 1$ mm.

Optimized topologies obtained from B-spline and NURBS hyper-surfaces, for both problems (18) and (28), are shown in Fig. 15, where the relative compliance, volume fraction, displacement at point I, and number of iterations are reported in the figure sub-captions. For both problems, the maximum number of iterations is $N_{\text{iter}}^{\max} = 200$.

As for BK2-2D, the final compliance for problem (18) is lower than that of problem (28) due to the additional displacement requirement at point I in the latter formulation, which is fulfilled at the end of the process for both B-spline and NURBS solutions. It is noteworthy that, for both problems (18) and (28) a lower value of the objective function is obtained with a NURBS hyper-surface with a number of CPs lower than that of the B-spline counterpart. Nevertheless, the B-spline hyper-surface is still characterized by a fewer number of design variables: 3200 for B-spline solution vs. 3600 for NURBS solution (because of symmetry).

4.2.4 Multi-Displacement Requirement on a Non-loaded Region

The benchmark problem BK3-3D aims at showing the influence of the displacement requirement, imposed on multiple non-loaded regions, on the final optimized topology. To this end, analogously to BK2-3D, both problems (18) and (28) are solved, and the resulting topologies are compared. A further constraint on the symmetry of the topology with respect to plane $x_2 = \frac{a_2}{2}$ is added to both problem formulations. CPs number and degrees of B-spline and NURBS entities are the same as in BK2-3D.

Firstly, problem (18) is solved with a volume fraction $\gamma = 0.3$ ($V_{\text{ref}} = 10^7$ mm^3). Secondly, problem (28) is solved by introducing a requirement on the displacement component along x_3 axis at point I, i.e. $u_{\text{I3,ref}} = 0.20$ mm, and on the displacement component along x_1 axis at point J, i.e. $u_{\text{I3,ref}} = 0.10$ mm. For both problems, the maximum number of iterations is $N_{\text{iter}}^{\max} = 200$.

Optimized topologies obtained from B-spline and NURBS hyper-surfaces, for both problems (18) and (28), are shown in Fig. 16, where the relative compliance, volume fraction, displacement components at points I and J and number of iterations are reported in the figure sub-captions. As far as numerical results are concerned, the same remarks already done for BK2-3D apply also in this case.

5 Conclusions

In this work, a general formulation of the multi-displacement requirement has been proposed in the framework of a density-based topology optimization algorithm making use of NURBS hyper-surfaces.

Some points of the proposed method deserve attention.

Firstly, unlike the classical SIMP approach, there is no need to define a further filter zone, as the NURBS local support establishes an implicit relationship among contiguous mesh elements. The size of this filter zone depends on the NURBS discrete parameters, which can be properly tuned in order to obtain a good compromise among performances, variables saving and smoothness of boundary.

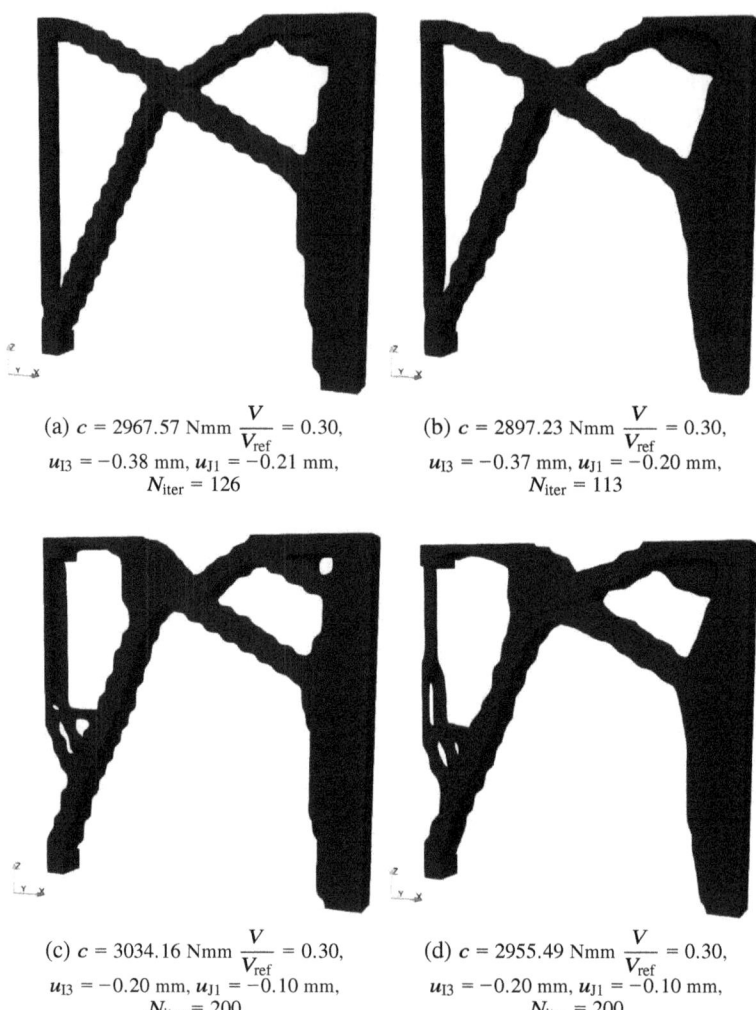

Fig. 16 Benchmark BK3-3D, optimized solutions of problem (18) for (**a**) B-spline and (**b**) NURBS hyper-surfaces and optimized solutions of problem (28) for (**c**) B-spline and (**d**) NURBS hyper-surfaces

Secondly, in all the considered benchmarks the role of NURBS weights has been assessed. In particular by considering the same number of CPs and the same degrees, the objective function of the NURBS solution is lower than the B-spline counterpart, and the boundary of the NURBS solution is always smoother than that of the B-spline solution.

Finally, the optimized topology does not depend upon the quality of the mesh of the FE model, but only upon the integer parameters of the NURBS entity (i.e. degrees of the blending functions and number of CPs). The FE model is used only to assess the physical responses of the problem at hand. The optimized topology can be easily reassembled at the end of the optimization process because it is described by means of a CAD-compatible entity. Moreover, the performances (in terms of volume fraction, displacement requirement, compliance, etc.) of the optimized solution reassembled in the CAD environment are better than those of the related density-map solution and the set of constraints is systematically fulfilled on both solutions.

As far as TO problems involving a requirement on structural displacements are concerned, two meaningful prospects can be identified. Firstly, the multi-displacement requirement should be properly formulated in the case of large generalized displacements and strains. Of course also the influence of the non-linear behaviour of the material should be considered in this case. Secondly, the geometrical wrinkling affecting the boundary of the optimized topology (especially for B-spline solutions) could be properly reduced by integrating the knot-vector components as design variables into the problem formulation. Research is ongoing on the aforementioned aspects.

Acknowledgments Thibaut Rodriguez is grateful to region Nouvelle-Aquitaine for supporting this research work through the OCEAN-ALM project.

References

1. Bendsoe, M., Kikuchi, N.: Generating optimal topologies in structural design using a homogenization method. Comput. Methods Appl. Mech. Eng. **71**, 197–224 (1988).
2. Suzuki, K., Kikuchi, N.: A homogenization method for shape and topology optimization. Comput. Methods Appl. Mech. Eng. **93**(3), 291–318 (1991)
3. Allaire, G., Bonnetier, E., Francfort, G., Jouve, F.: Shape optimization by the homogenization method. Numer. Math. **76**, 27–68 (1997)
4. Xie, Y.M., Steven, G.P.: A simple evolutionary procedure for structural optimization. Comput. Struct. **49**(5), 885–896 (1993)
5. Yang, X.Y., Xie, Y., Steven, G., Querin, O.: Bidirectional evolutionary method for stiffness optimization. AIAA J. **37**, 1483–1488 (1999)
6. Huang, Y.M., Xie, X.: Evolutionary topology optimization of continuum structures with an additional displacement constraint. Struct. Multidiscip. Optim. **40**, 409 (2009)
7. Huang, X., Xie, M.: Evolutionary Topology Optimization of Continuum Structures: Methods and Applications, pp. 121–150. Wiley, Hoboken (2010)
8. Bendsoe, M., Sigmund, O.: Topology Optimization - Theory, Methods and Applications. Springer, Berlin (2003)
9. Sigmund, O.: A 99 line topology optimization code written in Matlab. Struct. Multidiscip. Optim. **21**(2), 120–127 (2001)
10. Allaire, G., Jouve, F., Toader, A.-M.: Structural optimization using sensitivity analysis and a level-set method. J. Comput. Phys. **194**(1), 363–393 (2004)
11. Wang, M.Y., Wang, X., Guo, D.: A level set method for structural topology optimization. Comput. Methods Appl. Mech. Eng. **192**(1), 227–246 (2003)
12. de Ruiter, M.J., van Keulen, F.: Topology optimization using a topology description function. Struct. Multidiscip. Optim. **26**(6), 406–416 (2004)
13. van Dijk, N.P., Maute, K., Langelaar, M., van Keulen, F.: Level-set methods for structural topology optimization: a review. Struct. Multidiscip. Optim. **48**(3), 437–472 (2013)

14. HyperWorks: OptiStruct User's guide, version 23.0. In: Optistruct. Altair Engineering Inc., Troy MI, United States (2014)
15. Abaqus: ABAQUS/Standard User's Manual, Version R2021. Dassault Systèmes Simulia Corp, United States (2021)
16. Guest, J.K., Prévost, J.H., Belytschko, T.: Achieving minimum length scale in topology optimization using nodal design variables and projection functions. Int. J. Numer. Methods Eng. **61**(2), 238–254 (2003)
17. Wang, F., Lazarov, B.S., Sigmund, O.: On projection methods, convergence and robust formulations in topology optimization. Struct. Multidiscip. Optim. **43**(6), 767–784 (2011)
18. Costa, G., Montemurro, M., Pailhès, J.: A 2D topology optimisation algorithm in NURBS framework with geometric constraints. Int. J. Mech. Mater. Des. **14**(4), 669–696 (2018)
19. Costa, G., Montemurro, M., Pailhès, J.: NURBS hypersurfaces for 3D topology optimisation problems. Mech. Adv. Mater. Struct. **28**(7), 665–684 (2021)
20. Costa, G., Montemurro, M., Pailhès, J.: A general hybrid optimization strategy for curve fitting in the non-uniform rational basis spline framework. J. Optim. Theory Appl. **176**, 225–251 (2018)
21. Bertolino, G., Montemurro, M., Perry, N., Pourroy, F.: An efficient hybrid optimisation strategy for surface reconstruction. Comput. Graph. Forum. **40**(6), 215–241 (2021)
22. Costa, G., Montemurro, M., Pailhès, J.: Minimum length scale control in a NURBS-based SIMP method. Comput. Methods Appl. Mech. Eng. **354**, 963–989 (2019)
23. Costa, G., Montemurro, M., Pailhès, J., Perry, N.: Maximum length scale requirement in a topology optimisation method based on NURBS hyper-surfaces. CIRP Ann. **68**, 153–156 (2019)
24. Rodriguez, T., Montemurro, M., Le Texier, P., Pailhès, J.: Structural displacement requirement in a topology optimization algorithm based on isogeometric entities. J. Optim. Theory Appl. **184**, 250–276 (2020)
25. Costa, G., Montemurro, M.: Eigen-frequencies and harmonic responses in topology optimisation: a CAD-compatible algorithm. Eng. Struct. **214**, 110602 (2020)
26. Montemurro, M., Bertolino, G., Roiné, T.: A general multi-scale topology optimisation method for lightweight lattice structures obtained through additive manufacturing technology. Compos. Struct. **258**, 113360 (2021)
27. Roiné, T., Montemurro, M., Pailhès, J.: Stress-based topology optimisation through non-uniform rational basis spline hyper-surfaces. Mech. Adv. Mater. Struct. (2021). https://doi.org/10.1080/15376494.2021.1896822
28. Seo, Y.D., Kim, H.J., Youn, S.K.: Shape optimization and its extension to topological design based on isogeometric analysis. Int. J. Solids Struct. **47**(11–12), 1618–1640 (2010)
29. Xie, X., Wang, S., Xu, M., Wang, Y.: A new isogeometric topology optimization using moving morphable components based on r-functions and collocation schemes. Comput. Methods Appl. Mech. Eng. **339**, 61–90 (2018)
30. Zhang, W., Li, D., Kang, P., Guo, X., Youn, S.-K.: Explicit topology optimization using IGA-based moving morphable void (MMV) approach. Comput. Methods Appl. Mech. Eng. **360**, 112685 (2020)
31. Rong, J., Yi, J.H.: A structural topological optimization method for multi-displacement constraints and any initial topology configuration. Acta Mech. Sin. **26**, 735–744 (2010)
32. Yi, G.L., Sui, Y.K.: Different effects of economic and structural performance indexes on model construction of structural topology optimization. Acta Mech. Sin. **31**(5), 777–788 (2015)
33. Csébfalvi, A.: Volume minimization with displacement constraints in topology optimization of continuum structures. Int. J. Optim. Civil Eng. **6**, 447–453 (2016)
34. Yi, J., Zeng, T., Rong, J.: Topology optimization for continua considering global displacement constraint. Strojniski Vestnik **60**, 43–50 (2014)
35. Ye, H.-L., Dai, Z.-J., Wang, W.-W., Sui, Y.-K.: ICM method for topology optimization of multimaterial continuum structure with displacement constraint. Acta Mech. Sin. **35**(3), 552–562 (2019)

36. Kocvara, M.: Topology optimization with displacement constraints: a bilevel programming approach. Struct. Optim. **14**, 256–263 (1997)
37. Sigmund, O.: On the design of compliant mechanisms using topology optimization. Mech. Struct. Mach. **25**, 493–524 (1997)
38. Sigmund, O.: Manufacturing tolerant topology optimization. Acta Mech. Sin. **25**(2), 227–239 (2009)
39. Luo, Z., Tong, L.: A level set for shape and topology optimization of large-displacement compliant mechanisms. Int. J. Numer. Methods Eng. **76**, 862–892 (2008)
40. Rong, J.H., Liu, X.H., Yi, J.J., Yi, J.H.: An efficient structural topological optimization method for continuum structures with multiple displacement constraints. Finite Elements Anal. Des. **47**(8), 913–921 (2011)
41. Piegl, L., Tiller, W.: The NURBS Book. Springer, Berlin (1997)
42. Liu, Y., Zhao, G., Zavalnyi, O., Cao, X., Cheng, K., Xiao, W.: STEP-compliant CAD/CNC system for feature-oriented machining. Comput.-Aided Des. Appl. **16**, 358–368 (2019)
43. Errico, R.M.: What is an adjoint model? Bull. Am. Meteorol. Soc. **78**(11), 2577–2592 (1997)
44. Svanberg, K. (2002). A class of globally convergent optimization methods based on conservative convex separable approximations. SIAM J. Optim. **12**(2), 555–573 (2002)

Anti-plane Shear in Hyperelasticity

Jendrik Voss, Herbert Baaser, Robert J. Martin, and Patrizio Neff

1 Introduction

The anti-plane shear problem is considered one of the classical challenges in applied nonlinear elasticity theory [25–27]. The essence of this problem is to consider only a very special and simple deformation mode, the so-called anti-plane shear (**APS**), which allows for a reduction of the governing set of equations to an analytically more tractable form, in the compressible as well as the incompressible case. It has been traditional (although not mandatory) to interpret the anti-plane shear problem in a certain non-trivial sense: namely which nonlinear elastic formulations (with nonlinear energies) allow for solutions in APS-form provided only the boundary data is in APS-form [20, 22, 36, 37, 39].

In contrast to this established approach, Gao [13–16] has recently re-interpreted the APS-problem as a simple search for minimizers of the energy functional within the restricted class of APS-deformations. Obviously, the two approaches share some concepts but are, in general, distinct from each other. In this paper, we clarify the differences between both approaches, including numerical examples to highlight the unsuitability of the latter approach in the general case. For simplicity, we pose the APS-problem only for pure Dirichlet boundary data and restrict our considerations to the isotropic case.

We also give a counterexample to a recent statement from the 2015 Int. J. Eng. Sci. article "*On the determination of semi-inverse solutions of nonlinear*

J. Voss · R. J. Martin · P. Neff (✉)
University of Duisburg-Essen, Duisburg, Germany
e-mail: max.voss@uni-due.de; robert.martin@uni-due.de; patrizio.neff@uni-due.de

H. Baaser
University of Applied Sciences Bingen, Bingen am Rhein, Germany
e-mail: h.baaser@th-bingen.de

Cauchy elasticity: The not so simple case of anti-plane shear" by Pucci et al. [38], erroneously connecting ellipticity and the so-called empirical inequalities.

2 Anti-plane Shear Deformations

We employ the usual notion of an anti-plane shear deformation.

Definition 1 An *anti-plane shear deformation* (or **APS**-deformation) is a mapping $\varphi \colon \Omega \subset \mathbb{R}^3 \to \mathbb{R}^3$ of the form $\varphi(x_1, x_2, x_3) = (x_1, x_2, x_3 + u(x_1, x_2))$ with an arbitrary scalar-valued function $u \colon \Omega_{xy} \subset \mathbb{R}^2 \to \mathbb{R}$.

Due to its form, an APS-deformation of a cylinder-shaped body can be identified entirely by the displacement of the bottom or top with the scalar-valued heightfunction $u(x_1, x_2)$. Let $\alpha := u_{,x_1}$, $\beta := u_{,x_2}$, $\gamma := \|\nabla u\|$, and $\gamma^2 = \alpha^2 + \beta^2$. Then the deformation-gradient $F = \nabla \varphi$ and the left and right Cauchy-Green-deformation tensors $B = FF^T =: U^2$ and $C = F^T F =: V^2$ corresponding to an arbitrary APS-deformation are given by

$$F = \begin{pmatrix} 1 & 0 & 0 \\ 0 & 1 & 0 \\ \alpha & \beta & 1 \end{pmatrix}, \quad B = \begin{pmatrix} 1 & 0 & \alpha \\ 0 & 1 & \beta \\ \alpha & \beta & 1 + \gamma^2 \end{pmatrix}, \quad C = \begin{pmatrix} 1 + \alpha^2 & \alpha\beta & \alpha \\ \alpha\beta & 1 + \beta^2 & \beta \\ \alpha & \beta & 1 \end{pmatrix}. \quad (1)$$

In this case, the three isotropic matrix invariants of B (or, equivalently, C)

$$I_1 = \operatorname{tr}(B) = \|F\|^2, \quad I_2 = \frac{1}{2}[(\operatorname{tr} B)^2 - \operatorname{tr}(B^2)] = \operatorname{tr}(\operatorname{Cof} B) = \|\operatorname{Cof} F\|^2,$$

$$I_3 = \det(B) = (\det F)^2 \quad (2)$$

are given by $I_1 = I_2 = 3 + \gamma^2$ and $I_3 = 1$. In particular, APS-deformations always satisfy the condition $\det F = 1$ of incompressibility.

In this paper, we discuss the deformation φ of an elastic isotropic body Ω, cylinder-shaped in its natural state, induced by given boundary conditions. We assume a hyperelastic material behavior, i.e., that the resulting deformation φ is stationary with respect to the energy functional $I(\varphi) = \int_\Omega W(\nabla \varphi) \, dx$ for some nonlinear elastic energy potential $W(F)$. The given boundary conditions are assumed to be satisfiable by an APS-deformation of the whole body, which is the case if and only if the Dirichlet boundary conditions only contain consistent x_3-shifting.[1] Under such boundary conditions, we investigate whether or not an

---

[1] Possible boundary conditions are Dirichlet or Neumann boundary conditions which permit an APS-deformation of the surface $\partial \Omega$ of Ω. Here, we restrict our attention to Dirichlet boundary condition for simplicity of exposition.

APS-deformation of the whole body Ω exists which is stationary with respect to the energy functional $I(\varphi)$.

Of course, due to the involved nonlinearity, the existence of minimizers or solutions to the corresponding Euler-Lagrange equations is not guaranteed without further assumptions (like polyconvexity (cf. Sect. 4.1) and appropriate coercivity conditions) on the energy function. Furthermore, in the case of non-unique solutions (i.e., multiple stationary points), it has to be demonstrated that at least one critical point has the form of an APS-deformation. Therefore, we introduce a number of different terms describing the respective solutions.

- An **APS-equilibrium** is an APS-deformation which is stationary with respect to the restriction of the energy functional to the class of APS-deformations (cf. Sect. 5).
- A **global equilibrium** is an arbitrary deformation which is stationary without the restriction to the class of APS-deformations (cf. Sect. 3).
- A **global APS-equilibrium** is an APS-deformation which is a global equilibrium. Note that every global APS-equilibrium is an APS-equilibrium, but that it is not clear a priori whether the converse holds.
- We call an energy function W **APS-admissible** if *every* APS-equilibrium is also a global equilibrium, i.e., if every APS-equilibrium is a global APS-equilibrium. Note that APS-admissibility does not imply the existence of an APS-equilibrium. Furthermore, even for a non-APS-admissible energy, it is possible for some specific kind of APS-deformation to be a global APS-equilibrium, cf. Remark 3.1.

2.1 Minimizing with Additional Constraints

We can consider APS-deformations as an additional constraint regarding the general minimization problem

$$I(\varphi) = \int_\Omega W(\nabla \varphi(x))\,dx \to \min_{\varphi \in M}, \qquad (3)$$

where M includes some given boundary conditions. Thereby, we distinguish between a priori and a posteriori constraints.

In the former case, the whole variational problem (3) becomes restricted by adding further constraints to the class of allowed deformations M. A very prominent example is the incompressibility assumption $\det \nabla \varphi = 1$ for all $x \in \Omega$. For materials such as rubber, it is much easier to change their shape than their volume, e.g., the relation between infinitesimal shear modulus μ and infinitesimal bulk modulus κ is more extreme for rubber compared to steel or other elastic materials. By restricting the minimization problem to the class of incompressible deformations we simplify the whole model and its corresponding Euler-Lagrange equations to the class of deformations which represent a change in shape but do not alter the volume anywhere in the material. The resulting deformation is a priori incompressible and used as an approximation for rubber-like materials. In this case, incompressibility is an additional assumption for the mathematical model used to reduce its complexity, ideally without moving far away from the initial problem (no material is completely incompressible).

A posteriori constraints work quite differently. For instance, we consider the minimization problem of some arbitrary energy function W with specific boundary conditions and want to solve its Euler-Lagrange equations. Now, for some arbitrary reason, be it the boundary conditions or the energy function itself, we believe that in this specific setting the resulting deformation should be incompressible even though the general material model is compressible. In this case, we start without the incompressibility constraint and derive the full Euler-Lagrange equations.[2] Now, a posteriori, we add the additional incompressibility constraint as an ansatz to simplify the Euler-Lagrange equations. Note that the existence of a minimizer is not ensured for this ansatz of incompressibility; instead, we just try to solve the full Euler-Lagrange equations by testing this approach. The resulting deformation, if it exists at all, is incompressible by the a posteriori assumption and solves the original (compressible) equilibrium problem.

A priori and a posteriori minimization are not identical: A solution of the latter approach directly solves the former minimization problem but not vice versa. In general, a deformation of the a priori case does not solve the original (compressible) minimization problem and therefore it is not ensured that it is a solution of the a posteriori minimization problem, too. Only the a posteriori restriction to the class of incompressible deformations remains generally valid in the sense that the resulting incompressible solution is energetically optimal in the bigger class of compressible deformations. It is crucial to understand that even the raw number of Euler-Lagrange equations of these two problems differ.

We can visualize this issue with a simple two-dimensional function $I : \mathbb{R}^2 \to \mathbb{R}$ which is bounded below. In this example, we replace the constraint of incompressibility by the simple condition $x = y$. Now, in the a priori case, we consider the one-dimensional problem $I_1 : \mathbb{R} \to \mathbb{R}$ with $I_1(x) = I(x, x)$ instead which can be solved by the single equation $I_1'(x) = \frac{d}{dx} I(x, x) = 0$. In the a posteriori case, we start with the two-dimensional problem $\nabla I(x, y) = (0, 0)$ and add the assumption $x = y$ afterward. The latter approach has no solution in most cases while the former case does not necessarily minimize $I(x, y)$ (cf. Fig. 1).

To summarize the conflict, a minimum state in a subclass, like the class of incompressible deformations, does not have to be a global minimum of the general class of compressible deformations. A priori constraints are very strong assumptions for the elasticity model to reduce the analytical complexity significantly while a posteriori constraints are used as mathematical tools trying to solve the general model. Besides, the latter can also be used to validate energy functions regarding

---

[2] The solutions of the Euler-Lagrange equations are the stationary points of the original minimization problem and for the most part impossible to solve analytically.

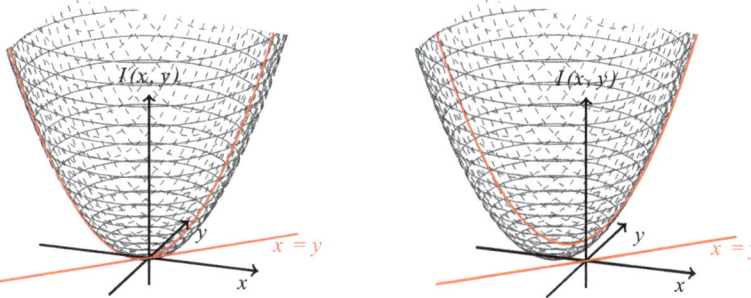

Fig. 1 Difference between a posteriori (black graph) and a priori minimization (red line) using a two-dimensional visualization $I: \mathbb{R}^2 \to \mathbb{R}$ and the constraint $x = y$. Left: Both methods end up with the same minimum, i.e., the a priori solution happens to solve the a posteriori minimization. Right: The a priori minimum is not a minimum of the a $I(x, y)$ while the a posteriori method has no solution at all, i.e., it exists no solution which minimizes $I(x, y)$ and also satisfies the constraint $x = y$

a desired physical behavior, e.g., the observation that every global minimum of the energy in Fig. 1 (Left) satisfies the constraint $x = y$ without any additional a priori assumption.

The same consideration applies to APS-deformations: An **APS-equilibrium** is a solution to the general minimization problem (3) with the a priori restriction of APS-deformations while a **global APS-equilibrium** solves the general problem using APS-deformation as an a posteriori constraint, i.e., only as an ansatz to reduce the full Euler-Lagrange equations.

3 The Classical Full Equilibrium Approach (A Posteriori)

We first discuss the full equilibrium approach to answer the following closely related questions:

- **Under which conditions is every APS-equilibrium a global APS-equilibrium?**
- **Under which conditions does a solution of the Euler-Lagrange equations have the form of an APS-deformation?**

In general, it is possible to obtain non-APS solutions to the equations of equilibrium which *do* have the form of an APS-deformation on the boundary of Ω, see the example in Figs. 3b, and 4b. It is therefore not sufficient to consider the equilibrium equations for APS-deformations exclusively, as described in Sect. 5, an approach followed, e.g., by Gao [14, 16].

The guiding questions were answered exhaustively by Knowles in 1976 [25, 26]. In the following, we elaborate on his work and set it in context with the notion of APS-convexity.

A *stationary deformation* or *equilibrium solution* φ is a deformation which satisfies the Euler-Lagrange equation

$$\text{Div}[DW(\nabla\varphi)] = 0 \qquad (4)$$

to the variational problem

$$I(\varphi) = \int_\Omega W(\nabla\varphi)\,dx \longrightarrow \min. \qquad (5)$$

In the following, we will assume that $W(F) \geq W(\mathbb{1})$ for all $F \in \text{GL}^+(3)$, i.e., that the natural state $F = \mathbb{1}$ is optimal.

Any isotropic energy function W can be represented in terms of the invariants I_1, I_2, I_3 of the left Cauchy-Green deformation tensor globally $B = FF^T$, see, e.g., Antman [3, Chapter 12.13]. Since the derivatives of these invariants with respect to F are given by

$$D_F I_1(FF^T).H = \langle 2F, H \rangle, \qquad D_F I_2(FF^T).H = \langle 2(I_1\mathbb{1} - B)F, H \rangle, \qquad (6)$$

$$D_F I_3(FF^T).H = \langle 2I_3 F^{-T}, H \rangle,$$

respectively, the derivative $DW(F)$ of the energy can be expressed in terms of I_1, I_2, and I_3 via

$$\begin{aligned}DW(F) &= \frac{\partial W}{\partial I_1} D_F I_1(FF^T) + \frac{\partial W}{\partial I_2} D_F I_2(FF^T) + \frac{\partial W}{\partial I_3} D_F I_3(FF^T) \\ &= 2\frac{\partial W}{\partial I_1} F + 2\frac{\partial W}{\partial I_2}(I_1\mathbb{1} - B)F + 2I_3\frac{\partial W}{\partial I_3} F^{-T}. \end{aligned} \qquad (7)$$

In the special case of APS-deformations (for which $I_3 = 1$), the (full) equations of equilibrium are therefore given by (cf. Knowles [26, eq.(10)])

$$\text{Div}\left(2\frac{\partial W}{\partial I_1} F + 2\frac{\partial W}{\partial I_2}(I_1\mathbb{1} - B)F + pF^{-T}\right) = 0 \qquad (8)$$

with $p(I_1, I_2) := 2\frac{\partial W}{\partial I_3}(I_1, I_2, 1)$.

Anti-plane Shear in Hyperelasticity

Using the general representation (1) of APS-deformations as an a posteriori constraint, Eq. (8) can be written more explicitly as

$$\text{Div} \begin{pmatrix} 2\frac{\partial W}{\partial I_1} + 2(2+\beta^2)\frac{\partial W}{\partial I_2} + p & -2\alpha\beta\frac{\partial W}{\partial I_2} & -2\alpha\left(\frac{\partial W}{\partial I_2} + p\right) \\ -2\alpha\beta\frac{\partial W}{\partial I_2} & 2\frac{\partial W}{\partial I_1} + 2(2+\alpha^2)\frac{\partial W}{\partial I_2} + p & -2\beta\left(\frac{\partial W}{\partial I_2} + p\right) \\ 2\alpha\left(\frac{\partial W}{\partial I_1} + \frac{\partial W}{\partial I_2}\right) & 2\beta\left(\frac{\partial W}{\partial I_1} + \frac{\partial W}{\partial I_2}\right) & 2\frac{\partial W}{\partial I_1} + 4\frac{\partial W}{\partial I_2} + p \end{pmatrix} = 0. \tag{9}$$

Since an APS-deformation is completely defined by a single scalar-valued function $u(x_1, x_2)$, the system (9) is *over-determined* by two equations. In order to ensure the existence of a solution $\bar{u}(x_1, x_2)$ to all three partial differential equations, the energy function W should therefore satisfy certain conditions so that two equations are omitted.

Since all occurring terms in (9) are independent of x_3, the last column of the matrix does not contribute to the divergence term. We therefore simplify the equations by replacing these entries by $\star$. Introducing the notation

$$G(I_1, I_2) := 2\frac{\partial W}{\partial I_2}(I_1, I_2, 1), \quad H(I_1, I_2) := 2\left[\frac{\partial W}{\partial I_1}(I_1, I_2, 1) + \frac{\partial W}{\partial I_2}(I_1, I_2, 1)\right],$$

$$q(I_1, I_2) := p(I_1, I_2) + 2\frac{\partial W}{\partial I_1}(I_1, I_2, 1) + 2(I_1 - 1)\frac{\partial W}{\partial I_2}(I_1, I_2, 1) \tag{10}$$

$$\text{with} \quad p(I_1, I_2) := 2\frac{\partial W}{\partial I_3}(I_1, I_2, 1),$$

we write the Euler-Lagrange equations for the general compressible case as

$$\text{Div}\begin{pmatrix} q - \alpha^2 G & -2\alpha\beta G & \star \\ -2\alpha\beta G & q - \beta^2 G & \star \\ \alpha H & \beta H & \star \end{pmatrix} = 0 \iff \begin{array}{ll} q_{,x_1} = (\alpha^2 G)_{,x_1} + (\alpha\beta G)_{,x_2}, & \text{(I)} \\ q_{,x_2} = (\alpha\beta G)_{,x_1} + (\beta^2 G)_{,x_2}, & \text{(II)} \\ 0 = (\alpha H)_{,x_1} + (\beta H)_{,x_2}. & \text{(III)} \end{array}$$

Our approach to the problem of over-determination consists of two steps: first, we consider under which circumstances Eq. (III) has a solution $\bar{u}(x_1, x_2)$; next, we find conditions under which this solution $\bar{u}(x_1, x_2)$ necessarily satisfies the other two Eqs. (I) and (II).

Remark 3.1 (Simple Plane Shear) Some classes of APS-deformations satisfy the above equations in a trivial way without further conditions on the energy function. Such deformations are known as *simple plane shear deformations*. These specific APS-deformations solve the full Euler-Lagrange equations (I)–(III) without the need of APS-admissibility.

The most simple example of a simple plane shear deformation is that of homogeneous shear: Since the Euler-Lagrange equations depend on $\gamma^2 = \|\nabla u\|^2$ via $I_1 = I_2 = 3 + \gamma^2$, every APS-deformation with constant γ satisfies (I)–(III) trivially.[3] For a detailed discussion of different types of simple plane shear, including axial-symmetric APS-deformations with $u(x_1, x_2) = \widetilde{u}(R)$, see [2, 8, 18, 38].

In the general case, however, APS-boundary conditions do not necessarily allow for simple plane shear deformations. The focus of this work is therefore to elaborate conditions (cf. Sect. 3.2) for APS-admissibility, i.e., conditions under which (I)–(III) can be satisfied for APS-deformations which are *not* simple plane shear deformations.

3.1 APS-Convexity

The third Euler-Lagrange equation, rewritten in divergence form

$$0 = \operatorname{div}\left[H\left(3 + \|\nabla u\|^2, 3 + \|\nabla u\|^2\right) \nabla u\right] \tag{11}$$

$$= 2 \operatorname{div}\left[\left(\frac{\partial W}{\partial I_1}(I_1, I_2, I_3) + \frac{\partial W}{\partial I_2}(I_1, I_2, I_3)\right) \nabla u\right],$$

can be represented as

$$\operatorname{div}\left(g'(\|\nabla u\|^2) \nabla u\right) = 0, \quad \text{with} \quad g(x) := W(3 + x, 3 + x, 1). \tag{12}$$

From the assumption that the natural state $F = \mathbb{1}$ is globally optimal for $W(F)$, we infer $g(x) \geq g(0)$. Thus Eq. (12) is the Euler-Lagrange equation corresponding to the scalar variational problem

$$\int_\Omega \frac{1}{2} g(\|\nabla u\|^2) \, dx \longrightarrow \min. \tag{13}$$

Of course, the simplest way of ensuring a solution to this equation is to require the convexity of the energy functional.[4] Therefore, the third Euler-Lagrange equation (III) of our original variation problem does have a solution if the mapping $(\alpha, \beta) \mapsto g(\|(\alpha, \beta)\|^2)$ is convex.

---

[3] For the homogeneous deformation $u(x_1, x_2) = c_1 x_1 + c_2 x_2 + c_3$ with constants $c_1, c_2, c_3 \in \mathbb{R}$, it follows directly from the linearity of u that $\alpha = u_{,x_1} = c_1$ and $\beta = u_{,x_2} = c_2$. This implies $I_1 = I_2 = 3 + \alpha^2 + \beta^2 = \text{const.}$, which shows that $G(I_1, I_2)$, $H(I_1, I_2)$, $p(I_1, I_2)$, $q(I_1, I_2) = \text{const.}$ Thus, all three Euler-Lagrange equations are trivially fulfilled.

[4] Convexity is clearly not necessary for the existence of a minimizer, see, e.g., [16], but it will turn out later that this convexity condition is not a particularly limiting property for most elastic energy functions.

Definition 2 (APS-convexity) We call an energy function $W: \mathbb{R}^{3\times 3} \longrightarrow \mathbb{R}$, $F \mapsto W(F)$ *anti-plane shear convex* (or *APS-convex*) if it is convex on the convex set

$$\mathcal{APS} = \left\{ \begin{pmatrix} 1 & 0 & 0 \\ 0 & 1 & 0 \\ \alpha & \beta & 1 \end{pmatrix} \, \Big| \, \alpha, \beta \in \mathbb{R} \right\}.$$

Remark 3.2 If the function W is expressed in terms of the matrix invariants I_1, I_2, and I_3, then the function is APS-convex if and only if the mapping $(\alpha, \beta) \mapsto W(3 + \gamma^2, 3 + \gamma^2, 1)$, where $\gamma^2 = \alpha^2 + \beta^2$, is convex. This equivalence results from the equalities $I_1 = I_2 = 3 + \|\nabla u\|^2$ and $I_3 = 1$ for APS-deformations.

In the following, we will consider explicit conditions for APS-convexity of isotropic energy functions W.

Lemma 1 *If $g : [0, \infty) \to \mathbb{R}$ satisfies $g(x) \geq g(0)$ for all x in $\mathbb{R}$, then convexity of g implies APS-convexity of $W(I_1, I_2, I_3)$.*

Proof If g is convex and minimal at 0, then g is monotone increasing on $[0, \infty)$. Then the mapping $(\alpha, \beta) \mapsto W(3 + \gamma^2, 3 + \gamma^2, 1) = g(\|(\alpha, \beta)\|^2)$ is convex as the composition of the convex mapping $\|\cdot\|^2$ and the convex and monotone increasing mapping g. □

Lemma 2 *If $W(I_1, I_2, I_3)$ is sufficiently smooth and has its global minimum in the natural state, then the condition*

$$\forall R > 0: \quad \mathcal{W}''(3 + R^2) \geq 0 \quad \text{with} \quad \mathcal{W}(I_1) := W(I_1, I_1, 1), \quad \text{(APS1)}$$

implies APS-convexity.

Proof Condition (APS1) is equivalent to the convexity of $\mathcal{W}$ on $[3, \infty)$, i.e., convexity of the mapping $x \mapsto W(3 + x, 3 + x, 1) = g(x)$ on $[0, \infty)$ which, due to Lemma 1, implies APS-convexity of W. □

The reverse of this implication does not hold in general. In order to obtain a condition equivalent to APS-convexity, we need to directly consider the convexity of the mapping $x \mapsto g(x^2)$ instead.

Theorem 1 *The condition*

$$\forall R > 0: \quad \frac{d^2}{dR^2} W(3 + R^2, 3 + R^2, 1) \geq 0 \quad (APS2)$$

is equivalent to APS-convexity of $W(F)$.

Proof Recall that APS-convexity is equivalent to the convexity of the mapping $(\alpha, \beta) \mapsto g(\|(\alpha, \beta)\|^2)$, which immediately implies the convexity of the mapping $R \mapsto g(\|(R, 0)\|^2) = g(R^2) = W(3 + R^2, 3 + R^2, 1)$ and thus (APS2).

If, on the other hand, (APS2) holds, then the mapping $R \mapsto g(R^2) = W(3 + R^2, 3 + R^2, 1)$ is convex and hence, due to the assumed minimality of the energy at the reference configuration, monotone increasing on $[0, \infty)$. Thus the mapping $(\alpha, \beta) \mapsto g(\|(\alpha, \beta)\|^2)$ is convex as the composition of the (convex) Euclidean norm with a monotone increasing, convex function. □

Remark 3.3 Under the assumption that W is minimal in $\mathbb{1}$ we have shown that the two statements

$$W(I_1, I_2, I_3) \text{ is APS-convex}: \quad (\alpha, \beta) \mapsto W(3 + \gamma^2, 3 + \gamma^2, 1),$$

$$\text{where} \quad \gamma^2 = \alpha^2 + \beta^2, \quad \text{is convex,}$$

$$(APS2): \quad \gamma \mapsto W(3 + \gamma^2, 3 + \gamma^2, 1) \text{ is convex on } [0, \infty)$$

are equivalent.

Remark 3.4 The so-called *ellipticity condition*

$$\forall R > 0: \frac{d}{dR}\left[R\left(\frac{\partial W}{\partial I_1}(I_1, I_2, 1) + \frac{\partial W}{\partial I_2}(I_1, I_2, 1)\right)\bigg|_{I_1 = I_2 = 3 + R^2}\right] \geq 0, \quad (APS3)$$

given by Knowles [26, eq.(19)] is equivalent to APS-convexity of $W(F)$.

Proof $(APS3) \iff \frac{d}{dR}\left[Rg'(R^2)\right] \geq 0 \iff \frac{d^2}{dR^2} g(R^2) \geq 0 \iff (APS2)$.

□

The following implication was pointed out by Fosdick et al. [10, 11].

Lemma 3 *APS-convexity* (APS3) *implies*

$$\forall R > 0: \quad \mathcal{W}'(3 + R^2) > 0 \quad \text{with} \quad \mathcal{W}(I_1) := W(I_1, I_1, 1). \tag{14}$$

Proof Let $f(R) = R\mathcal{W}'(3 + R^2)$. Then (APS3) implies $f'(R) > 0$ for all $R > 0$, i.e., monotonicity of f, thus $0 = f(0) < f(R) = R\mathcal{W}'(3 + R^2)$ and hence $\mathcal{W}'(3 + R^2) > 0$ for all $R > 0$. □

3.2 Energy Function Admissibility Conditions

We introduced APS-convexity as a sufficient condition for the existence of a solution $\bar{u}(x_1, x_2)$ to Eq. (III) and, by means of (APS2), derived a simple criterion for this condition. Another way to obtain such a solution $\bar{u}(x_1, x_2)$ without requiring APS-convexity is discussed in Gao [13, Theorem 5], cf. Sect. 5.

In the following, we consider under which circumstances this solution also satisfies the other two Eqs. (I) and (II) so that the APS-deformation induced by $\bar{u}(x_1, x_2)$ is an overall solution of the full equilibrium equations (I)–(III). The following theorem was obtained by Knowles [26, eq. (21)]; here, we want to elaborate on his proof.

Recall that an energy function is APS-admissible if every APS-equilibrium (solution of equation (III)) is also a global equilibrium (solves Eqs. (I)–(III)).

Theorem 2 (Compressible Case) *Let $W(I_1, I_2, I_3)$ be an isotropic, elastic energy function. Then W is APS-admissible if and only if the following conditions are satisfied:*

$$\exists b \in \mathbb{R}: \forall I_1 = I_2 \geq 3, I_3 = 1: \ b\frac{\partial W}{\partial I_1}(I_1, I_2, I_3) + (b-1)\frac{\partial W}{\partial I_2}(I_1, I_2, I_3) = 0, \tag{K1}$$

$$\frac{\partial^2 W}{\partial I_1^2} + I_1 \frac{\partial^2 W}{\partial I_1 \partial I_2} + \frac{\partial^2 W}{\partial I_1 \partial I_3} + (I_1 - 1)\frac{\partial^2 W}{\partial I_2^2} + \frac{\partial^2 W}{\partial I_2 \partial I_3} + \frac{1}{2}\frac{\partial W}{\partial I_2} = 0. \tag{K2}$$

Proof Recall from Sect. 3 that for an APS-deformation, the three equations of equilibrium are given by

$$q_{,x_1} = (\alpha^2 G)_{,x_1} + (\alpha\beta G)_{,x_2}, \qquad \text{(I)}$$

$$q_{,x_2} = (\alpha\beta G)_{,x_1} + (\beta^2 G)_{,x_2}, \qquad \text{(II)} \qquad (15)$$

$$0 = (\alpha H)_{,x_1} + (\beta H)_{,x_2}. \qquad \text{(III)}$$

Let $\bar{u}(x_1, x_2)$ be an arbitrary solution of equation (III), i.e., an APS-equilibrium. We want to derive Eqs. (K1) and (K2) as conditions on $W(I_1, I_2, I_3)$ for the other two Euler-Lagrange equations to be necessarily satisfied for $\bar{u}(x_1, x_2)$.

If relation (K1) holds,[5] we can simplify Eq. (I) to read[6]

$$q_{,x_1} = (\alpha^2 G)_{,x_1} + (\alpha\beta G)_{,x_2} = \alpha(\alpha G)_{,x_1} + \alpha_{,x_1}\alpha G + \alpha(\beta G)_{,x_2} + \alpha_{,x_2}\beta G$$

$$= \alpha \operatorname{div}(G\,\nabla u) + \alpha_{,x_1}\alpha G + \alpha_{,x_2}\beta G = G(\alpha_{,x_1}\alpha + \alpha_{,x_2}\beta)$$

$$= G(\alpha\alpha_{,x_1} + \beta\beta_{,x_1}) = G\frac{\partial}{\partial x_1}\left[\frac{1}{2}\gamma^2\right] = G\gamma\gamma_{,x_1}, \qquad (16)$$

where $\alpha_{,x_2} = u_{,x_1 x_2} = u_{,x_2 x_1} = \beta_{,x_1}$.

By utilizing the fact that the invariants[7] depend only on $u(x_1, x_2)$, the term $q(I_1, I_2)$ can be expressed as

$$u(x_1, x_2) \mapsto q(3 + \gamma^2, 3 + \gamma^2) := \tilde{q}(\gamma^2)$$

$$\text{with} \quad \frac{\partial q}{\partial x_1}(3 + \gamma^2, 3 + \gamma^2) = \tilde{q}'(\gamma^2) 2\gamma\gamma_{,x_1}. \qquad (17)$$

Therefore, we can transform (II) and similarly (I) to

$$\left(\tilde{q}'(\gamma^2) - \frac{1}{2}G(3+\gamma^2, 3+\gamma^2)\right) 2\gamma\gamma_{,x_1} = 0,$$

$$\left(\tilde{q}'(\gamma^2) - \frac{1}{2}G(3+\gamma^2, 3+\gamma^2)\right) 2\gamma\gamma_{,x_2} = 0, \qquad (18)$$

respectively. As a result, the Euler-Lagrange equations are simplified by condition (K1) to the system of equations

$$\left[\tilde{q}'(\gamma^2) - \frac{\partial W}{\partial I_2}(3+\gamma^2, 3+\gamma^2, 1)\right] 2\gamma\gamma_{,x_1} = 0, \qquad \text{(I)}$$

$$\left[\tilde{q}'(\gamma^2) - \frac{\partial W}{\partial I_2}(3+\gamma^2, 3+\gamma^2, 1)\right] 2\gamma\gamma_{,x_2} = 0, \qquad \text{(II)} \qquad (19)$$

$$\left[\alpha\frac{\partial W}{\partial I_2}(3+\gamma^2, 3+\gamma^2, 1)\right]_{,x_1} + \left[\beta\frac{\partial W}{\partial I_2}(3+\gamma^2, 3+\gamma^2, 1)\right]_{,x_2} = 0. \qquad \text{(III)}$$

---

[5] For the necessity of (K1), see Knowles [25, eq.(3.22)].

[6] With the notation from (10), we can restate (K1) as $bH(I_1, I_2) = G(I_1, I_2)$ with constant $b \in \mathbb{R}$. Therefore, the relationship $\operatorname{div}(H\,\nabla u) = 0$ together with $bH(I_1, I_2) = G(I_1, I_2)$ yields $\operatorname{div}(G\,\nabla u) = 0$.

[7] Note again that $I_1 = I_2 = 3 + \gamma^2 = 3 + \|\nabla u\|^2$.

Note that Eqs. (I) and (II) are trivially satisfied if $\gamma = \|\nabla u\|^2$ is constant, i.e., if φ is a simple plane shear deformation. In the general case of arbitrary APS-deformations, however, (I) and (II) are satisfied if and only if the equation

$$\tilde{q}'(R^2) = \frac{\partial W}{\partial I_2}(3 + R^2, 3 + R^2, 1) \tag{20}$$

holds for all $R \in \mathbb{R}$. Thus, Eqs. (I) and (II) are reduced to a single new Eq. (20) by (K1). The system of equations is still over-determined by one equation. Therefore, we need to show that the last Eq. (20) is equivalent to the energy function compatibility condition (K2):

$$\tilde{q}'(R^2) = 2\frac{d}{dR^2}\frac{\partial W}{\partial I_3}(3 + R^2, 3 + R^2, 1) + 2\frac{d}{dR^2}\frac{\partial W}{\partial I_1}(3 + R^2, 3 + R^2, 1)$$

$$+ 2\frac{d}{dR^2}\left[(2 + R^2)\frac{\partial W}{\partial I_2}(3 + R^2, 3 + R^2, 1)\right]$$

$$= 2\left(\frac{\partial^2 W}{\partial I_3 \partial I_1} 1 + \frac{\partial^2 W}{\partial I_3 \partial I_2} 1\right) + 2\left(\frac{\partial^2 W}{\partial I_1^2} 1 + \frac{\partial^2 W}{\partial I_1 \partial I_2} 1\right) + 2\frac{\partial W}{\partial I_2} \tag{21}$$

$$+ 2(2 + R^2)\left(\frac{\partial^2 W}{\partial I_2 \partial I_1} 1 + \frac{\partial^2 W}{\partial I_2^2} 1\right)$$

$$= 2\left[\frac{\partial^2 W}{\partial I_1 \partial I_3} + \frac{\partial^2 W}{\partial I_2 \partial I_3} + \frac{\partial^2 W}{\partial I_1^2} + I_1 \frac{\partial^2 W}{\partial I_1 \partial I_2} + (I_1 - 1)\frac{\partial^2 W}{\partial I_2^2} + \frac{\partial W}{\partial I_2}\right].$$

Thus (20) and (K2) are, in fact, equivalent in this case.

Altogether, under the two conditions (K1) and (K2), the Euler-Lagrange equations for a compressible energy function are always simplified such that Eqs. (I) and (II) can be omitted for any solution of equation (III). □

In the case of incompressible nonlinear elasticity, energy functions are only defined on the special linear group of isochoric deformations with $I_3 = 1$, thus condition (K2) is not well defined. However, since APS-deformations belong to the class of isochoric deformations, the problem of APS-admissibility can be considered in the incompressible case as well. It should be expected that in the incompressible case, less restricting requirements than the conditions (K1) and (K2) are needed to ensure APS-admissibility.

The concept of APS-convexity remains the same for incompressible and compressible energy functions, starting with the variational problem

$$\min_{\det \nabla \varphi = 1} \int_\Omega W(\nabla \varphi)\, dx \implies \min \int_\Omega W(\nabla \varphi) + p(x)\,(\det(\nabla \varphi) - 1)\, dx\,, \tag{22}$$

where $p(x_1, x_2, x_3) \in C^1(\Omega)$ is now the Lagrange multiplier for the constraint $\det \nabla \varphi = 1$ of incompressibility. With the same notation as before, the Euler-Lagrange equations are simplified to

$$\text{Div}\,[DW(F) + p(x)\,\text{Cof}(F)] = 0$$

with $\text{Cof}(F) = (\det F)\, F^{-T} = F^{-T}$ by incompressibility. We obtain the same formal equation as in the compressible case (8):

$$\text{Div}\left(2\frac{\partial W}{\partial I_1}F + 2\frac{\partial W}{\partial I_2}(I_1\mathbb{1} - B)F + pF^{-T}\right) = 0. \tag{23}$$

Here, however, $p \in C^1(\Omega, \mathbb{R})$ is the Lagrange multiplier and not a fixed term given by the energy function $W(F)$. This yields the same equilibrium system of three coupled partial differential equations, but this time in two scalar-valued functions $u(x_1, x_2)$ and $p(x_1, x_2, x_3)$. Therefore, the equilibrium system is only over-determined by one equation, which means that although the system still does not have a solution in general, only one condition on the energy function is required for APS-admissibility.

Theorem 3 (Incompressible Case) *Let $W(I_1, I_2)$ be an isotropic and **incompressible** elastic energy function. The function W is APS-admissible if and only if*

$$\exists b \in \mathbb{R}: \forall I_1 = I_2 \geq 3: \quad b\frac{\partial W}{\partial I_1}(I_1, I_2) + (b-1)\frac{\partial W}{\partial I_2}(I_1, I_2) = 0. \tag{K1}$$

Proof Analogously to the proof of Theorem 2, the Euler-Lagrange equations can be reduced with the condition (K1) by one equation. Therefore, we can remove one of the first two Euler-Lagrange equations and leave two equations to determine $u(x_1, x_2)$ and $p(x_1, x_2, x_3)$. The system of equations is therefore no longer over-determined under the assumption of (K1). Moreover, it is possible to compute the Lagrange multiplier $p(x_1, x_2, x_3)$ for a given solution $\bar{u}(x_1, x_2)$.[8] □

Remark 3.5 For $\frac{\partial W}{\partial I_1} = c_1$ and $\frac{\partial W}{\partial I_2} = c_2$ with arbitrary constants $c_1, c_2 > 0$, condition (K1) is automatically satisfied with $b = \frac{c_2}{c_1 + c_2}$ and the energy function is APS-convex (APS3).

Remark 3.6 In linear elasticity, the energy function $W_{\text{lin}}(\varepsilon) = \mu \,\|\varepsilon\|^2 + \frac{\lambda}{2}\,\text{tr}(\varepsilon)^2$ with $\varepsilon = \text{sym}\,\nabla u$ is automatically APS-admissible and APS-convex [43]. Therefore, any linear elasticity solution constrained by APS-boundary conditions is

---

[8] For detailed calculations, see [43].

automatically an APS-deformation. Thus APS-admissibility is an inherently non-linear concept.

4 Connections to Constitutive Requirements in Nonlinear Elasticity

The concept of APS-convexity can be extended to the class of APS$^+$-deformations $\varphi : \Omega \subset \mathbb{R}^3 \to \mathbb{R}^3$,

$$\varphi(x_1, x_2, x_3) = (x_1, \ x_2, \ x_3 + u(x_1, x_2, x_3)) \quad \text{with} \quad \varphi \in C^1(\Omega). \tag{24}$$

We call convexity of this type of functions **APS$^+$-convexity**. Note that APS$^+$-convexity immediately implies APS-convexity.

4.1 Convexity

The following lemma shows that an energy function W is APS$^+$-convex (and thus APS-convex) if it is *polyconvex*, i.e., if [5, eq.(0.8)]

$$W(F) = P(F, \ \text{Cof}(F), \ \det(F)) \quad \text{with} \quad P : \mathbb{R}^{3\times 3} \times \mathbb{R}^{3\times 3} \times \mathbb{R} \cong \mathbb{R}^{19} \longrightarrow \mathbb{R} \ \text{convex}.$$

Lemma 4 *Every polyconvex energy function $W(F)$ is APS^+-convex.*

Proof For APS$^+$-convexity of W in $F = \nabla \varphi$ we have to show that

$$W(t \, \nabla\varphi_1 + (1-t) \, \nabla\varphi_2) \le t \, W(\nabla\varphi_1) + (1-t) \, W(\nabla\varphi_2), \quad t \in [0, 1]$$

holds for arbitrary APS$^+$-deformations φ_1, φ_2 (24). In this case, the minors of $F = \nabla\varphi$ are given by

$$F = \begin{pmatrix} 1 & 0 & 0 \\ 0 & 1 & 0 \\ u_{,x_1} & u_{,x_2} & 1+u_{,x_3} \end{pmatrix}, \ \text{Cof}(F) = \begin{pmatrix} 1+u_{,x_3} & 0 & -u_{,x_1} \\ 0 & 1+u_{,x_3} & -u_{,x_2} \\ 0 & 0 & 1 \end{pmatrix}, \ \det(F) = 1 + u_{,x_3}.$$

$$\tag{25}$$

Due to the affine linearity of the above terms, we find for $\varphi = t \, \varphi_1 + (1-t) \, \varphi_2$:

$$F = t \, F_1 + (1-t) \, F_2,$$
$$\text{Cof}(t \, F_1 + (1-t) \, F_2) = t \, \text{Cof}(F_1) + (1-t) \, \text{Cof}(F_2), \tag{26}$$
$$\det(t \, F_1 + (1-t) \, F_2) = t \, \det(F_1) + (1-t) \, \det(F_2),$$

where $F = \nabla \varphi$. If P is convex, then

$$\begin{aligned}
&W(t\, F_1 + (1-t)\, F_2) \\
&= P(t\, F_1 + (1-t)\, F_2,\, \operatorname{Cof}(t\, F_1 + (1-t)\, F_2),\, \det(t\, F_1 + (1-t)\, F_2)) \\
&= P(t\, F_1 + (1-t)\, F_2,\, \operatorname{Cof}(F_1) + (1-t)\operatorname{Cof}(F_2),\, t \det(F_1) + (1-t)\det(F_2)) \\
&\leq t\, P(F_1, \operatorname{Cof}(F_1), \det(F_1)) + (1-t)\, P(F_2, \operatorname{Cof}(F_2), \det(F_2)) \\
&= t\, W(F_1) + (1-t)\, W(F_2),
\end{aligned}$$

which concludes the proof. □

We now want to reduce the requirement of polyconvexity to that of rank-one convexity. An energy function $W(F)$ is called *rank-one convex* if the mapping $t \mapsto W(F + t\, \xi \otimes \eta)$ is convex on $[0, 1]$ for all $F \in \mathbb{R}^{3 \times 3}$ and all $\xi, \eta \in \mathbb{R}^3$.

Lemma 5 *Every rank-one convex energy function $W(F)$ is APS^+-convex.*

Proof Again, we need to show that the mapping

$$t \mapsto W(t\, F_1 + (1-t)\, F_2) = W(F_2 + t(F_1 - F_2))$$

is convex on $[0, 1]$ for all F_1, F_2 of the form $(25)_1$. However, this convexity property follows directly from the rank-one convexity since $F_1 - F_2$ is of the form

$$F_1 - F_2 = \begin{pmatrix} 0 & 0 & 0 \\ 0 & 0 & 0 \\ u_{,x_1} - v_{,x_1} & u_{,x_2} - v_{,x_2} & u_{,x_3} - v_{,x_3} \end{pmatrix} = \begin{pmatrix} 0 \\ 0 \\ 1 \end{pmatrix} \otimes \begin{pmatrix} u_{,x_1} - v_{,x_1} \\ u_{,x_2} - v_{,x_2} \\ u_{,x_3} - v_{,x_3} \end{pmatrix},$$

which concludes the proof. □

Remark 4.1 The above proof also shows that W is APS^+-convex if and only if the mapping $t \mapsto W(F + t\, (0, 0, 1)^T \otimes \eta)$ is convex on $[0, 1]$ for all F of the form $(25)_1$ and all $\eta \in \mathbb{R}^3$.

Corollary 1 *If $W(F)$ is strictly rank-one convex and APS-admissible, then the antiplane shear solution (APS-equilibrium) is a unique APS-equilibrium and minimal in the class of APS-deformations, due to APS-convexity.*

Remark 4.2 As demonstrated by Lemma 5, APS-convexity is not a highly restrictive condition for physically viable elastic energy functions. Moreover, it is remarkable that APS-convexity is equivalent to the monotonicity of the Cauchy shear stress in simple shear [41], see Lemma 8 in Appendix.

Remark 4.3 In a recent article by Pucci et al. [38, eq.(7.1)] it is claimed that in the compressible case, Knowles' "*ellipticity condition* [...] *is a consequence of the empirical inequalities and* [compatibility with linear elasticity]", i.e., that the so-called *empirical inequalities* [3, 29, 42]

$$\beta_0 := \frac{2}{\sqrt{I_3}}\left(I_2\frac{\partial W}{\partial I_2}+I_3\frac{\partial W}{\partial I_3}\right) \leq 0, \qquad \beta_1 := \frac{2}{\sqrt{I_3}}\frac{\partial W}{\partial I_1} > 0,$$

$$\beta_{-1} := -2\sqrt{I_3}\frac{\partial W}{\partial I_2} \leq 0 \qquad (27)$$

together with the condition of a stress-free reference configuration imply Knowles' ellipticity condition (Remark 3.4), cf. Remark 7.1 in Appendix. We show here that for large deformations, this statement is erroneous: Consider the energy function

$$W(F) = \frac{3\mu}{4}\alpha\underbrace{[\log(I_1)+\log(I_2)-\log(I_3)}_{=\log(\|U\|^2)+\log(\|U^{-1}\|^2)}-2\log(3)] + \frac{\mu}{2}(1-\alpha)\left[I_1+\frac{2}{\sqrt{I_3}}-5\right], \qquad (28)$$

with $\mu > 0$ and $0 < \alpha < 1$. The first term is isochoric (i.e., invariant with respect to volume change) and therefore has bulk modulus $\kappa = 0$, the second term ensures positive bulk modulus in the reference state. The empirical inequalities

$$\beta_0 = \frac{3\mu}{4}\alpha\left[\frac{2}{\sqrt{I_3}}\left(I_2\frac{1}{I_2}-I_3\frac{1}{I_3}\right)\right]+\frac{\mu}{2}(1-\alpha)\left[\frac{2}{\sqrt{I_3}}\left(0-I_3\frac{1}{I_3^{3/2}}\right)\right] = \frac{\mu(1-\alpha)}{I_3} < 0,$$

$$\beta_1 = \frac{3\mu}{4}\alpha\frac{2}{\sqrt{I_3}}\frac{1}{I_1}+\frac{\mu}{2}(1-\alpha)\frac{2}{\sqrt{I_3}}\cdot 1 > 0, \qquad \beta_{-1} = -\frac{3\mu}{4}\alpha\sqrt{I_3}\frac{2}{I_2} \leq 0 \qquad (29)$$

are satisfied. Moreover, the energy function is stress-free in the reference configuration $F = \mathbb{1}$, since

$$\left[\frac{\partial W}{\partial I_1}+2\frac{\partial W}{\partial I_2}+\frac{\partial W}{\partial I_3}\right]_{F=\mathbb{1}} = \frac{3\mu}{4}\alpha\left[\frac{1}{3}+\frac{2}{3}-\frac{1}{1}\right]+\frac{\mu}{2}(1-\alpha)\left[1-\frac{1}{1}\right] = 0 \qquad (30)$$

and the generated infinitesimal shear modulus can be determined from

$$(\beta_1 - \beta_{-1})_{F=\mathbb{1}} = \frac{3\mu}{4}\alpha\left[\frac{2}{3}+\frac{2}{3}\right]+\frac{\mu}{2}(1-\alpha)\left[\frac{2}{1}\right] = \mu. \qquad (31)$$

Recall from Lemma 3.4 that Knowles' ellipticity condition is equivalent to the condition (APS2) of APS-convexity which, in this case, reads

$$0 \leq \frac{d^2}{dR^2}W(3+R^2, 3+R^2, 1) = \frac{3\mu}{4}\alpha\frac{d^2}{dR^2}\left[2\log(3+R^2)\right]+\frac{\mu}{2}(1-\alpha)\frac{d^2}{dR^2}\left[R^2\right]$$

$$= \frac{3\mu}{2}\alpha\frac{d}{dR}\left[\frac{2R}{3+R^2}\right]+\frac{\mu}{2}(1-\alpha)\frac{d}{dR}[2R] = 3\mu\alpha\left[\frac{3-R^2}{(3+R^2)^2}\right]+\mu(1-\alpha). \qquad (32)$$

However, if $\frac{8}{9} < \alpha < 1$, then there exists an interval where APS-convexity is violated. Therefore, the energy function (28) with $\frac{8}{9} < \alpha$ is compatible with linear elasticity and satisfies the empirical inequalities (27) as well as the condition of a stress-free reference configuration but does not satisfy Knowles' ellipticity condition, in contradiction to the claim by Pucci et al. [38].

4.2 Tension-Compression Symmetry

Table 1 shows a number of elastic energy potentials used in nonlinear elasticity theory and their properties regarding APS-convexity. The detailed calculations can be found in [43].

Note that an APS-admissible energy in the incompressible case only has to satisfy condition (K1), whereas an APS-admissible energy for the general compressible case must also fulfill condition (K2). A still unsolved problem is to find a compressible viable energy function which is APS-admissible but depends nonlinearly on I_2. It is noticeable in Table 1 that many energy functions satisfy condition (K1) with $b = 0$ or $b = \frac{1}{2}$; the former case can be easily explained by the independence from the second invariant.

Lemma 6 *Every isotropic energy function $W(F)$ which can be expressed in the form $W(F) = W(I_1, I_3)$, i.e., which does not depend on the second invariant I_2, satisfies condition* (K1) *with $b = 0$.*

Proof Condition (K1) with $b = 0$ can be simplified to $\frac{\partial W}{\partial I_2} = 0$, which is trivially fulfilled for every isotropic energy function of the type $W(F) = W(I_1, I_3)$. □

The special case $b = \frac{1}{2}$, on the other hand, shows a more interesting relation to the so-called *tension-compression symmetry* of an energy.

Definition 3 An energy function $W(F)$ is called *tension-compression symmetric* if $W(F) = W(F^{-1})$ for all $F \in \text{GL}^+(3)$.

Lemma 7 *An isotropic tension-compression symmetric energy function W is invariant under permutation of the two invariants I_1 and I_2 under the constraint of incompressibility, i.e., $W(I_1, I_2, 1) = W(I_2, I_1, 1)$, see also [4].*

Proof Let $I_1' = I_1(B^{-1})$, $I_2' = I_2(B^{-1})$, $I_3' = I_3(B^{-1})$. Then

$$I_1' = \text{tr}(B^{-1}) = \text{tr}\left(\frac{\det(B)}{\det(B)} B^{-1}\right) = \frac{1}{\det(B)} \text{tr}(\det(B)\, B^{-T}) = \frac{\text{tr}(\text{Cof}(B))}{\det(B)} = \frac{I_2}{I_3},$$

$$I_2' = \text{tr}(\text{Cof}(B^{-1})) = \text{tr}(\det(B^{-1})\, (B^{-1})^{-T}) = \det(B^{-1})\, \text{tr}(B^T) = \frac{\text{tr}(B)}{\det(B)} = \frac{I_1}{I_3},$$

$$I_3' = \det(B^{-1}) = \frac{1}{\det(B)} = \frac{1}{I_3}. \tag{33}$$

Table 1 An overview of APS-related properties for several important energy functions

Name	Energy expression	Rank-1 convex	APS-convex	K1 incomp.	K2 compr.
Neo-Hooke [35]	$W(F) = \frac{\mu}{2}(I_1 I_3^{-\frac{1}{3}} - 3) + h(I_3)$	Yes	Yes	$b = 0$	No
Mooney-Rivlin [35]	$W(F) = \frac{\mu}{2}\alpha \left(I_1 I_3^{-\frac{1}{3}} - 3 \right) + \frac{\mu}{2}(1-\alpha)\left(I_2 I_3^{-\frac{2}{3}} - 3 \right) + h(I_3)$	Yes	Yes	$b = 1 - \alpha$	No
Blatz-Ko [21]	$W(F) = \frac{\mu}{2}\left(I_1 + \frac{2}{\sqrt{I_3}} - 5 \right)$	Yes	Yes	$b = 0$	Yes[a]
Veronda-Westman [34]	$W(F) = \mu\left(\frac{e^{\gamma(I_1-3)}-1}{\gamma} - \frac{I_2-3}{2} \right) + h(I_3)$	No	Yes	No	No
Mihai-Neff [28, 31]	$W(F) = \frac{\mu}{2}\left(I_1 I_3^{-\frac{1}{3}} - 3 \right) + \frac{\tilde{\mu}}{4}(I_1-3)^2 + \frac{\kappa}{2}\left(I_3^{\frac{1}{2}} - 1 \right)^2$	No	Yes	$b = 0$	$\tilde{\mu} = \frac{\mu}{3}$
Knowles	$W(F) = \frac{\mu}{2b}\left(\left[1 + \frac{b}{n}\left(I_1 I_3^{-\frac{1}{3}} - 3 \right) \right]^n - 1 \right) + \frac{1}{D_1}\left(I_3^{\frac{1}{2}} - 1 \right)^2$	?	Yes	$b = 0$	No
Bazant	$W(F) = \|B - B^{-1}\|^2$	No	Yes	$b = \frac{1}{2}$	No
Ciarlet [9]	$W(F) = \frac{c_1}{2} I_1 + \frac{c_2}{2} I_2 + h(\sqrt{I_3})$	Yes	Yes	$b = \frac{c_2}{c_1+c_2}$	$c_2 = 0$
SVK [9]	$W(F) = \frac{\mu}{4}\|C - \mathbb{1}\|^2 + \frac{\lambda}{8} \text{tr}(C-\mathbb{1})^2$	No	Yes	No	—
4th Order	$W(F) = \mu \, \text{tr}(E^2) + \frac{A}{3}\text{tr}(E^3) + D\,\text{tr}(E^2)^2$	No	Yes	No	—
Hencky [17, 33]	$W(F) = \mu \|\text{dev}\log V\|^2 + \frac{\kappa}{2}(\text{tr}(\log V))^2$	No	No	$b = \frac{1}{2}$	No
exp-Hencky [32]	$W(F) = \frac{\mu}{k} e^{k\|\text{dev}\log V\|^2} + \frac{\kappa}{2\hat{k}} e^{\hat{k}(\text{tr}(\log V))^2}$	No	Yes	$b = \frac{1}{2}$	No
Martin-Neff	$W(F) = \frac{\|F\|^3}{\det(F)} + \det(F)\|F^{-1}\|^3$	Yes	Yes	$b = \frac{1}{2}$	No
Model [43]	$W(F) = c_1 \left(\sqrt{I_1} + \sqrt{I_2} + \frac{\sqrt{3}}{\sqrt{I_3}} - 3\sqrt{3} \right)$ No approximation to linear elasticity in $F = \mathbb{1}$.	Yes	Yes	$b = \frac{1}{2}$	Yes
Becker [7]	$W(F) = 2\mu \langle U, \log U - \mathbb{1} \rangle$	No	No	No	Yes

[a] A general class of APS-admissible energy functions $W(I_1, I_3)$ can be found in [23]

Therefore, tension-compression-symmetry implies $W(I_1, I_2, I_3) = W(I_1', I_2', I_3') = W(\frac{I_2}{I_3}, \frac{I_1}{I_3}, \frac{1}{I_3})$ and thus, in particular, $W(I_1, I_2, 1) = W\left(\frac{I_2}{1}, \frac{I_1}{1}, \frac{1}{1}\right) = W(I_2, I_1, 1)$. □

Theorem 4 *Every isotropic tension-compression-symmetric energy function $W(F)$ satisfies condition (K1) with $b = \frac{1}{2}$.*

Proof The condition (K1) with $b = \frac{1}{2}$ can be restated as

$$\frac{1}{2}\frac{\partial W}{\partial I_1}(I_1, I_1, 1) + \left(\frac{1}{2} - 1\right)\frac{\partial W}{\partial I_2}(I_1, I_1, 1) = 0 \iff \frac{\partial W}{\partial I_1}(I_1, I_1, 1) = \frac{\partial W}{\partial I_2}(I_1, I_1, 1)$$

for all $I_1 \geq 3$, and for tension-compression symmetric W we find

$$\frac{\partial W}{\partial I_1}(I_1, I_1, 1) = \frac{d}{dt}W(t, I_1, 1)\bigg|_{t=I_1} = \frac{d}{dt}W(I_1, t, 1)\bigg|_{t=I_1} = \frac{\partial W}{\partial I_2}(I_1, I_1, 1)$$

due to Lemma 7. □

Coming back to Table 1, we observe that no energy function which satisfies condition (K1) with $b = \frac{1}{2}$ also fulfills the second condition (K2). Therefore, we hypothesize that APS-admissibility is not a reasonable characteristic for physically motivated compressible energy functions.

5 The Constrained Equilibrium Approach (A Priori)

By testing several examples, we are led to believe that most viable energy functions in compressible nonlinear elasticity are *not* APS-admissible. Therefore, in general, APS-boundary conditions do *not* necessarily lead to an APS-deformation of the whole body. Nevertheless, it is possible to compute the energetically optimal APS-deformation by minimization only over the class of APS-deformations, i.e., APS-deformation as an a priori constraint (cf. Sect. 5):

$$I(\varphi) = \int_\Omega W(\nabla\varphi)\, dx \longrightarrow \min_{\varphi \in \mathcal{APS}}. \qquad (34)$$

An equilibrium of the corresponding Euler-Lagrange equations of (34) (with respect to the restriction of the energy functional to the class of APS-deformations) is called APS-equilibrium and does not have to be stationary in the global sense (5). As emphasized by Saccomandi [39], this approach was chosen by Gao [13–16]:[9] starting with

$$I(u) = \int_\Omega \mathcal{W}(3 + \|\nabla u\|^2)\, dx \longrightarrow \min, \qquad (35)$$

---

[9] Gao [12]:" [...] the equilibrium equation [...] has just one non-trivial component [namely equation (III)]." Gao claims that Knowles' condition (K1) is automatically satisfied for every elastic energy function with $b = 0$, which is clearly not the case (Table 1).

where we employ the same notation[10] $W(I_1, I_1, 1) = \mathcal{W}(I_1)$ as before, we obtain the Euler-Lagrange equation $\mathrm{div}(\mathcal{W}'(3+\|\nabla u\|^2)\nabla u) = 0$ for stationarity within the class of APS-deformations, which is equivalent to Eq. (III) from the full equilibrium approach.

Corollary 2 *APS-Convexity of the energy function $W(F)$ ensures the existence of a unique APS-equilibrium which is a global energy minimizer (among the class of APS-deformations).*

Contrary to Theorem 5 in [16], we see in Lemma 5 that strict rank-one convexity implies strict APS-convexity which, in turn, implies uniqueness of the APS-equilibrium.

Gao [16] prominently discusses the case where $g(\|u\|^2) = \mathcal{W}(3 + \|u\|^2) = W(3+\|u\|^2, 3+\|u\|^2, 1)$ is *not* convex. In this case, the existence of a solution to the minimization problem is not clear due to the loss of APS-convexity (see Lemma 1), and one needs to resort to just solutions of the Euler-Lagrange equation (III); of course, while such solutions may exist, it is by no means obvious why they should satisfy the general equations of equilibrium.

Remark 5.1 If an energy function is APS-admissible (satisfies (K1) for incompressible material behavior or (K1) and (K2) in the compressible case), then the full and the constrained equilibrium approach provide the same solution.

6 Finite Element Simulations

We consider the deformation of a unit cube Ω with APS-type Dirichlet boundary conditions on the four lateral sides of the cube, see Fig. 2. In order to compare the APS-computations for different constitutive laws, we perform numerical simulations using the finite element system ABAQUS [1], which supports the use of internal models (e.g., the compressible Neo-Hooke or the compressible Mooney-Rivlin model) as well as the implementation of custom hyperelastic models via the provided user subroutine uhyper, which requires the user to provide the energy function $W(I_1, I_2, I_3)$ in terms of the invariants as well as its first, second, and third derivatives.

In the latter case for the energy expression of Becker [7], which is formulated in terms of the right stretch tensor U (and its logarithms) instead of the deformation invariants, we implement the constitutive relations via the umat user subroutine into the ABAQUS environment following [30] for strain–energies based on principal stretches or their logarithms, see also [19]. Here, we obtain the analytical derivations, e.g., $\frac{d^2 W}{d(\log \lambda_i)d(\log \lambda_j)}$, by Matlab [40] following [24] and transform the resulting

---

[10] For APS-deformations, $I_1 = I_2 = 3 + \|\nabla u\|^2$ and $I_3 = 1$.

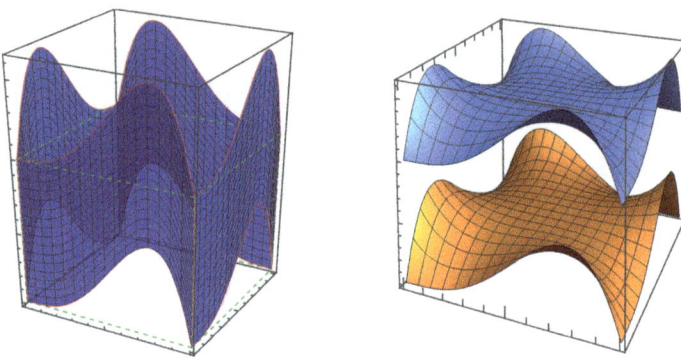

Fig. 2 (Left) Prescribed APS-boundary conditions. (Right) possible APS-deformation of the top and bottom side of the cube

material tangent modulus C into the typical mixed configuration used by ABAQUS, see [30], for the Jaumann stress rate.

For our numerical calculations, we use a grid of $21 \times 21 \times 21$ nodes. The considered unit cube is discretized by 8-noded linear brick elements with hybrid formulation (C3D8H) in order to get better approximations for the (quasi-)incompressible hyperelastic models.

The APS-boundary conditions as shown in Fig. 2 are realized by the `disp` subroutine which enables the user to prescribe values for selected node sets and their addressed degree of freedom (DOF) for each iteration increment. Here, we apply the functional value depending on the nodal x_1, x_2-position onto the boundary nodes of the unit cube.

In the following, we want to visualize the difference between APS-admissible energy functions in the general compressible case, APS-admissibility only for the constraint of incompressibility and an energy function that satisfies neither condition. We start with the incompressible case and choose the Mooney-Rivlin and Veronda-Westman energy functions (see Table 1). Both are APS-convex, but only the Mooney-Rivlin energy satisfies the condition (K1) which implies APS-admissibility in the incompressible case. An exact APS-deformation is characterized by an exclusive displacement in e_3-direction for every node of the whole body Ω. Therefore, the e_1-e_2-plane grid-structure of the nodes in the undeformed body Ω has to be maintained by any deformation in equilibrium for an APS-admissible energy function. We introduce the measure $u_\delta = \sqrt{(\varphi_1(x) - x_1)^2 + (\varphi_1(x) - x_2)^2}$ of deviation from an APS-deformation.

The graphics in Fig. 3a visualize the deformation induced by the Mooney-Rivlin energy, which is APS-admissible in the incompressible case. The slice of the inside of the cube shows perfect APS-behavior, maintaining the original grid-structure. The deformation induced by the non-APS-admissible Veronda-Westman energy function is shown in Fig. 3b. The deviation to the original grid-structure is more distinct and affects the whole body Ω.

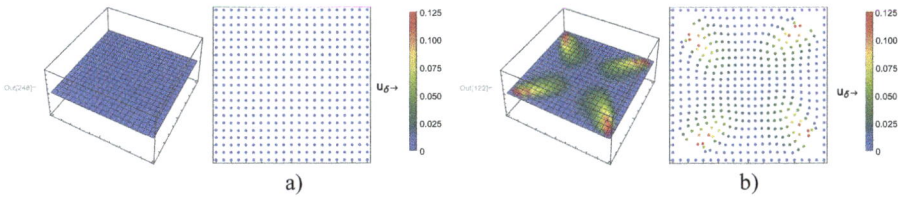

Fig. 3 Visualization of a slice (at $x_3 = 0.5$) of the deformed square with APS-boundary condition (Fig. 2) for the Mooney-Rivlin (**a**) and the Veronda-Westman (**b**) energy in the quasi-incompressible case (bulk modulus $K \sim 10^5 \mu$ shear modulus). The color shows the displacement u_δ in x_1- and x_2-direction

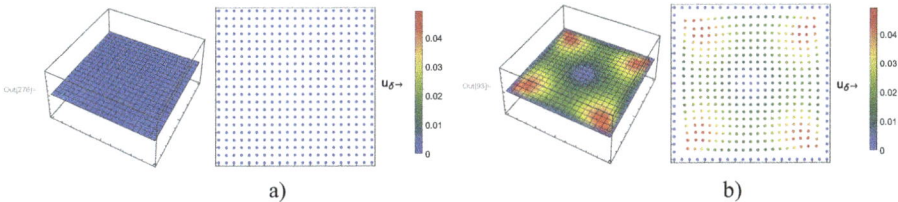

Fig. 4 Visualization of a slice ($x_3 = 0.5$) of the deformed square with APS-boundary condition (Fig. 2) with the Blatz-Ko (**a**) and the Mooney-Rivlin (**b**) energy in the compressible case (bulk modulus $K \sim \mu$ shear modulus). The color shows the displacement u_δ in x_1- and x_2-direction

For the visualization of APS-admissibility in the compressible case, we again use the Mooney-Rivlin energy and compare it to the APS-admissible Blatz-Ko model (cf. Table 1); note that the Mooney-Rivlin energy is not APS-admissible in the compressible case. Similar to our observation of the quasi-incompressible case, the equilibrium solution for the APS-admissible energy function (Blatz-Ko, Fig. 4a) shows perfect APS-behavior inside the cube. The deformation induced by the Mooney-Rivlin energy, on the other hand, shows more distinguished deviations from an APS-deformation throughout the whole body.

Last, we repeat our quasi-incompressible calculations for the Becker-energy $\widehat{W}(F) = W(F) + \frac{K}{2}(\det F - 1)^2$ with $K \gg \mu$ (cf. Fig. 5). The Becker model is not APS-admissible, i.e., condition (K1) is not satisfied but still shows almost perfect APS-behavior at the middle height ($x_3 \approx 0.5$) of the cube for the APS-boundary conditions used here, while the grid-structure of the deformed configuration gradually becomes more convoluted near the top and bottom sides of the cube. In addition, the lack of APS-convexity of the Becker-energy leads to the loss of convergence in the compressible case. As mentioned above, these two versions of the model (compressible and quasi-incompressible) are implemented in ABAQUS via the umat user interface.

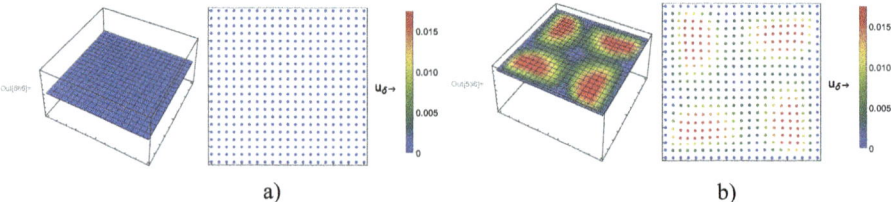

Fig. 5 Visualization of a slice (**a**) at $x_3 = 0.5$ and (**b**) at $x_3 = 0.9$ of the deformed square with APS-boundary condition (Fig. 2) for the Becker-energy in the quasi-incompressible case (bulk modulus $K \sim 10^5 \mu$ shear modulus). The color shows the displacement u_δ in x_1- and x_2-direction

7 Conclusion

This work elaborates on Knowles' paper [26] and the difference between Knowles' (full) and Gao's (constrained) approach. The two conditions (K1) and (K2) as discovered by Knowles were derived directly from the Euler-Lagrange equations of the energy function W. The required ellipticity condition [26, eq.(19)] was identified with the introduced APS-convexity and inferred from the important concepts of polyconvexity and rank-one convexity. Since the latter condition is a highly desirable property in nonlinear elasticity from a mathematical point of view, the requirement of APS-convexity does not further restrict the class of viable energy functions. Moreover, investigating different elastic energy functions revealed that even a number of commonly used non-rank-one convex energy functions are in fact still APS-convex due to its equivalence to the monotonicity of the Cauchy shear stress in simple shear.

Furthermore, it was shown that, contrary to expectations, (K1) is fulfilled by almost all investigated energy functions; indeed, this condition follows from the physically reasonable requirement of tension-compression-symmetry. Therefore, it is to be expected for incompressible elastic materials to exhibit APS-deformations under given APS-boundary conditions.

However, within the context of nonlinear hyperelasticity, the additional APS-admissibility condition (K2) in the compressible case appears to be satisfied only in the trivial case of energies which do not depend on the second invariant I_2. Note that the only energy function listed in Table 1 that satisfies condition (K2) and is not independent of I_2 is a function without consistency to linear elasticity, unsuitable for mechanical application.

By numerical simulations, we were able to visualize the difference between the deformations under APS-type boundary conditions induced by APS-admissible and non-APS-admissible energy functions in the incompressible as well as the compressible case.

Acknowledgment We thank Giuseppe Saccomandi (University of Perugia) and Roger Fosdick (University of Minnesota) for helpful discussions.

Appendix

Recall that in the isotropic case, the Cauchy-stress tensor can always be expressed in the form

$$\sigma = \beta_0 \mathbb{1} + \beta_1 B + \beta_{-1} B^{-1} \qquad (36)$$

with scalar-valued functions β_i depending on the invariants of B. In the hyperelastic isotropic case, β_0, β_1, and β_{-1} are given by

$$\beta_0 = \frac{2}{\sqrt{I_3}}\left(I_2 \frac{\partial W}{\partial I_2} + I_3 \frac{\partial W}{\partial I_3}\right), \quad \beta_1 = \frac{2}{\sqrt{I_3}}\frac{\partial W}{\partial I_1}, \quad \beta_{-1} = -2\sqrt{I_3}\frac{\partial W}{\partial I_2}. \qquad (37)$$

Lemma 8 *Let $\varphi: \Omega \to \mathbb{R}$, $\varphi(x) = (x_1 + \gamma x_2, x_2, x_3)$ be a simple shear deformation, with $\gamma \in \mathbb{R}$ denoting the amount of shear. Then the Cauchy shear stress σ_{12} of an arbitrary isotropic energy function $W(I_1, I_2, I_3)$ is monotone as a scalar-valued function depending on the amount of shear for positive γ if and only if W is APS-convex.*

Proof We consider the Cauchy-stress tensor for an arbitrary material which is stress-free in the reference configuration:

$$\sigma = \beta_0 \mathbb{1} + \beta_1 B + \beta_{-1} B^{-1}. \qquad (38)$$

In the case of simple shear we compute [6, p.41]

$$\nabla\varphi = \begin{pmatrix} 1 & \gamma & 0 \\ 0 & 1 & 0 \\ 0 & 0 & 1 \end{pmatrix}, \; B = FF^T = \begin{pmatrix} 1+\gamma^2 & \gamma & 0 \\ \gamma & 1 & 0 \\ 0 & 0 & 1 \end{pmatrix}, \; B^{-1} = \begin{pmatrix} 1 & -\gamma & 0 \\ -\gamma & 1+\gamma^2 & 0 \\ 0 & 0 & 1 \end{pmatrix}, \qquad (39)$$

$$I_1 = \mathrm{tr}\, B = 3+\gamma^2, \; I_2 = \mathrm{tr}(\mathrm{Cof}\, B) = \mathrm{tr}\begin{pmatrix} 1 & -\gamma & 0 \\ -\gamma & 1+\gamma^2 & 0 \\ 0 & 0 & 1 \end{pmatrix} = 3+\gamma^2, \; I_3 = \det B = 1,$$

$$\implies \sigma = (\beta_0 + \beta_1 + \beta_{-1})\mathbb{1} + \begin{pmatrix} \beta_1\gamma^2 & (\beta_1-\beta_{-1})\gamma & 0 \\ (\beta_1-\beta_{-1})\gamma & \beta_{-1}\gamma^2 & 0 \\ 0 & 0 & 0 \end{pmatrix}. \qquad (40)$$

Therefore, the Cauchy shear stress component σ_{12} is a scalar-valued function depending on the amount of shear γ, given by

$$\sigma_{12}(\gamma) = (\beta_1 - \beta_{-1})\gamma = \gamma \, \frac{2}{\sqrt{I_3}}\left(\frac{\partial W}{\partial I_1} + I_3 \frac{\partial W}{\partial I_2}\right)\bigg|_{I_1=I_2=3+\gamma^2, I_3=1}$$

$$= 2\gamma \left(\frac{\partial W}{\partial I_1} + \frac{\partial W}{\partial I_2}\right)\bigg|_{I_1=I_2=3+\gamma^2, I_3=1} = \frac{d}{d\gamma} W(3+\gamma^2, 3+\gamma^2, 1). \qquad (41)$$

The positivity of the Cauchy shear stress is already implied by the (weak) empirical inequalities $\beta_1 > 0$, $\beta_{-1} \leq 0$. The condition for shear-monotonicity is given by

$$\frac{d}{d\gamma}\sigma_{12}(\gamma) = \frac{d^2}{d\gamma^2}W(3+\gamma^2, 3+\gamma^2, 1) > 0 \qquad \forall \gamma \geq 0, \tag{42}$$

which is equivalent to APS-convexity condition (APS2) of the energy function $W(I_1, I_2, I_3)$. □

Remark 7.1 The empirical inequalities (27) state that $\beta_0 \leq 0$, $\beta_1 > 0$, $\beta_{-1} \leq 0$. In the case of APS-deformations ($I_1 = I_2 = 3+\gamma^2$, $I_3 = 1$), Pucci et al. [38, eq.(4.3)] obtain the inequality

$$(I_1 - 3)^p (h^*)' + q^2 h^* > 0, \quad \forall I_1 \geq 3 \quad \text{with } h^*(I_1) = \beta_1 - \beta_{-1}|_{I_1=I_2, I_3=1}, \tag{43}$$

"where p, q are real numbers such that $p > 0$ and $q \neq 0$", by a "simple manipulation of the empirical inequalities (27) and [the stress-free reference configuration]". In [38, Remark III], it is pointed out correctly that in the case of $p = 1$, $q^2 = \frac{1}{2}$ (they erroneously use $q = 1$) the resulting constitutive inequality

$$0 < 2\left((I_1 - 3)(h^*)'(3+\gamma^2) + \frac{1}{2}h^*(3+\gamma^2)\right)$$

$$= (h^*)'(3+\gamma^2) \cdot 2\gamma^2 + h^*(3+\gamma^2) = \frac{d}{d\gamma}\left[\gamma h^*(3+\gamma^2)\right] \tag{44}$$

is equivalent to APS-convexity by Eq. (APS3) with

$$\left(\frac{\partial W}{\partial I_1} + \frac{\partial W}{\partial I_2}\right)\bigg|_{I_1=I_2=3+\gamma^2, I_3=1} \stackrel{(37)}{=} \beta_1 - \beta_{-1}|_{I_1=I_2=3+\gamma^2, I_3=1} = h^*(3+\gamma^2). \tag{45}$$

We are, however, not able to reproduce a proof of inequality (43), see also the counterexample in Remark 4.3.

Lemma 9 *Let W be a sufficiently smooth isotropic energy function such that the induced Cauchy-stress response satisfies the (weak) empirical inequalities. Then for sufficiently small shear deformations (i.e., within a neighborhood of the identity $\mathbb{1}$), the Cauchy shear stress is a* monotone *function of the amount of shear.*

Proof In Lemma 8, we already computed the Cauchy shear stress corresponding to a simple shear to be $\sigma_{12}(\gamma) = (\beta_1 - \beta_{-1})\gamma$, with $\gamma \in \mathbb{R}$ denoting the amount of shear. The monotonicity of this mapping is equivalent to

$$0 < \frac{d}{d\gamma}\sigma_{12}(\gamma) = \frac{d}{d\gamma}\left[(\beta_1(3+\gamma^2) - \beta_{-1}(3+\gamma^2))\gamma\right]$$
$$= \left(\beta_1'(3+\gamma^2) - \beta_{-1}'(3+\gamma^2)\right)2\gamma^2 + \beta_1(3+\gamma^2) - \beta_{-1}(3+\gamma^2). \qquad (46)$$

According to the (weak) empirical inequalities, $\beta_1(3) - \beta_{-1}(3) =: \mu > 0$. Therefore, $\beta_1(3+\gamma^2) - \beta_{-1}(3+\gamma^2) \geq \varepsilon > 0$ for sufficiently small $\gamma \in \mathbb{R}$. If W and thus β_1, β_{-1} are sufficiently smooth, then $\beta_1' - \beta_2'$ is locally Lipschitz-continuous, and thus within a compact neighborhood of $\mathbb{1}$,

$$\frac{d}{d\gamma}\sigma_{12}(\gamma) = \underbrace{\beta_1(3+\gamma^2) - \beta_{-1}(3+\gamma^2)}_{\geq \varepsilon} + \underbrace{\left(\beta_1'(3+\gamma^2) - \beta_{-1}'(3+\gamma^2)\right)}_{\leq \text{const.}} \cdot 2\gamma^2 > 0$$

for every sufficiently small shear deformation, i.e., sufficiently small γ. □

References

1. ABAQUS/Standard User's Manual. Simulia, Providence, RI (2017)
2. Agarwal, V.: On finite anti-plane shear for compressible elastic circular tube. J. Elast. **9**(3), 311–319 (1979)
3. Antman, S.S.: Nonlinear Problems of Elasticity, 2nd edn. Springer, New York (2005)
4. Baaser, H., Hopmann, C., Schobel, A.: Reformulation of strain invariants at incompressibility. Arch. Appl. Mech. **83**(2), 273–280 (2013)
5. Ball, J.M.: Convexity conditions and existence theorems in nonlinear elasticity. Arch. Ration. Mech. Anal. **63**(4), 337–403 (1976)
6. Beatty, M.F.: Seven lectures on finite elasticity. In: Topics in Finite Elasticity, pp. 31–93. Springer, Berlin (2001)
7. Becker, G.F.: The finite elastic stress-strain function. Am. J. Sci. **46**, 337–356 (1893). Newly typeset version available at https://www.uni-due.de/imperia/md/content/mathematik/ag_neff/becker_latex_new1893.pdf
8. Carroll, M.M., Hayes, M.A.: Nonlinear Effects in Fluids and Solids, vol. 45. Springer, New York (2012)
9. Ciarlet, P.G.: Mathematical Elasticity. Volume I: Three-Dimensional Elasticity, vol. 20. Studies in Mathematics and Its Applications. Elsevier Science, North-Holland (1988)
10. Fosdick, R., Kao, B.: Transverse deformations associated with rectilinear shear in elastic solids. J. Elast. **8**(2), 117–142 (1978)
11. Fosdick, R., Serrin, J.: Rectilinear steady flow of simple fluids. Proc. R. Soc. Lond. A Math. Phys. Eng. Sci. **332**(1590). The Royal Society, pp. 311–333 (1973)
12. Gao, D.Y.: On analytical solutions to general anti-plane shear problems in finite elasticity. Preprint, arXiv:1402.6025v1 (2014)
13. Gao, D.Y.: Remarks on anti-plane shear problem and ellipticity condition in finite elasticity. Preprint, arXiv:1507.08748 (2015)
14. Gao, D.Y.: Duality in G. Saccomandi's challenge on analytical solutions to anti-plane shear problem in finite elasticity. Preprint, arXiv:1511.03374 (2015)
15. Gao, D.Y.: Analytical solutions to general anti-plane shear problems in finite elasticity. Contin. Mech. Thermodyn. **28**(1–2), 175–194 (2016)

16. Gao, D.Y.: Remarks on analytic solutions and ellipticity in anti-plane shear problems of nonlinear elasticity. In: Canonical Duality Theory, pp. 89–103. Springer, New York (2017)
17. Hencky, H.: Welche Umstände bedingen die Verfestigung bei der bildsamen Verformung von festen isotropen Körpern? Z. Phys. **55**, 145–155 (1929). https://www.uni-due.de/imperia/md/content/mathematik/ag_neff/hencky1929.pdf
18. Hill, J.M.: Generalized shear deformations for isotropic incompressible hyperelastic materials. J. Aust. Math. Soc. Ser. B Appl. Math. **20**(02), 129–141 (1977)
19. Holzapfel, G.A.: Nonlinear Solid Mechanics. Wiley, Chichester (2000)
20. Horgan, C.: Anti-plane shear deformations in linear and nonlinear solid mechanics. SIAM Rev. **37**(1), 53–81 (1995)
21. Horgan, C.: Remarks on ellipticity for the generalized Blatz-Ko constitutive model for a compressible nonlinearly elastic solid. J. Elast. **42**(2), 165–176 (1996)
22. Horgan, C., Miller, K.: Antiplane shear deformations for homogeneous and inhomogeneous anisotropic linearly elastic solids. J. Appl. Mech. **61**(1), 23–29 (1994)
23. Jiang, C., Cheung, Y.: An exact solution for the three-phase piezoelectric cylinder model under antiplane shear and its applications to piezoelectric composites. Int. J. Solids Struct. **38**(28), 4777–4796 (2001)
24. Klein, D., Baaser, H.: FEM Implementation of elastomeric stress softening. KGK Kautsch. Gummi Kunst. **71**(7–8), 41–45 (2018)
25. Knowles, J.K.: On finite anti-plane shear for incompressible elastic materials. J. Aust. Math. Soc. Ser. B Appl. Math. **19**(04), 400–415 (1976)
26. Knowles, J.K.: A note on anti-plane shear for compressible materials in finite elastostatics. J. Aust. Math. Soc. Ser. B Appl. Math. **20**(01), 1–7 (1977)
27. Knowles, J.K.: The finite anti-plane shear field near the tip of a crack for a class of incompressible elastic solids. Int. J. Fract. **13**(5), 611–639 (1977)
28. Mihai, L.A., Neff, P.: Hyperelastic bodies under homogeneous Cauchy stress induced by non-homogeneous finite deformations. Int. J. Non-Linear Mech. **89**, 93–100 (2017)
29. Moon, H., Truesdell, C.: Interpretation of adscititious inequalities through the effects pure shear stress produces upon an isotropic elastic solid. Arch. Ration. Mech. Anal. **55**(1), 1–17 (1974)
30. Nedjar, B., Baaser, H., Martin, R.J., Neff, P.: A finite element implementation of the isotropic exponentiated Hencky-logarithmic model and simulation of the eversion of elastic tubes. Comput. Mech. **62**(4), 635–654 (2018). arXiv:1705.08381
31. Neff, P., Mihai, L.A.: Injectivity of the Cauchy-stress tensor along rank-one connected lines under strict rank-one convexity condition. J. Elast. **127**(2), 309–315 (2017)
32. Neff, P., Ghiba, I.-D., Lankeit, J.: The exponentiated Hencky-logarithmic strain energy Part I: constitutive issues and rank-one convexity. J. Elast. **121**(2), 143–234 (2015)
33. Neff, P., Eidel, B., Martin, R.J.: Geometry of logarithmic strain measures in solid mechanics. Arch. Ration. Mech. Anal. **222**(2), 507–572 (2016)
34. Oberai, A.A., Gokhale, N.H., Goenezen, S., Barbone, P.E., Hall, T.J., Sommer, A.M., Jiang, J.: Linear and nonlinear elasticity imaging of soft tissue in vivo: demonstration of feasibility. Phys. Med. Biol. **54**(5), 1191 (2009)
35. Ogden, R.W.: Non-Linear Elastic Deformations, 1st edn. Mathematics and Its Applications. Ellis Horwood, Chichester (1983)
36. Pucci, E., Saccomandi, G.: A note on antiplane motions in nonlinear elastodynamics. Atti Accad. Peloritana Pericolanti-Cl. Sci. Fis. Mat. Nat. **91**(S1) (2013)
37. Pucci, E., Saccomandi, G.: The anti-plane shear problem in nonlinear elasticity revisited. J. Elast. **113**(2), 167–177 (2013)
38. Pucci, E., Rajagopal, K., Saccomandi, G.: On the determination of semi-inverse solutions of nonlinear Cauchy elasticity: the not so simple case of anti-plane shear. Int. J. Eng. Sci. **88**, 3–14 (2015)
39. Saccomandi, G.: DY Gao: Analytical solutions to general anti-plane shear problems in finite elasticity. Contin. Mech. Thermodyn. **28**(3), 915–918 (2016)

40. The MathWorks, Inc. Symbolic Math Toolbox, Natick, MA (2019). https://www.mathworks.com/help/symbolic/
41. Thiel, C., Voss, J., Martin, R.J., Neff, P.: Shear pure and simple. Int. J. Non-Linear Mech. **112**, 57–72 (2019)
42. Truesdell, C.: Mechanical foundations of elasticity and fluid dynamics. J. Ration. Mech. Anal. **1**, 125–300 (1952)
43. Voss, M.J.: Anti-plane shear deformation. MA thesis, Universität Duisburg-Essen (2017)

Identification of Diffusion Properties of Polymer-Matrix Composite Materials with Complex Texture

Marianne Beringhier, Marco Gigliotti, and Paolo Vannucci

1 Introduction

Structural parts made of polymer-matrix composite materials exposed to the environment may suffer from material degradation related to species (water, oxygen ...) diffusion-reaction phenomena within the materials substrate. Species concentration may affect material properties and give rise to internal stresses, leading to aging and durability issues. In order to predict the durability of materials and structures it is therefore of paramount importance to identify the materials diffusion behavior. In an industrial context the availability of rapid identification procedures is also desirable.

According to the Thermodynamics of Irreversible Processes, the diffusion behavior can be isotropic or orthotropic (we will see below, in fact, that the diffusion of a species is ruled by a 2nd rank tensor): for many materials, due to the complexity of the microscopic texture, the principal directions of orthotropy are not known a priori and are part of the identification procedure.

From the experimental point of view, differently from thermal conduction—where temperature spatial fields can be measured (for instance, by thermocouples or by infrared thermography)—in the case of chemical diffusion (diffusion of species molecules bounded to the material substrate driven by concentration gradients) the species concentration fields cannot be directly measured if not qualitatively with the aid of physical-chemical characterization techniques (Fourier-transform infrared spectroscopy—FTIR, Nuclear Magnetic Resonance—NMR, Raman Anal-

---

M. Beringhier · M. Gigliotti (✉)
PPRIME Institute, CNRS - ISAE-ENSMA - University of Poitiers, Poitiers, France
e-mail: marianne.beringhier@ensma.fr; marco.gigliotti@ensma.fr

P. Vannucci
LMV - UMR8100 CNRS and University of Versailles Saint Quentin, Versailles, France
e-mail: paolo.vannucci@uvsq.fr

ysis ... [22]). Therefore the diffusion behavior (isotropic, orthotropic) cannot be assessed explicitly.

A popular experimental technique for diffusion identification involves carrying out gravimetric tests. In this technique, material samples (usually of parallelepipedic shape) are exposed to a controlled environment (with fixed temperature and relative humidity) and are periodically weighted. If diffusion is Fickian, the sample mass uptake is linear with respect to the square root of time in the first times of the conditioning, then reaches saturation. The diffusion behavior can be then inferred indirectly by exploiting the gravimetric curve.

A great deal of literature studies concern the identification of the diffusion properties of materials whose diffusion behavior (isotropic, orthotropic) is known a priori, for instance, through the knowledge of the material microstructure. This is the case, for instance, of moisture diffusion in bulk polymer materials (no privileged diffusion direction—isotropic behavior), polymeric materials with randomly dispersed reinforcing particles (the random nature of the microstructures generates isotropic behavior) or reinforced by hydrophobic long continuum carbon fibers (orthotropic behavior—principal directions of orthotropy are along the fibers and perpendicularly to fibers). In these cases, the identification of the diffusive parameters is usually carried out by inverse analysis of the isotropic and orthotropic Fick's law by minimizing the error between the simulated and the experimental curves (1 gravimetric curve for the isotropic case, 3 gravimetric curves for the orthotropic case). The simulated curves can be obtained by means of analytical methods [14], Finite Element Method [3, 21, 30], more recently by methods based on Proper Generalized Decomposition techniques [4, 5].

Shen and Springer [24] have proposed a method for rapid identification of the diffusion coefficients in the isotropic and orthotropic cases. The method (also called the *slope method*) involves measuring the initial slope of the gravimetric curve then using the analytical expression of the slope, which corresponds to a short time approximation of the Fick's equation.

As early as 1987, Arhonime et al. [2] presented gravimetric curves for moisture absorbing Kevlar-epoxy laminated unidirectional composite parallelepipedic thin samples with identical dimensions and different fiber orientations, showing that mass uptake curves were significantly affected by the fiber orientation of samples.

Beringhier et al. [6] simulated this behavior by using an anisotropic Fick diffusion model in which anisotropy was generated by rotating the sample axes with respect to the fiber direction (the orientation of the orthotropic reference frame).[1] They also showed that the sample apparent anisotropic behavior can be enhanced by the sample shape and dimensions.

---

[1] We have already mentioned that two are the possible cases for the behavior concerning the diffusion properties: isotropic or orthotropic. However, for the sake of shortness we denote here and in the following as *anisotropic* an orthotropic case when the orthotropy axes do not coincide with the reference frame. In such a case, as well known, the property is still orthotropic but the tensor representing it in the reference frame looks like that of a completely anisotropic case [29].

Furthermore, by adapting the Shen and Springer slope method [24] to the anisotropic case, Beringhier et al. [6] exploited the experimental results of Arhonime et al. [2] in inverse manner, allowing identifying the principal coefficients of diffusion and the orientation of the orthotropy reference frame. In the 2D planar case, for fixed sample dimensions, the slope of the mass uptake curves is affected by the 2 in-plane principal values of the diffusion coefficients (along the axis of the principal orthotropic frame) and the 1 in-plane rotation angle between the sample and the orthotropic frame. The inverse problem consists in the solution of a 3×3 system of nonlinear algebraic equations with 3 unknowns. A protocol procedure was proposed but it was only applied for a 2D planar case as the identification procedure requires for 3D case the use of a more robust algorithm than the Gauss-Newton ones as discussed in this paper.

The present paper focuses in particular on the identification of the diffusion behavior of materials with complex texture in the full 3D case framework. In this case, for samples with fixed shape and dimensions, the issue resides in the identification of 6 unknown quantities, the 3 principal values of the diffusion coefficients (along the axis of the principal orthotropic frame) and the 3 rotation angles between the sample and the orthotropic frame. The inverse problem consists in the solution of a 6×6 nonlinear algebraic equations with 6 unknowns. Because of the high nonlinearity of these equations, the method followed for the identification is the transformation of it into an optimization problem: a distance is introduced in the space of the physical parameters and then the solution is searched as the point, in the physical space, that minimizes, i.e. reduces to zero, such a distance. A dedicated protocol and appropriate optimization algorithms for achieving this goal are presented. The methodology is here developed mainly for polymer-matrix composite materials with complex texture exhibiting Fickian behavior; however, in the future it can be extended to other kind of material textures and to other materials showing different diffusion behavior.

The paper is organized as follows. Section 2 presents the theoretical background, with an overview of the Thermodynamics of Irreversible Processes (TIP) approach, the illustration of the structure of a diffuso-mechanical model, the general structure of the diffusion tensor. Section 3 presents the identification of the diffusion properties in the orthotropic case, based on the Shen and Springer slope method. Section 4 concerns the anisotropic diffusion through the short time approximation. Section 5 concerns the identification in the anisotropic case, a review of the full 3D case, with details on the numerical procedure (Adaptive Local Evolution-PSO) used for solving the minimization problem and a case study. Finally, Sect. 6 presents conclusions and perspectives of the present research.

2 Theoretical Background and Experimental Premises

Within the framework of the Thermodynamics of Irreversible Processes with Internal Variables (TIP/TIV approaches, see, for instance, [11, 13, 18, 19, 23]), the evolution and the state equations of the state variables (for instance, the species concentration, c and the strain $\mathbf{E}$) or their dual (the chemical potential of the diffusing species, μ, the Cauchy stress, $\mathbf{T}$) in a weakly coupled diffuso-mechanical elastic problem can be calculated by the dissipation and state potentials and are given by

$$\mathbf{j} = -\mathbf{D}\nabla c, \tag{1}$$

$$\mathbf{T} = \mathbf{C}_E(c)(\mathbf{E} - \mathbf{E}^H) = \mathbf{C}_E(c)(\mathbf{E} - \beta \Delta c), \tag{2}$$

where $\mathbf{j}$ is the mass flux of the diffusing species, $\mathbf{D}$ the diffusivity tensor, $\mathbf{C}_E(c)$ the elasticity tensor which may depend on c, and $\mathbf{E}^H$ is the swelling strain, linearly related to a change of concentration Δc (with respect to a reference concentration, c_0, $\Delta c = c - c_0$) through the tensor of hygroscopic expansion β. Equations (1) and (2) are in fact constitutive equations which must be accompanied by proper balance equations, for instance, the mass balance of the diffusing species

$$\frac{\partial c}{\partial t} = -\nabla \cdot \mathbf{j} + \sigma, \tag{3}$$

in which σ represents the production of species mass (due, for instance, to chemical reaction). By exploiting Eq. (1), Eq. (3) becomes

$$\frac{\partial c}{\partial t} = \nabla \cdot (\mathbf{D}\nabla c) + \sigma, \tag{4}$$

which is analogous to the thermal conduction equation in solids, including a diffusive term and a reaction term. Other relevant balance equations are the solid mass balance, the momentum balance, and the internal energy balance. It can be seen that even in the case of pure diffusive behavior (non-chemical reactions, $\sigma = 0$) the solution of a diffuso-mechanical problem passes through the solution of Eq. (4), with appropriate initial and boundary conditions, which is essential for the identification of the concentration dependent elasticity tensor and for the calculation of the hygroscopic swelling strain (see [30] for more details). In turn, the solution of Eq. (4) passes through the knowledge of the diffusivity tensor $\mathbf{D}$, which must be identified though proper experiments.

An important note on the symmetry of the diffusivity tensor, $\mathbf{D}$, has been compiled by Powers [23] within the framework of TIP, in the analogous case of thermal conduction (in thermal conduction the analogous of $\mathbf{D}$ is the conductivity tensor $\mathbf{K}$).

By taking the tensor $\mathbf{D}$ as the sum of a symmetric, $\mathbf{D}_s$, and an unsymmetric, $\mathbf{D}_u$, part

$$\mathbf{D} = \mathbf{D}_s + \mathbf{D}_u. \tag{5}$$

The Onsager conjecture [19] states

$$\mathbf{D}_u = \mathbf{0}. \tag{6}$$

The condition $\mathbf{D}_s \nabla c \cdot \nabla c \geq 0$ is satisfied if and only if $\mathbf{D}_s$ is positive semi-definite. Therefore eigenvalues of $\mathbf{D}_s$ are positive, that is, invariants of $\mathbf{D}_s$ are positive.

In this context, the (3) eigenvalues of $\mathbf{D}$, D_1, D_2, D_3, can be identical, $D_1 = D_2 = D_3 = D$ or different $D_1 \neq D_2 \neq D_3$. In the first case the diffusive behavior is isotropic and $\mathbf{D} = D\mathbf{I}$. In the second case the diffusive behavior is orthotropic and $\mathbf{D}$ can be represented by a (3×3) diagonal matrix in the principal reference frame $(1, 2, 3)$

$$\mathbf{D} = \begin{bmatrix} D_1 & 0 & 0 \\ 0 & D_2 & 0 \\ 0 & 0 & D_3 \end{bmatrix}. \tag{7}$$

In a generalized reference frame (x, y, z) rotated with respect to $(1, 2, 3)$ the diffusivity matrix attached to $\mathbf{D}$ is not diagonal. This last case can be considered by misuse of language as anisotropic due the fact that non-diagonal terms appear in the $\mathbf{D}$ matrix due to a frame rotation, see Note 1. This fact is extremely important when dealing with samples of material having a given spatial orientation. The identification of $\mathbf{D}$ consists in the identification of 1 parameter (D) in the isotropic case and of 3 parameters $(D_1, D_2,$ and $D_3)$ in the orthotropic case. However, it is not possible to establish a priori whether the behavior is isotropic or orthotropic. As stressed in the introduction, for many materials, due to the complexity of the microscopic texture, the principal directions of orthotropy are not known a priori and are part of the identification procedure.

3 Identification of the Diffusion Properties in Orthotropic Case

More details about the description of the Shen and Springer Slope Based approach can be found in [6]. Shen and Springer [24] have proposed an approximated short time analytical solution for the mass gain $M(t)$ as a function of time for the isotropic and orthotropic case: the solution is based on the exact solution for Fickian diffusion into a semi-infinite domain with constant and homogeneous fixed concentration $(c = c_\infty)$ on the external boundary (see Crank [9] and Carslaw and Jaeger [7], for the analogous thermal conduction problem). In the 3D case, Shen and Springer assume that each of the six surfaces of the parallelepipedic domain behaves as a semi-infinite media, the mass gain is given by

$$M(t) = \frac{4M_\infty}{\sqrt{\pi}} \left[\frac{\sqrt{D_1}}{L_x} + \frac{\sqrt{D_2}}{L_y} + \frac{\sqrt{D_3}}{L_z} \right] \sqrt{t}, \qquad (8)$$

where D_1, D_2, and D_3 are the three diffusion coefficients in the orthotropic reference frame $(1, 2, 3)$ L_x, L_y, L_z are the sample dimensions along x, y and z, respectively and M_∞ is the mass gain at saturation. According to the short time solution, the mass evolves linearly with respect to $\sqrt{t}$, the short time solution represents the initial slope of the analytical mass gain curve, giving a satisfactory description of the analytical solution in a $\frac{M(t)}{M_\infty}$ range between 0 and 0.2–0.4. This solution is applicable to initially dry samples exposed to a regulated humidity/temperature environment in which temperature and relative humidity values are fixed and stay constant. The weight of the sample (that is, the weight of absorbed humidity-water, since the initially dry sample weight is known) is periodically measured by a precision balance and reported. In the presence of Fickian behavior, the gravimetric curve (water mass uptake) is approximately linear as a function of square root of time, for short times.

A discussion concerning the validity of this approach can be found in [1, 6] and the references therein.

The notable point in the approach by Shen and Springer is that it allows to relate the identification of the diffusion coefficients to the measurement of the slope, S, of the mass gain curve $\frac{M(t)}{M_\infty}$ with respect to $\sqrt{t}$ as

$$S = \frac{M(t)}{M_\infty \sqrt{t}} = \frac{4}{\sqrt{\pi}} \left[\frac{\sqrt{D_1}}{L_x} + \frac{\sqrt{D_2}}{L_y} + \frac{\sqrt{D_3}}{L_z} \right]. \qquad (9)$$

This approach also referred as slope method is adapted to the determination of the diffusion properties as it allows to drastically decrease the test time. It has to be noted, however, that even when employing the slope method the value of mass gain at saturation, M_∞, has to be known, since it enters the slope equation (Eq. (9)); therefore at least one gravimetric test has to be carried out up to saturation.

The protocol for rapid identification in the orthotropic (isotropic) diffusion behavior based on the use of the Shen and Springer short time solution—slope method—can be here resumed (see also [24])

1. Identification of M_∞ by 1 gravimetric test up to saturation.
2. Realization of 3 (1) short time gravimetric tests to obtain 3 (1) distinct slopes.
3. Measurement of the 3 (1) slopes.
4. Use of the slope equations (Eq. (9)) to solve a 3×3 (1) linear algebraic system of 3 (1) unknowns (D_1, D_2, D_3 (D)).

An example of identification based on this approach is fully illustrated in [6] the reader can find more details therein. In that case, $D_3 = D_2$ so that only two parameters (namely D_1 and D_2) need to be identified through the employment of two separated slopes. Two unknown parameters D_1 and D_2 can be calculated by using the Shen and Springer short time solution (Eq. (8)) with $D_3 = D_2$. The two experimental slopes, for instance, denoted by S_1 and S_2 allow computing D_1 and D_2 by solving the linear system with respect to $\sqrt{D_1}$ and $\sqrt{D_2}$

$$\frac{4}{\sqrt{\pi}} \begin{bmatrix} \frac{1}{L_x} & \frac{L_y+L_z}{L_y L_z} \\ \frac{1}{L_y} & \frac{L_x+L_z}{L_x L_z} \end{bmatrix} \begin{bmatrix} \sqrt{D_1} \\ \sqrt{D_2} \end{bmatrix} = \begin{bmatrix} S_1 \\ S_2 \end{bmatrix}. \tag{10}$$

It is essential to note that since the gravimetric curve is stopped at $\frac{M(t)}{M_\infty}$ equal to 0.4 by the slope method the time gain with respect to a gravimetric test up to saturation can approximately span from 3 to 5, depending on the time to reach saturation.

4 Anisotropic Diffusion and Short Time Approximation

We recall that anisotropy is here generated by rotating the sample axes with respect to the fiber direction (the orientation of the orthotropic reference frame). In this case, the second-order diffusivity tensor written in a generalized reference frame (x, y, z) can be represented by the following matrix

$$\mathbf{D} = \begin{bmatrix} D_{xx} & D_{xy} & D_{xz} \\ D_{yx} & D_{yy} & D_{yz} \\ D_{zx} & D_{zy} & D_{zz} \end{bmatrix}. \tag{11}$$

The relationship between the anisotropic diffusivity matrix given by Eq. (11) in the (x, y, z) reference frame and the orthotropic diffusivity matrix in the $(1, 2, 3)$ reference frame is given by the algebraic relations between the 2 coordinate systems (orthotropic and anisotropic or generalized) by the rotation angles as follows

$$\begin{bmatrix} D_1 & 0 & 0 \\ 0 & D_2 & 0 \\ 0 & 0 & D_3 \end{bmatrix} = \mathbf{Q}^T \begin{bmatrix} D_{xx} & D_{xy} & D_{xz} \\ D_{xy} & D_{yy} & D_{yz} \\ D_{xz} & D_{yz} & D_{zz} \end{bmatrix} \mathbf{Q}, \tag{12}$$

where $\mathbf{Q}$ is the operator moving from the anisotropic (generalized) axes to the principal axes. The $\mathbf{Q}$ matrix can be expressed under the form

$$\mathbf{Q} = \mathbf{R}_\theta \, \mathbf{R}_\phi \, \mathbf{R}_\psi, \tag{13}$$

where

- $\mathbf{R}_\theta$ is the rotation matrix around the x-axis of an angle θ

$$\mathbf{R}_\theta = \begin{bmatrix} 1 & 0 & 0 \\ 0 & \cos(\theta) & -\sin(\theta) \\ 0 & \sin(\theta) & \cos(\theta) \end{bmatrix}; \tag{14}$$

- $\mathbf{R}_\phi$ is the rotation matrix around the y-axis of an angle ϕ

$$\mathbf{R}_\phi = \begin{bmatrix} \cos(\phi) & 0 & \sin(\phi) \\ 0 & 1 & 0 \\ -\sin(\phi) & 0 & \cos(\phi) \end{bmatrix}; \quad (15)$$

- $\mathbf{R}_\psi$ is the rotation matrix around the z-axis of an angle ψ

$$\mathbf{R}_\psi = \begin{bmatrix} \cos(\psi) & -\sin(\psi) & 0 \\ \sin(\psi) & \cos(\psi) & 0 \\ 0 & 0 & 1 \end{bmatrix}. \quad (16)$$

We then deduce the following relationships between the anisotropic diffusion coefficients, D_{xx}, D_{yy}, D_{zz}, and the diagonal diffusion coefficients D_1, D_2, D_3, by the rotation angles θ, ϕ, ψ:

$$\begin{aligned}
D_{xx} &= D_1 \hat{c}^2(\phi)\hat{c}^2(\psi) + D_2 \hat{c}^2(\phi)\hat{s}^2(\psi) + D_3 \hat{s}^2(\phi), \\
D_{yy} &= D_1(\hat{s}(\theta)\hat{s}(\phi)\hat{c}(\psi) + \hat{c}(\theta)\hat{s}(\psi))^2 + D_2(\hat{c}(\theta)\hat{c}(\psi) - \\
&\quad \hat{s}(\theta)\hat{s}(\phi)\hat{s}(\psi))^2 + D_3 \hat{s}^2(\theta)\hat{c}^2(\phi), \\
D_{zz} &= D_1(\hat{s}(\theta)\hat{s}(\psi) - \hat{c}(\theta)\hat{s}(\phi)\hat{c}(\psi))^2 + D_2(\hat{s}(\theta)\hat{c}(\psi) + \\
&\quad \hat{c}(\theta)\hat{s}(\phi)\hat{s}(\psi))^2 + D_3 \hat{c}^2(\theta)\hat{c}^2(\phi), \\
D_{xy} &= D_1 \hat{c}(\phi)\hat{c}(\psi)(\hat{c}(\theta)\hat{s}(\psi) + \hat{c}(\psi)\hat{s}(\phi)\hat{s}(\theta)) - \\
&\quad D_2 \hat{c}(\phi)\hat{s}(\psi)(\hat{c}(\psi)\hat{c}(\theta) - \hat{s}(\phi)\hat{s}(\psi)\hat{s}(\theta)) - D_3 \hat{c}(\phi)\hat{s}(\phi)\hat{s}(\theta), \\
D_{xz} &= D_1 \hat{c}(\phi)\hat{c}(\psi)(\hat{s}(\psi)\hat{s}(\theta) - \hat{c}(\psi)\hat{c}(\theta)\hat{s}(\phi)) - \\
&\quad D_2 \hat{c}(\phi)\hat{s}(\psi)(\hat{c}(\psi)\hat{s}(\theta) + \hat{c}(\theta)\hat{s}(\phi)\hat{s}(\psi)) + D_3 \hat{c}(\phi)\hat{c}(\theta)\hat{s}(\phi), \\
D_{yz} &= -D_3 \hat{c}(\theta)\hat{s}(\theta)\hat{c}^2(\phi) + D_1(\hat{c}(\theta)\hat{s}(\psi) + \\
&\quad \hat{c}(\psi)\hat{s}(\phi)\hat{s}(\theta))(\hat{s}(\psi)\hat{s}(\theta) - \hat{c}(\psi)\hat{c}(\theta)\hat{s}(\phi)) + \\
&\quad D_2(\hat{c}(\psi)\hat{s}(\theta) + \hat{c}(\theta)\hat{s}(\phi)\hat{s}(\psi))(\hat{c}(\psi)\hat{c}(\theta) - \hat{s}(\phi)\hat{s}(\psi)\hat{s}(\theta)),
\end{aligned} \quad (17)$$

where $\hat{c}$ and $\hat{s}$ stand for the cosine and the sine functions, respectively. Let us note that no analytical solution is available in the anisotropic case.

Due to the interest of short time analysis for the identification and the lack of analytical solution, the Shen and Springer orthotropic (isotropic) slope expression (Eq. (8)) can be generalized to the anisotropic case by directly replacing the coefficients D_1, D_2, and D_3 by their related values D_{xx}, D_{yy}, and D_{zz} (see [6] for more details and in-depth discussion on the matter)

$$M(t) = \frac{4M_\infty}{\sqrt{\pi}} \left[\frac{\sqrt{D_{xx}}}{L_x} + \frac{\sqrt{D_{yy}}}{L_y} + \frac{\sqrt{D_{zz}}}{L_z} \right] \sqrt{t}. \quad (18)$$

The extradiagonal coefficients of the anisotropy tensor are assumed to have no influence on the diffusive behavior for short times (see again [2] for in-depth discussion about this assumption).

Identification of Diffusion Properties of PMC Materials

By replacing D_{xx}, D_{yy}, and D_{zz} by their expressions Eq. (17) the slope, S, of the curve $\frac{M(t)}{M_\infty}$ with respect to $\sqrt{t}$ is finally expressed by

$$S = \frac{4}{\sqrt{\pi}} \left[\frac{\sqrt{D_{xx}}}{L_x} + \frac{\sqrt{D_{yy}}}{L_y} + \frac{\sqrt{D_{zz}}}{L_z} \right]. \tag{19}$$

In the case of planar anisotropic diffusion in the (x, y) plane, there exists only a rotation around the z-axis, denoted by ψ. The diffusivity matrix has the form

$$\mathbf{D} = \begin{bmatrix} D_{xx} & D_{xy} & 0 \\ D_{xy} & D_{yy} & 0 \\ 0 & 0 & D_{zz} \end{bmatrix} \tag{20}$$

with

$$\begin{aligned}
D_{xx} &= D_1 \hat{c}^2(\psi) + D_2 \hat{s}^2(\psi) \\
D_{yy} &= D_1 \hat{s}^2(\psi) + D_2 \hat{c}^2(\psi) \\
D_{zz} &= D_3 \\
D_{xy} &= D_1 \hat{c}(\psi) \hat{s}(\psi) - D_2 \hat{s}(\psi) \hat{c}(\psi) \\
D_{xz} &= D_{yz} = 0,
\end{aligned} \tag{21}$$

where $\hat{c}$ and $\hat{s}$ stand for the cosine and the sine functions, respectively. By considering the particular case $D_3 = D_2$, the slope can be expressed under the form

$$S = \frac{4}{\sqrt{\pi}} \left[\frac{\sqrt{D_1 \hat{c}^2(\psi) + D_2 \hat{s}^2(\psi)}}{L_x} + \frac{\sqrt{D_1 \hat{s}^2(\psi) + D_2 \hat{c}^2(\psi)}}{L_y} + \frac{\sqrt{D_2}}{L_z} \right], \tag{22}$$

where $\hat{c}$ and $\hat{s}$ stand for the cosine and the sine functions, respectively. The study of planar anisotropic diffusion can be also dealt within the polar method [27, 29]. The diffusion coefficients have thus the form

$$\begin{aligned}
D_{xx} &= T + \text{sgn}(D_1 - D_2) \, R \, \hat{c}(2\psi) \\
D_{yy} &= T - \text{sgn}(D_1 - D_2) \, R \, \hat{c}(2\psi) \\
D_{zz} &= T - \text{sgn}(D_1 - D_2) \, R \\
D_{xy} &= \text{sgn}(D_1 - D_2) \, 2R \, \hat{c}(\psi) \hat{s}(\psi),
\end{aligned} \tag{23}$$

where $\hat{c}$ and $\hat{s}$ stand for the cosine and the sine functions, respectively, sgn stands for the sign function and T and R are the polar invariants defined by

$$T = \frac{D_1 + D_2}{2}$$

$$R = \frac{|D_1 - D_2|}{2}.$$

(24)

Within the polar representation, the slope is given by

$$S = \frac{4}{\sqrt{\pi}} \left[\frac{\sqrt{T + \text{sgn}(D_1 - D_2) \, R \, \hat{c}(2\psi)}}{L_x} + \frac{\sqrt{T - \text{sgn}(D_1 - D_2) \, R \, \hat{c}(2\psi)}}{L_y} + \frac{\sqrt{T - \text{sgn}(D_1 - D_2) \, R}}{L_z} \right],$$

(25)

where $\hat{c}$ stands for the cosine function and sgn stands for the sign function.

5 Identification of the Diffusion Properties in the Anisotropic Case

5.1 Generalities

The identification problem in the 3D anisotropic case is related to three angles of rotation around the x, y, and z-axis, respectively, denoted by θ, ϕ, and ψ, respectively. The change from the current frame (frame attached to the initial configuration as, for example, the axes of the plate where the tests samples are cut) to the principal orthotropic frame can be represented by the operator $\mathbf{Q} = \mathbf{R}_\theta \mathbf{R}_\phi \mathbf{R}_\psi$. To identify these three angles, the sample is rotated from his current configuration by three angles around the x, y, and z-axis, respectively, denoted by θ_t, ϕ_t, and ψ_t, respectively: this can be represented by the operator $\mathbf{Q}_t = \mathbf{R}_{\theta_t} \mathbf{R}_{\phi_t} \mathbf{R}_{\psi_t}$. The change from the rotated frame to the principal frame can be represented by the operator $\mathbf{Q}_{tm} = \mathbf{Q}_t^T \mathbf{Q}$ which describes the composition of 3D rotations as depicted in Fig. 1. Let us note that the rotated axes represent the generalized axes, or the axes of the samples for which the gravimetric tests are carried out. The expression of the anisotropic diffusion coefficients depends on:

- the 3 principal diffusion coefficients, D_1, D_2, and D_3, the 3 angles θ, ϕ, and ψ expressing the rotation from the initial frame to the principal (orthotropic) axis—these 6 coefficients are unknown,
- the angles θ_t, ϕ_t, and ψ_t expressing the rotation from the rotated frame to the initial frame—these three coefficients are known,

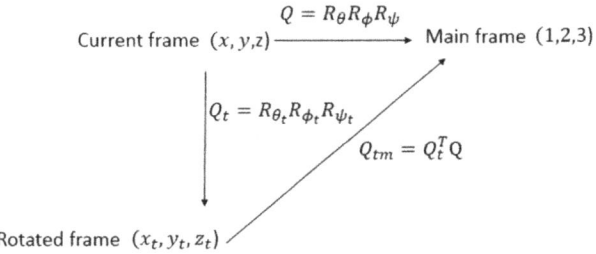

Fig. 1 Bases changes: current frame, main frame, and rotated frame (figure from Ref. [6])

and the expression is the following

$$\begin{bmatrix} D_{xx} & D_{xy} & D_{xz} \\ D_{yx} & D_{yy} & D_{yz} \\ D_{zx} & D_{zy} & D_{zz} \end{bmatrix} = \mathbf{Q}_{tm}(\theta, \phi, \psi, \theta_t, \phi_t, \psi_t) \begin{bmatrix} D_1 & 0 & 0 \\ 0 & D_2 & 0 \\ 0 & 0 & D_3 \end{bmatrix}$$
$$\times \mathbf{Q}_{tm}(\theta, \phi, \psi, \theta_t, \phi_t, \psi_t)^T . \quad (26)$$

D_{xx}, D_{yy}, and D_{zz} are then used in the slope analytical expression. By considering 6 different 3-uplet $(\theta_t, \phi_t, \psi_t)$, 6 slopes can be calculated. The system to determine the 6 unknowns D_1, D_2, D_3, θ, ϕ, and ψ is a 6×6 nonlinear algebraic system. This system can be simplified by considering the planar anisotropic diffusion as in this case the combination of two rotations of a same axis is investigated leading to simplified relations.

5.2 Full 3D Diffusion

In the 3D anisotropic diffusion case, the rotations around the three different axes (x, y, and z, respectively) have to be considered: θ, ϕ, and ψ from the current axis to the orthotropic axis, θ_t, ϕ_t, and ψ_t from the initial axis (configuration of the initial plate) to the rotated axis (configuration of the test sample). The operator $\mathbf{Q}_{tm}$ depends on the six angles θ, ϕ, ψ, θ_t, ϕ_t, and ψ_t. By using Eq. (26), the expression of the coefficients of the anisotropic diffusivity tensor are deduced and depend on D_1, D_2, D_3 (the 3 principal diffusion coefficients) and θ, ϕ, ψ, θ_t, ϕ_t, and ψ_t. They allow to determine the expression of the slope S given by Eq. (19) which depends on D_1, D_2, D_3, θ, ϕ, ψ, θ_t, ϕ_t, ψ_t, L_x, L_y, L_z. Let us recall L_x, L_y, and L_z are the sample dimensions along x, y, and z, respectively. The six unknowns D_1, D_2, D_3, θ, ϕ, ψ can be determined by considering six different values for the 3-uplet $(\theta_t, \phi_t, \psi_t)$ and the sample size (L_x, L_y, and L_z) leading to six different expressions for the slopes. Finally a 6×6 nonlinear algebraic system has to be solved. The expression of the slope can be written under the form

$$f_i(D_1, D_2, D_3, \theta, \phi, \psi) = 0 \qquad (27)$$

for a fixed value of the 3-uplet $(\theta_t, \phi_t, \psi_t)_i$ and the sample size $(L_x, L_y, L_z)_i$ where the exponent i denotes the configuration. More precisely, f_i is the difference between the predicted value of S_i deduced from Eq. (19) and the experimental value of S_i calculated from a gravimetric test (slope of the gravimetric curve at short time). The dependence of the f_i with respect to their arguments is nonlinear. Six unknowns having to be determined, six values of $(\theta_t, \phi_t, \psi_t)_i$ and their associated sample size $(L_x, L_y, L_z)_i$ and slope values have to be considered. We then obtain a system of six equations:

$$f_i(D_1, D_2, D_3, \theta, \phi, \psi) = 0 \quad \text{for } i = 1\ldots 6. \qquad (28)$$

This system can be numerically solved by turning into a constrained minimization problem

$$\min_{D_1,D_2,D_3,\theta,\phi,\psi} \left(\sum_{i=1}^{6} f_i^2(D_1, D_2, D_3, \theta, \phi, \psi) \right) \qquad (29)$$

subject to

$$D_1 \geq 0, D_2 \geq 0, D_3 \geq 0,$$
$$\theta \in \left[-\frac{\pi}{2}, \frac{\pi}{2}\right], \phi \in \left[-\frac{\pi}{2}, \frac{\pi}{2}\right], \psi \in \left[-\frac{\pi}{2}, \frac{\pi}{2}\right]. \qquad (30)$$

5.3 Numerical Procedure for Identification

As seen in the previous sections, we identify the quantities D_1, D_2, D_3 through a minimization procedure: we look for the minimum of the objective function specified in Eq. (29). Because such a function is non-convex, the search for a minimum by the classical descent methods, sensitive to the starting point, is unsuited: a bad choice of the starting point could lead to a local minimum. In the problem at hand, there are no indications to choose a suitable initial guess, so the use of such classical descent methods is not recommended.

That is why we have decided to make use of a metaheuristic; as well known, metaheuristics are optimization strategies inspired by some heuristics mimicking a real-world phenomenon: simulated annealing is inspired by a metallurgic process for decreasing the internal energy in metal alloys, and colonies refer to the social behavior of ants, genetic algorithms mimic the Darwinian selection in nature and so on. A general characteristic of metaheuristics is that they work on a population of individuals rather than on only a single element; in such a way, the exploration of the feasible domain is much more effective and by consequence the probability to be trapped in a local minimum decreases.

For the problem at hand, we have used an algorithm for Particle Swarm Optimization (PSO). This is a metaheuristic introduced by Eberhart and Kennedy [12] to solve complex optimization problems. It is inspired by and it mimics in an algebraic way the social behavior and dynamics of groups of individuals (particles), such as the flocks of birds, whose groups displacements are not imposed by a leader: the same overall behavior of the flock guides itself.

In this analogy, each particle composing the swarm is actually a vector of $\mathbb{R}^n$, with n the number of design variables, who is a candidate to the solution of the minimization problem under consideration. A true advantage of a PSO algorithm is its true simplicity: a PSO is a zeroth order algorithm, i.e. it just needs the evaluation of the objective function: no calculation of derivatives is needed, so that it is able to tackle problems ruled by discrete variables too. In addition, the standard algorithm is very simple and short, so that calculations are usually very quick.

In particular, we have used the PSO algorithm ALE-PSO (Adaptive Local Evolution-PSO) [28]; this is an algorithm having the possibility to tune all the parameters of a standard PSO algorithm and to let evolve the main numerical coefficients along the computation steps through a power law specified by the user. Each coefficient can be updated independently by the others, so as to optimize the trade-off between the exploration and exploitation capabilities of the numerical procedure with respect to a given objective function. The algorithm can also handle constrained problems, by a technique which is essentially a barrier method.

Actually, as the most part of numerical methods, also PSO algorithms depend upon some parameters to be chosen by the user to guarantee convergence, rapidity, and robustness of the algorithm itself. Tuning these parameters, i.e. the best choice of them, is one of the major issues with PSO algorithms. Generally speaking, tuning of parameters depends mostly upon the objective function; so, a set of well suited parameters for a given problem can be a bad choice in another case. Tuning of parameters is mostly based upon experience, and in some cases several tests can be necessary to find a good tuning. For this reason, some attempts have been made to establish, on the one hand, the dynamics of the swarm, though often on very simplified models, to obtain some conditions for the convergence of the swarm towards a final point of the design space, see, for instance, the works of Shi and Eberhart [25], Ozcan and Mohan [20], Clerc and Kennedy [8], Trelea [26], Liu et al. [17], Jiang et al. [16], and, on the other hand, to propose adaptive PSO algorithms, i.e. algorithms which tune the parameters during the calculation, see, for instance, the papers of Hu and Eberhart [15], Yasuda et al. [32], Yamaguchi et al. [31], DeBao and ChunXia [10].

The general scheme of a PSO algorithm can be condensed in the following updating rule: in a swarm composed by m particles, the new position of the particle k

in the feasible space, i.e. the vector x_{t+1}^k containing the value of the design variables represented by such a particle at the iteration $t+1$, $1 \leq t \leq s$, is given by

$$x_{t+1}^k = x_t^k + u_{t+1}^k, \qquad (31)$$

with the displacement u_{t+1}^k updated as

$$u_{t+1}^k = r_0 c_0 u_t^k + r_1 c_1 \left(p_t^k - x_t^k \right) + r_2 c_2 \left(p_t^g - x_t^k \right). \qquad (32)$$

In the above equations:

- p_t^k : vector recording the best position occupied so far by the kth particle (*personal best position*); in a minimization problem for the objective function $f(x)$, p_t^k is updated as follows

$$p_{t+1}^k = \begin{cases} p_t^k, & \text{if } f(x_{t+1}^k) \geq f(p_t^k), \\ x_{t+1}^k, & \text{if } f(x_{t+1}^k) < f(p_t^k); \end{cases} \qquad (33)$$

- p_t^g : vector recording the best position occupied so far by any particle in the swarm (*global best position*); in a minimization problem for the objective function $f(x)$, p_t^g is updated as

$$p_{t+1}^g \in \left\{ p_{t+1}^k, k = 1, \ldots, m \right\} \text{ such that } f(p_{t+1}^g) = \min_{k=1,\ldots,m} f(p_{t+1}^k); \qquad (34)$$

- r_0, r_1, r_2: independent random coefficients, uniformly distributed in the range $[0, 1]$;
- c_0, c_1, c_2: real coefficients called, respectively, *inertial*, *cognition*, and *social* parameters.

The use of the random coefficients r_0, r_1, r_2, together with the fact that usually the original swarm is randomly generated, gives a stochastic nature to the algorithm. Anyway, this stochastic nature is weighted, in some way, by the presence of the coefficients c_0, c_1, c_2: their value has a paramount importance in the convergence, stability, and search domain exploration of the algorithm. The choice of these coefficients is of the greatest importance. Unfortunately, as said above, the best choice of these parameters is function-dependent and sometimes it is difficult to found a suitable set of parameters. For this reason, one can think to update also the coefficients c_0, c_1, and c_2, so as to improve the convergence of the algorithm.

Normally, high values of c_0 and c_1 improve exploration, while increasing values of c_2 increases stability and rapidity of convergence. Hence, a good strategy can be the following one: to begin the computation with high values of c_0 and c_1 and a low value of c_2, for better exploring the feasible domain. Once, hopefully, the search concentrated in a good region of the feasible domain, to decrease the values of c_0 and c_1 and to increase the value of c_2. Nevertheless, some relations among these parameters must be respected in order to ensure stability and convergence. In ALE-PSO, it has been chosen to update the coefficients $c_j, j = 0, 1, 2$ according to a power law ruled by three parameters, $c_{j,0}, c_{j,s}$, and e_j:

$$c_{j,t} = c_{j,s} + (c_{j,0} - c_{j,s}) \left(\frac{s-t}{s-1}\right)^{e_j}, \quad j \in \{0, 1, 2\}. \tag{35}$$

In the above equation $c_{j,t}$ is the value of the coefficient c_j at iteration t, $c_{j,s}$ at the end of the iterations ($t = s$) and $c_{j,0}$ at the beginning ($t = 0$). The user has to choose the coefficients $c_{j,0}$ and $c_{j,s}$ besides the power e_j. This last modulates the variation of the coefficient c_j along the iterations, giving the possibility to the user to increase the rapidity of the variation at the beginning or at the end of the iterations. The use of such an updating rule is motivated by the need to adapt the coefficients to the dynamics of the storm. This is affected by the problem itself and also by the storm size, so there is not an optimal general strategy for the updating of the coefficients c_js. That is why disposing of an updating rule that can be tailored to a specific problem is suitable. The law (35), depending upon 3 parameters, has proven to be rather effective to this purpose, the reader is referred to [28] for more details.

In ALE-PSO, the frequency of updating the random parameters r_0, r_1, and r_2 can be fixed by the user; three are the possibilities:

1. r_0, r_1, and r_2 are updated at each iteration (all the particles receive the same value of r_0, r_1, and r_2);
2. r_0, r_1, and r_2 are updated at each particle (all the components of u_{t+1}^k receive the same value of r_0, r_1, and r_2);
3. r_0, r_1, and r_2 are updated at each component of each particle (no repeated values, in principle).

Normally, the third option ensures a better exploration of the feasible domain, but it is more time consuming; option 1 is quickest but recommended only for simple problems, as it lowers too much the search capacity of the algorithm, while option 2 is a good compromise for many problems.

If the new position of a particle increases the value of $f(x)$, ALE-PSO can, optionally, refuse to update the position; this option tends to concentrate the search on good regions, though it increases the probability of convergence towards non global minima.

5.4 Identification Test

We have applied ALE-PSO to the identification of the D_1, D_2, D_3 and θ, ϕ, ψ for the case whose data are shown in Table 1 where a set of six samples with different values of $(\theta_t, \phi_t, \psi_t)_i$, size $(L_x, L_y, L_z)_i$ and slopes are considered.

Table 1 Configurations (rotated angles and sample size) and computed slopes

Configuration i	$(\theta_t, \phi_t, \psi_t)_i$ (°)	$(L_x, L_y, L_z)_i$ (mm)	Slope S_i (h$^{-1/2}$)
1	(0, 0, 0)	(110, 10, 1.7)	0.0275
2	(45, 0, 0)	(110, 10, 1.7)	0.0471
3	(0, 0, 0)	(1.7, 110, 10)	0.0166
4	(0, 45, 0)	(1.7, 110, 10)	0.0135
5	(0, 0, 0)	(10, 1.7, 110)	0.0793
6	(0, 0, 45)	(10, 1.7, 110)	0.0141

The parameters used for ALE-PSO are:

- inertial parameter: $c_{0,0} = 1, c_{0,s} = 0.5, e_0 = 0.5$;
- cognition parameter: $c_{1,0} = 5, c_{1,s} = 1.5, e_1 = 0.5$;
- social parameter: $c_{2,0} = 1, c_{2,s} = 1.8, e_2 = 2$;
- update of the coefficients: at each variable (option iii);
- size of the swarm: $m = 50$;
- number of iterations: $s = 200$;
- coefficient r_0: randomly determined.

The result of the search is:

$$D_1 = 0.015 \text{ mm}^2.\text{h}^{-1}, \ D_2 = 0.0015 \text{ mm}^2.\text{h}^{-1}, \ D_3 = 0.00015 \text{ mm}^2.\text{h}^{-1},$$

$$\theta = \phi = \psi = 0°.$$

The value of f for the best solution, i.e. the residual (we recall that, for the formulation used for the minimum problem, the solution corresponds to a zero-valued objective function), found at iteration 45, is $f = 0.12 \times 10^{-30}$, while the best average of the residual, denoting the best swarm, is 0.37×10^{-8}, at iteration 48. We show in Fig. 2 the diagrams of the objective function versus the iterations and in Fig. 3 the dynamics of the swarm during iterations: randomly generated, at the beginning the swarm occupies almost uniformly the design space; then, it moves more and more towards the solution and at the end it practically coincides with the best particle.

6 Proposal for a Rapid Identification Protocol Based on the Slope Method in the Anisotropic Case

A protocol for a rapid identification based on the slope method in the anisotropic case can be proposed and consists in:

1. Identification of mass (concentration) at saturation by 1 gravimetric test up to saturation

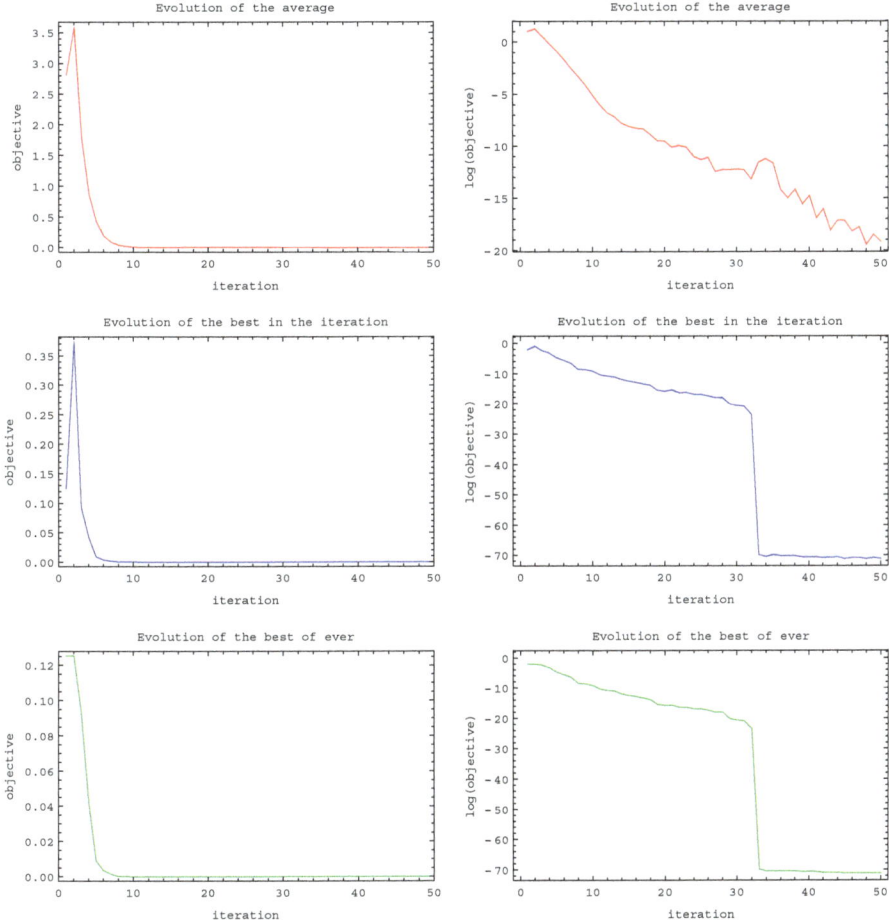

Fig. 2 The objective function f along iterations; left column: diagrams of f; right column: diagrams of $\log(f)$

2. Realization of at least 6 short time gravimetric tests to obtain 6 separated slopes. The samples must be realized by performing 6 distinct appropriate rotations about a given reference frame (for instance, that of the as-received material) or 2 distinct appropriate rotations for 3 distinct appropriate sample sizes (as, for instance, presented in Table 1).
3. Measurement of the slopes.
4. Use of the slopes equations to solve a 6×6 nonlinear algebraic system of 6 unknowns ($D_1, D_2, D_3, \theta, \phi$ and ψ). A robust optimization algorithm has to be used to determine the six unknowns: being the problem non-convex, it is better to use metaheuristics; in this case, we have used a PSO algorithm with dynamically changing coefficients, ALE-PSO.

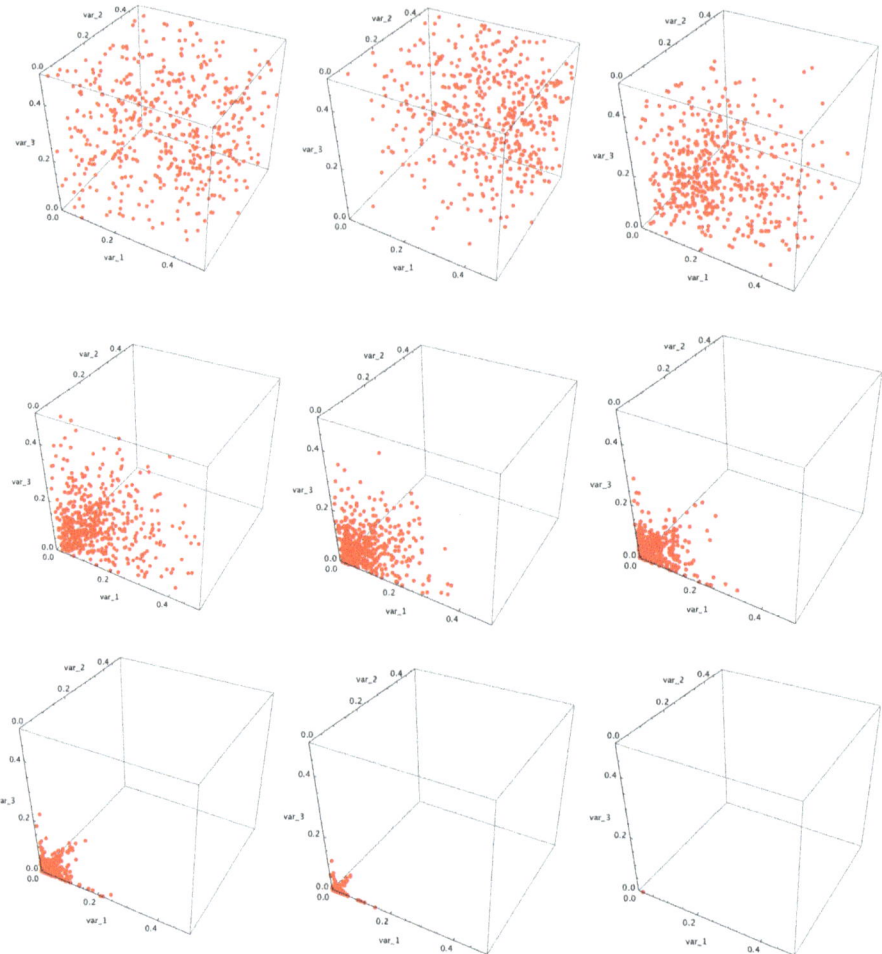

Fig. 3 The evolution of the swarm during iterations in the space D_1, D_2, D_3 (=var 1, var 2, var 3, respectively). The initial swarm, randomly generated, converges more and more towards the solution; at the end, i.e. after 50 iterations, the entire swarm practically coincides with the best particle

For planar anisotropic diffusion, the number of unknowns reduces to 3, with the need of measuring 3 distinct slopes on samples realized by performing 3 distinct appropriate rotations about a given reference frame (for instance, that of the as-received material).

Again we note that since the gravimetric curve is stopped at $\frac{M(t)}{M_\infty}$ equal to 0.4 by the slope method the time gain with respect to a gravimetric test up to saturation can approximately span from 3 to 5, depending on the time to reach saturation.

We stress moreover the fact that in order to separate slopes by at least 5% to take into proper account the experimental scatter, one can either change sample orientation in order to enhance the effect of one particular coefficient D_i with respect to another either choose proper sample dimensions to enhance diffusion along one particular direction. The optimization of the generalized parameters $\frac{D_i}{L_i^2}$ may generally not be straightforward: the knowledge of indicative values (order of magnitude) for the diffusion coefficients from step 1 may help deciding proper sample dimensions (sample rotations) for step 2.

7 Conclusions

The paper has been focused on the identification of 3D anisotropic diffusion properties of materials with complex texture, based on the exploitation of short time gravimetric tests. The paper has addressed the development of an experimental protocol and an identification algorithm for the full 3D diffusion case, aiming at establishing the 3 coefficients of diffusion along the principal directions of orthotropy and the orientation of the orthotropic reference frame with respect to the sample frame. The identification of the physical properties has been carried out through the minimization of a distance in the space of the physical parameters. The problem being non-convex, the chosen numerical strategy used for the search of the global minimum has been a Particle Swarm Optimization—PSO, the code ALE-PSO (Adaptive Local Evolution-PSO) with adaptive coefficients, proven to be more robust with respect to Gauss-Newton algorithms. Because of the high nonlinearity of the equations involved in the studied physical problem, possible research perspectives include the use of invariant properties for the expression of the material diffusivity and the use of quaternions for the expression of the rotations with respect to the orthotropic frame. The methodology here used has been mainly developed for polymer-matrix composite materials with complex texture exhibiting Fickian behavior; however, in the future it will be extended to other kind of material textures and to other materials showing different diffusion behavior. These topics are currently under investigation.

Acknowledgments This work pertains to the French Government programs "Investissements d'Avenir" LABEX INTERACTIFS (reference ANR-11-LABX-0017-01).

References

1. Aktas, L., Hamidi, Y.K., Altan, M.C.: Combined edge and anisotropy effects on fickian mass diffusion in polymer composites. J. Eng. Mater. Technol. **126**, 427–435 (2004)
2. Arhonime, M.T., Neuman, S., Marom, G.: The anisotropic diffusion of water in kevlar-epoxy composites. J. Mater. Sci. **22**, 2435–2446 (1987)
3. Arnold, J., Alston, S., Korkees, F.: An assessment of methods to determine the directional moisture diffusion coefficients of composite materials. Compos. Part A Appl. Sci. Manuf. **55**, 120–128 (2013)

4. Beringhier, M., Gigliotti, M.: A novel methodology for the rapid identification of the water diffusion coefficients of composite materials. Compos. Part A Appl. Sci. Manuf. **68**, 212–218 (2015)
5. Beringhier, M., Simar, A., Gigliotti, M., Grandidier, J.C., Ammar-Khodja, I.: Identification of the orthotropic diffusion properties of RTM textile composite for aircraft applications. Compos. Struct. **137**, 33–43 (2016)
6. Beringhier, M., Djato, A., Maida, Gigliotti, M.: A novel protocol for rapid identification of anisotropic diffusion properties of polymer matrix composite materials with complex texture. Compos. Struct. **201**, 1088–1096 (2018)
7. Carslaw, H.S., Jaeger, J.: Conduction of Heat in Solids. Oxford University Press, Oxford (1959)
8. Clerc, M., Kennedy, J.: The particle swarm: explosion, stability and convergence in a multidimensional complex space. IEEE Trans. Evol. Comput. **6**, 58–73 (2002)
9. Crank, J.: The Mathematics of Diffusion. Oxford University Press, Oxford (1975)
10. DeBao, C., ChunXia, Z.: Particle swarm optimization with adaptive population size and its application. Appl. Soft. Comput. **9**, 39–48 (2009)
11. De Groot, D.R., Mazur, P.: Thermodynamics of Irreversible Processes. North-Holland, Amsterdam (1962)
12. Eberhart, R.C., Kennedy, J.: Particles swarm optimization. In: Proceedings of the IEEE International Conference on Neural Networks, Piscataway (1995)
13. Glansdorff, P., Prigogine, I.: Structure, Stability and Fluctuations. Wiley, New York (1971)
14. Grace, L., Altan, M.: Characterization of anisotropic moisture absorption in polymeric composites using hindered diffusion model. Compos. Part A Appl. Sci. Manuf. **43**(8), 1187–1196 (2012)
15. Hu, X., Eberhart, R.C.: Adaptive particle swarm optimization: detection and response to dynamic systems. In: Proceedings of IEEE Congress on Evolutionary Computation, Hawaii (2002)
16. Jiang, M., Luo, Y.P., Yang, S.Y.: Stochastic convergence analysis and parameter selection of the standard particle swarm optimization algorithm. Inf. Proc Lett. **102**, 8–16 (2007)
17. Liu, H., Abraham, A., Clerc, M.: Chaotic dynamic characteristics in swarm intelligence. Appl. Soft. Comput. **7**, 1019–1026 (2007)
18. Muller, I.: Thermodynamik. Bertelsmann, Dusseldorf (1973)
19. Onsager, L.: Reciprocal Relations in Irreversible Processes. I. Phys. Rev. **37**, 405–426 (1931)
20. Ozcan, E., Mohan, C.K.: Particle swarm optimization: surfing the waves. In: Proceedings of IEEE Congress on Evolutionary Computation, Washington (1999)
21. Pierron, F., Poirette, Y., Vautrin, A.: A novel procedure for identification of 3D moisture diffusion parameters on thick composites: theory, validation and experimental results. J. Compos. Mater. **36**(19), 2219–2243 (2002)
22. Popineau, S., Rondeau-Mouro, C., Sulpice-Gaillet, C., Shanahan, M.: Free/bound water absorption in an epoxy adhesive. Polymer **46**, 10733–10740 (2005)
23. Powers, J.M.: On the necessity of positive semi-definite conductivity and Onsager reciprocity in modeling heat conduction in anisotropic media. J. Heat Transf. 126, 670–675 (2004)
24. Shen, C., Springer, G.: Moisture absorption and desorption of composite materials. J. Compos. Mater. **10**, 2–20 (1976)
25. Shi, Y., Eberhart, R.C.: Parameter selection in particle swarm optimization. In: Proceedings of the 7th Annual conference on Evolution Computation, New York (1998)
26. Trelea, I.C.: The particle swarm optimization algorithm: convergence analysis and parameter selection. Inf. Proc Lett. **85**, 317–325 (2003)
27. Vannucci, P: The polar analysis of a third order piezoelectricity-like plane tensor. Int. J. Solids Struct. **44**, 7803–7815 (2007)
28. Vannucci, P.: ALE-PSO: an adaptive swarm algorithm to solve design problems of laminates. Algorithms **2**, 710–734 (2009)
29. Vannucci, P.: Anisotropic Elasticity. Springer, Singapore (2018)
30. Weitsman, Y.: Fluid Effects in Polymers and Polymeric Composites. Springer, Berlin (2012)

31. Yamaguchi, T., Iwasaki, N., Yasuda, K.: Adaptive particle swarm optimization using information about global best. Electr. Eng. Japan **159**, 270–276 (2007)
32. Yasuda, K., Ide, A., Iwasaki, N.: Adaptive particle swarm optimization using velocity information of swarm. In: Proceedings of IEEE International Conference on Systems, Man & Cybernetics, The Hague (2004)

Correction to: Variational Views in Mechanics

Paolo Maria Mariano

Correction to:
P. M. Mariano (ed.), *Variational Views in Mechanics*, Advances in Continuum Mechanics 46,
https://doi.org/10.1007/978-3-030-90051-9

The original version of the book was inadvertently published without a volume number. This has now been amended in the book with volume number 46.

The updated online version of the book can be found at
https://doi.org/10.1007/978-3-030-90051-9

© The Author(s), under exclusive license to Springer Nature Switzerland AG 2022
P. M. Mariano (ed.), *Variational Views in Mechanics*, Advances in Continuum Mechanics 46, https://doi.org/10.1007/978-3-030-90051-9_12

GPSR Compliance

The European Union's (EU) General Product Safety Regulation (GPSR) is a set of rules that requires consumer products to be safe and our obligations to ensure this.

If you have any concerns about our products, you can contact us on

ProductSafety@springernature.com

In case Publisher is established outside the EU, the EU authorized representative is:

Springer Nature Customer Service Center GmbH
Europaplatz 3
69115 Heidelberg, Germany

www.ingramcontent.com/pod-product-compliance
Ingram Content Group UK Ltd.
Pitfield, Milton Keynes, MK11 3LW, UK
UKHW021446190426
11946UKWH00022B/53